Making Hands

A History of Prosthetic Arms

Making Hands

A History of Prosthetic Arms

Peter Kyberd
School of Energy and Electronic Engineering
Portsmouth University
Portsmouth, United Kingdom

Academic Press is an imprint of Elsevier
125 London Wall, London EC2Y 5AS, United Kingdom
525 B Street, Suite 1650, San Diego, CA 92101, United States
50 Hampshire Street, 5th Floor, Cambridge, MA 02139, United States
The Boulevard, Langford Lane, Kidlington, Oxford OX5 1GB, United Kingdom

Library of Congress Cataloging-in-Publication Data
A catalog record for this book is available from the Library of Congress

British Library Cataloguing-in-Publication Data
A catalogue record for this book is available from the British Library

ISBN: 978-0-12-820544-0

For information on all Academic Press publications
visit our website at https://www.elsevier.com/books-and-journals

Publisher: Mara Conner
Acquisitions Editor: Sonnini R. Yura
Editorial Project Manager: John Leonard
Production Project Manager: Nirmala Arumugam
Designer: Matthew Limbert

Typeset by VTeX

Dedication

I dedicate this book to the wearers and users I have encountered during my time in the field, from George and Mark to Eric and Ted, (the first to the most recent), and all those in between. Thank you for all your help. It is their insight and persistence that has taught me and all of us in the field humility in front of their efforts to overcome problems. Problems often created by a society that does not consider sufficiently the needs of those who have restrictions to movement or comprehension. It is the users who also inform us what we should know about thriving with fewer than two hands. It is for them that we once again return to the challenges of making hands.

Contents

List of figures

Biography

Peter Kyberd

Was born on the day that at a board meeting for the pharmaceutical company Grünenthal determined, despite evidence that the drug Thalidomide was having terrible teratogenic effects on foetuses, they would continue to market the drug. Growing up it seemed to Peter that the children affected by the drug were often on the television and it planted the idea that there was an unsolved problem to create good prosthetic limbs. After a degree in General Science from Durham University he obtained a masters in Electronics from the University of Southampton before embarking on a PhD that led to the first microprocessor controlled prosthesis to be worn by a user in the field. After his PhD he worked at the Oxford Orthopaedic Engineering Centre on many aspects of engineering in orthopaedics and he was part of the team that produced the first prosthetic limb that used a network of microprocessors. In 2003 he took up a Canada Research Chair at the Institute of Biomedical Engineering in Fredericton, Canada, where he worked with the team on applications of the technology to prosthetics and learned to become a ski instructor, including for disabled skiers. In 2015 he returned to the UK and in 2018 he took up a position as the head of the school of Energy and Electronic Engineering at the University of Portsmouth. He has chaired the popular upper limb conferences; both MEC in Canada and TIPS in the UK.

Preface

It is difficult to overestimate the importance of our hands. They are involved in every aspect of our lives; they are used both in activity and expression. They are distributed throughout our language and society. When labourers are referred to in the abstract, they are ‘hands’ not ‘backs’ or ‘legs’. Hands hold a fascination for many. They are the parts of our own bodies we see most, more even than our faces. This is perhaps why designing their replacement has become a popular past-time for the maker and hobbyist as well as the professional engineer. Indeed, hands have held a fascination for me for decades and have been the focus of much of my career. While at high school I even considered how I could design a prosthetic hand mechanism; it never got beyond this because the tools of easy home production, such as 3D printing, were more than thirty years away. The focus of the book is a biography of the field of prosthetics, and is the book I wish I had been able to read when I started my PhD back in 1985.

The genesis of this book was a PhD study by David Foord of University of New Brunswick. He was studying the process of research and development and used the history of myoelectric hands as his area of focus. This involved interviewing many of the most influential players in the field from the previous half century. I was one of the many people who contributed to his academic supervision. I realised that he had interviewed a generation of pioneers who were retiring and taking with them the knowledge of the development of an important tool for personal independence. I thought that we should capture something of their stories for the future, before they were lost. So I suggested that all this belonged in a book. David’s response was simple: “Why don’t you write it?”. Having worked in the field since my PhD, I knew many of the players and so was in a position to extend the book to cover the entire field. Thus I rose to the challenge. During the long gestation of the text, the impression that it was a disappearing cohort was supported when sadly, a number of these significant players either died or suffered strokes. David was available for consultation and has looked at many of the drafts and without his input this book would not be what it is.

A subject, discipline, or area of research can live on, as the community that makes up the effort changes and alters. It should, however, learn from its past. Arm prosthetics attracted a group of characters that remained active for years, some were well known and influential inside and outside this field. Despite this change of players, the subject remains vibrant today. Based on the number of papers published and companies formed recently, it is far larger than it was at any time in the previous half century. This book is a snapshot of some of those who worked in the field from the 1960s up to 2020.

Limb Prosthetics is a subject with a long history, its story is that of the users who needed the prostheses and the innovators who attempt to make the devices to help enable those users. Not all this help was needed or requested by the limb different. Some of the help was misguided, there were as many missteps and blind alleys in prosthetics, as any other human endeavour, but we can learn from our mistakes. From the perspective of someone working in field, it is the involvement of the users in the process that makes it rewarding and motivates us to work a little harder when we hit the inevitable brick wall.

No matter how comprehensive a book is there will always be gaps. Not only in information contained, but of the contributors and laboratories. I know that some labs and organisations have been only mentioned in passing. The important role of the Italian engineers and clinicians in this tale are only briefly mentioned, perhaps a story for another book. Similarly I realised that Dr Levi Hargrove has hardly been mentioned. Levi was a graduate of the University of New Brunswick where we both worked, before he joined Todd Kuiken. Levi's role in the development of the systems and processes in Chicago in the years after TMR was introduced is inestimable. For this and other omissions, I apologise.

I am indebted to a very great many people, colleagues, friends and strangers for their help, especially to the people we interviewed for the book. I thank them for their time and willingness to check what I wrote, plus the permission to use many of the illustrations from many of the manufacturers. Thanks also go to my very good friend Oyvind "D8" Stavdahl and my father, Bill, for their drawings. As customary, any mistakes in detail or tone or omissions are mine and mine alone. I naturally thank my family for their support especially Ann for patiently reading the drafts and setting me straight in many aspects of language and grammar, and also for enhancing my life with everything she does.

Peter Kyberd
Lockdown
April 2021

Acronyms

ACMC	The Assessment for the Capacity of Myoelectric Control
ACPOC	Association of Children's Prosthetic-Orthotic Clinics
AM	Additive Manufacture
AMI	Agonist/Antagonist Myoneural Interface
ANNs	Artificial Neural Networks
APRL	Army Prosthetic Research Laboratory
AWRE	Atomic Weapons Research Establishment
BBC	British Broadcasting Corporation
BP	Body Power
CAB	Chas A Blatchfords and Sons
CAN bus	Controller Area Network
CAPP	Child Amputee Prosthetic Project
CAT	Centre for Alternative Technology
CBM	Center for Bionic Medicine
CDMRP	Congressionally Directed Medical Research Programs
CEO	Chief Executive Officer
CIBC	Canadian Imperial Bank of Commerce
CNS	Central Nervous System
COO	Chief Operations Officer
CPRD	Committee on Prosthetics Research and Development
DARPA	Defense Advanced Research Projects Agency
DHSS	Department of Health and Social Security
DIPO	Delft Institute of Prosthetics and Orthotics
DoF	Degrees of Freedom
ECG	Electrocardiogram
EEG	Electroencephalograph
EJ	ENER Joint
EMAS	Edinburgh Modular Arm System
EMGs	Electromyogram - Myoelectric signal
ENG	Electroneurograph
EP	External power
EPP	Extended Physiological Proprioception
ETD/ETD2	Electric Terminal Device
FDA	US Food and Drug Administration
FES	Functional Electrical Stimulation
FLAG	Force Limiting Auto Grasp
fMRI	functional Magnetic Resonance Imaging
FSR	Force Sensitive Resistor
GUI	Graphical User Interface
HO	Heterotopic Ossification
IBM	International Business Machines Corporation
IBME	Institute of Biomedical Engineering
INAIL	Instituto Nazionale per l'Assicurazione control gli infortuni sul Lavoro, (National Institute for Insurance Against Accidents at Work)
IP	Internet Protocol
IT	Information Technology
ISPO	International Society of Prosthetics and Orthotics

ISS	International Space Station
ITAP	Intraosseous Transcutaneous Amputation Prosthesis
JHU	Johns Hopkins University Applied Physics Lab
LTI	Liberating Technologies Incorporated
MEC	MyoElectric Controls Symposium
MIT	Massachusetts Institute of Technology
NHS	National Health Service
NRC	National Research Council
NSERC	Natural Science and Engineering Research council of Canada
NTNU	Norwegian University of Science and Technology
NTO	Norwegian Technical Orthopaedics AS
NUPRL	Northwestern University Prosthetics Research Laboratory
OBU	Orthopaedic BioEngineering Unit
OCCC	Ontario Centre for Crippled Children
OSRD	Veterans Administration Office of Scientific Research and Development
OT	Occupational Therapist
OTA	US Congress Office of Technology Assessment
PGT	Prosthesis Guided Training
PLA	Polylactic Acid (plastic)
PMR	Princess Margaret Rose, Orthopaedic Hospital
PPU	Powered Prosthetic Unit
PR	Pattern Recognition
PRTU	Prosthetic Research and Training Unit - OCCC
PT	Physiotherapist
PWM	Pulse Width Modulation
QDW	Quick Disconnect Wrist
RFiD	Radio Frequency iDentification tags
RIC	Rehabilitation Institute of Chicago
RIM	Rehabilitation Institute of Montreal
RPNI	Regenerative Peripheral Nerve Interfaces
SAMS	Southampton Adaptive Manipulation Scheme
SHAP	Southampton Hand Assessment Procedure
SHIL	Scottish Health Innovations Limited
SLS	Selective Laser Sintering
SMA	Shape Memory Alloy
SMART	South East Mobility and Rehabilitation Technology centre
TAG	Technology Applications Group
TARGPR	Technical Assistance and Research Group for Physical Rehabilitation
TD	Terminal device
TIPS	Trent International Prosthetics Symposium
TMR	Targeted Muscle Reinnervation
ToMPAW	TOtally Modular Prosthetic Arm with high Workability
TPS	Trent Prosthetics Symposium
TRS	Therapeutic Recreation Systems
TU	Technical University in Delft, the Netherlands
UCLA	University of California, Los Angeles
UK	United Kingdom
ULPOM	Upper Limb Prosthesis Outcome Measures group
UNB	University of New Brunswick
USA	United States of America
VA	Veterans' Affairs
VASI	Variety Abilities Systems Incorporated
VF	Virtual Finger

VOIC	Voluntary Opening Involuntary Closing
VOVC	Voluntary Opening Voluntary Closing
WAK	Wearable Artificial Kidney
WHO-ICF	World Health Organisation's - International Classification of Functioning, Disability and Health

CHAPTER

Introduction

1

Imagine a young child, on their back, in the sunshine. They are holding their hands in front of their face staring at their shape and the way that the fingers move, exploring the manner in which they work. See them flexing their fingers, bending the thumb, getting the thumb tip to touch the tip of each digit in turn. Or imagine an engineer, on a sunny Sunday morning. Laying in bed, doing the same thing, exploring the finesse of a fine pinch and the range of motion of each digit, feeling the texture of the tip of the opposed finger, spreading the fingers then making a fist, then rolling the tip of the index finger round the end of the thumb. Both are marvelling at just how intricate the hand is.

The hand holds a similar fascination for many of us. Its adaptability is especially remarkable when one knows that the origins of the hand are a fin of the earliest fish. This fin evolved for walking, it was subsequently employed in climbing, and it was finally adapted to be used to hold, caress and create [11]. For the majority, hands are taken for granted, especially until injured or lost. When faced with these challenges for themselves or others, people often return to the child-like wonder of just what an incredible implement has been created by natural selection.

The hand has been described as the gateway to the mind. Focillon wrote *The mind makes the hand and the hand make the mind* [12]. It is often more expressive, more precise and more subconscious in use, than even the spoken word. Indeed, in cases where words let you down, gesture may be the only way you have to get your meaning across. When people lack effective hearing they invent new languages and talk with their hands. The most developed sign languages are not dumb-show imitations of the spoken language, they are at least as alive, real and expressive as any spoken tongue, perhaps (given the extra dimensions of space and movement) *more* expressive [13].

To be born without a hand is to be born with a challenge. To lose a hand in life can seem a terrible disaster. It is a testament to human ingenuity that the loss of a single hand, below the elbow, is often little more than an inconvenience. Many find ways around the lack of a hand in much the same way that a person with two functioning hands may, for a job requiring more than two things to be held, use their mouth, jam an object between knees or between body and arm to hold it. Another method to get round a two-handed task is to find an approach that requires only one manipulator, or the person may simply take longer to complete the task. Performing it slowly and methodically. For example, when I observed a colleague at a prosthetics conference, a woman who was born without a hand, was untangling the lanyards for the attendees. She did not need help, and indeed she refused the offers. She laid the lanyards out in front of her on the bench, draping them on the forearm of her shorter limb and unteasing the tangles with the fingers of her single hand. It took longer, but it was completed effectively.

Some people with an absent hand will get on with their lives without any prosthetic assistance, especially if they were born with a single flesh and blood hand. This is because these individuals are not incomplete, just differently equipped. For others, especially when there has been a loss later in life,

Making Hands. https://doi.org/10.1016/B978-0-12-820544-0.00010-X

this absence is important to them and they need some form of replacement. What they actually miss can vary. It could be the holding capability, or maybe the appearance, or perhaps it's the balance. In any eventuality, they seek to acquire a prosthetic limb from the prosthetics industry. It is for these people that the work detailed in this book was conducted.

When I say 'hand' here, I mean any device that can serve duty performing the same tasks as a natural hand. It may have one finger or many, be able to grasp or not, and look like a flesh and blood hand or a clamp; it depends on the wishes of the potential user.

The design of a prosthetic limb is the combination of many aspects, from engineering to biology, psychology to physics. That is its charm, and challenge. To be a real prosthesis it has to be useful to at least one person with a limb difference. It has to give them an advantage over not using anything at all and it must do it for more than an afternoon. There are many really clever and elaborate ideas, but unless they can give a user some benefit they are just that, a neat idea. Real prosthetic hands have to pass many different hurdles. They must be light enough to be worn, easy enough to be used quickly and without thought and cheap enough to be bought or prescribed for persons who may not have much money, discrete enough not attract the attention of other people, and look acceptable, if they are noticed. This, however, appears to be changing and some prostheses are now being designed to be noticed.

Prostheses must solve a problem in the user's life, such as holding a hand of cards. They do not need to solve every holding problem, just the ones the person cannot do without the device. If the prosthesis passes all these tests, it also has to be tough enough to survive being used for many activities it was not designed for, from hammer to clamp, from signal to probe.

The breadth and vagueness of the above criteria reflect the fact that natural hands are very adaptable and prosthetic limbs need to be right for every individual. There cannot be one device for all potential users. Like many other devices a person might want or use, the criteria for use change regularly with the hour, the day, the season and the stage in their life.

Prosthetic limbs have always been one subject that has engaged the ingenuity of the general population, as well as the limb absent themselves. The literature and the industry have many stories of amputees, who, frustrated with what they could get, (or not get) from others, made one of their own. Some of these user-innovators went on to make devices for others, and a few formed successful companies. This book aims to introduce the reader to some of these pioneers. For example two men, John Hanger and Bob Radocy, fit the profile of the dissatisfied amputee who became prosthetic entrepreneurs. Other companies were formed by people like David Gow who wanted to help, and had an idea that they believed could make a difference. Yet more companies were formed by prosthetists like Otto Bock, who wanted to supply a service and believed he could do better than the current providers.

Prosthetic limbs have a longer history than many other sorts of rehabilitation. People with amputations have survived in greater numbers than those with other forms of physical loss. As a consequence prostheses have a longer history and richer stories than most of the more recent rehabilitation devices, except perhaps braces (orthoses). Orthoses are more common than prosthetic limbs, but seemingly more neglected in research and literature. A discussion of why this is the case, is a topic for a different book.

This book offers an introduction to the field of upper-limb prostheses, as well as a popular history of the scientific research and technological innovation in upper-limb prostheses. It examines the history of significant technologies and devices in the field. It uses stories from Canada, Germany, Sweden, the Netherlands, the United States and the United Kingdom. The stories highlight the commonalities and differences in approaches to research and innovation. Most of these are tales of the movement of ideas from the laboratory to the clinic. They bring engineers into contact with device users, and into collaboration with the professional services providers such as occupational therapists, prosthetists, and physicians, but these contributions often overlook user shaping of devices. Advancements have not been steady and ever increasing, as some commentators suggest; instead there were bursts of progress followed by times of neglect. These changes in pace were triggered by everything from war to social progress and only sometimes were they dependent on technological advances. One aim of this book is to chart the changes and the influences.

The histories in this book do not operate in a clockwork fashion where, once set in motion, the laws of nature dictate every subsequent movement. Rather, they show that culture and ideas influenced technical outcomes. Although technology from other fields was instrumental, the devices and systems that were developed were not inevitable. Many bodies and organisations played their part. Commentators describe an idea that progresses from engineers in the lab to entrepreneurs who take it over and make a product based on the idea. More often ideas are tried and some progress is made before something prevents the idea being taken any further. Harold Sears of Motion Control, one of the pioneers of modern prosthetics, put forward an analogy from the early days of powered flight: It is like early aircraft that unsteadily and briefly left the ground. The ideas for new prosthetic limbs failed due to some underlying flaw in idea or execution, and like the early planes they *crashed on the hillside* [14]. Only later, with experience and newer ideas or materials, could the planes fly greater distances and pilots survive. In a similar way, basic prosthetic ideas were improved and the devices eventually became useful tools of daily life.

After a flowering of research following the Second World War, there was a period of reduced funding from the mid-1970s to 2000s which slowed research and development. Instead of just acting on their governments' agendas, researchers found ways to create their own options. For those of us working in the field during that time, it became necessary to identify other areas to study that might result in progress. We were striving to find the time and the resources to work on our ideas and try them with users. The innovations that did happen resulted from work in other investigations. The topics were perhaps not identified as prosthetic in nature, but the research created ideas that would be useful to prosthetics, such as advancements in electronics and batteries, which were driven by other markets and 'borrowed' for prosthetics.

By the end the 20th Century the technology of prosthetic limbs routinely used in the field was seen as primitive and old fashioned by anyone immersed in the technology of the day. In 1994 when I spoke to a group of engineering students in New Zealand, I was confronted with an opinion from the Kiwis: That the prostheses they saw in New Zealand had inferior technology, because their country were a very long way away from any manufacturers and was always last in the developed world to see technological innovations. This was untrue. The prostheses in New Zealand might have been scarcer and more expensive, but they were precisely the same devices as those available in Europe or North America. In the 1980s and 1990s, while technology was advancing, prosthetics seemed to be stuck in the 1970s. Then in the 2000s, the explosion of new technology changed the way advanced prosthetics were developed. Ironically, many of the new prostheses and software that has appeared in the 2000s, were based on ideas and technology that had existed for many years.

This book is the story of the individual engineers, scientists, entrepreneurs, users and institutions that contributed to the field of arm prosthetics. It is based on interviews with members from three generations of pioneers of this world of prosthetics. The aim was to capture something of what it was like to work in this field from the second half of the twentieth century to the present, and also to record elements of the personalities of those who participated in the story. The interviews were conducted over a number of years as I wished to preserve the knowledge and character of those involved. Over the course of the time of the research for the book, a number of the participants have sadly died. Without their work and inspiration, the field of prosthetics would not have advanced to where it is today.

Powered prosthetics did not begin with a particular war or end with a specific innovation. Nor has the innovation ended. There is still more progress to be made, more ideas to try and probably many more hillsides to crash upon. We are at a point where the way that prosthetics research and development has been conducted is changing again. New surgical techniques and much more fundamental methods are opening up. We can study in greater detail the way that people control their own limbs and their prosthetic replacements. There is a chance for new solutions to be created based on the new knowledge being gained. Here the new research agendas are about understanding human control of their limbs and not just about making better hands, so it is with a hint of the future developments I chose to stop telling the story. One source of change throughout the book is the names of companies, centres and institutes involved, making the tale potentially more confusing. I have tended to use the name that was current at the time the event took place.

Today, news broadcasts and the internet are filled with stories of cheap rapid prototyping machines, opening up the possibilities for many more people outside universities and companies to try ideas and to design prosthetic hands. It is fascinating that while 3D printing (also known as *additive manufacturing*) can allow many different objects to be built, and paths to be followed, by people outside the traditional areas of engineering and innovation. It is the production of prosthetic hands seems to be one of the most popular interests, and it seems every month or so there is another story of a group of interested citizens producing a hand. How novel any of these ideas are is a moot point. Many hand designs are repeated versions of the same mechanisms tried any time before 1919 [2]. Since then few of the new hands have passed the stated criteria of what makes a useful prosthetic hand. This is likely to be true for the majority of the new designs, but among them, a few designers might find a way to make a more useful hand. The important questions are: Are the devices being built better versions of the old ideas, and will the new way of making an old design produce hands that are more functional and less frustrating to use? If not then they will not survive as a prosthesis. If they do, then there is a chance they will make the difference their designers aim for.

What has changed in engineering from a decade ago today is the tools we have to turn ideas into devices. When designing any machine, it has always been possible to think of wonderful objects one cannot actually make. This inability could be because the shape cannot be cut on a conventional machine, or it uses materials with properties that do not yet exist. As part of the development of products, Harold Sears talked of ideas for products that needed 'unobtainium', the impossible alloy [14]. This would have wonderful material properties that would allow a designer to create the ideal prosthesis: Light, strong and cheaply formable into a new device. The potential advantage of 3D printing is that it *can* allow designers to design and make objects that cannot be physically realised any other way, and so solve problems in completely new ways. Are these new citizen designers exploiting the real novel capabilities of the materials to make hands that could never have been made before? Few of the new designs have yet to pass this test.

One aspect where 3D printing has the potential to make an impact is in the customisation of the appearance of a device to the precise wishes of the user, such as those produced by Open Bionics (see 14.1.4). While a prosthetist has always been able to customise the socket to the wearer, a completely crafted prosthesis has only been available to the richer users. Now this is possible for many more, although the majority of 3D printing techniques that make this process affordable cannot yet make devices robust enough for routine use.

Why is it that prosthetics has attracted as much interest as it has from the citizen innovators? What seems to be happening is that many more people are allowing their fascination with the hand to get them to think about the problem of designing a useful prosthetic limb. More people are joining the quest that a small number of us, those mentioned in this book, were swept up in, and spent our lives devoted to its resolution. With more people engaged there is more chance there will be someone who is inspired to make the next breakthrough. They could come from any walk of life or interest, very much as it has been in the past, like the people whose stories form the bulk of this book.

What this wave of interest shows is the fascination that the hand and its abilities hold for very many people around the world. This fascination can lead us to scale new heights of innovation and work in careers aimed at helping people, but it can also take us back to that morning, when lying on our backs in the sunshine, we rediscovered how amazing our hands are.

CHAPTER

Artificial arms

2

Prosthetic limbs are what people generally think of when they consider the creation of a functional replacement for an injury or absence. They are not the only way to solve the problem of limb difference. While in the past fewer people would have survived an amputation, the limb different have always survived in sufficient numbers to allow some techniques to be tried, forgotten and repeatedly rediscovered over many centuries. This makes for a long and varied history. This book is specifically about artificial arms, prosthetic legs are only mentioned in passing.

The functional differences between limbs lead to very different devices. At their most basic, prosthetic legs are necessary to restore equal limb length. A simple pylon can provide function without additional sophistication and can work on rough ground. An alternative, the wheelchair, is much newer invention that requires a much smoother and more level world to drive upon [15]. So while there are other roles for the feet and legs, a very simple device can provide a great deal of function for the wearer.

In contrast, arms and hands have many more roles. Arms are used equally for balance and expression as well as grasping. While an absence limits a person's abilities, some people choose not to use a prosthesis. This is hardly surprising. No prosthesis feels as light or comfortable as the natural arm. Only in a limited number of circumstances does it feel part of the person. In any respect, a person can achieve the overwhelming majority of tasks in everyday life with a missing hand and a short forearm. It is only when the replacement gives the person something extra that they may opt for wearing one.

A prosthesis usually consists of a socket, custom made to fit that person and designed to hold the prosthesis on to the wearer's residual limb (Fig. 2.1). Over the limb is a forearm section that positions the terminal device (TD) at the correct length relative to the shoulder or elbow, and the opposite hand, to make two handed manipulation possible. The socket may hold the controlling electrodes or the harness in the correct place to drive the device. In a powered hand, the forearm also contains the batteries. The socket is customised to the size and shape of the wearer and the outside is decorated to the user's wishes. Traditionally the forearm was 'skin colour', i.e. approximating the wearer's natural skin tone. These days, sockets are often decorated with colours or designs to match the wearer's choice.

Prostheses aim to restore/create function for the user. However, what that function is depends on the person and what they need. One function is simply to make clothes hang properly on the person, or restore some appearance of 'normality' so that the person may pass in society without drawing too much attention.

2.1 The language of prosthetics

This is a good point at which to consider the idea of 'normality'. This is not straightforward. Humans vary a great deal in a continuous range from one extreme to another, irrespective of the measure.

Making Hands. https://doi.org/10.1016/B978-0-12-820544-0.00011-1

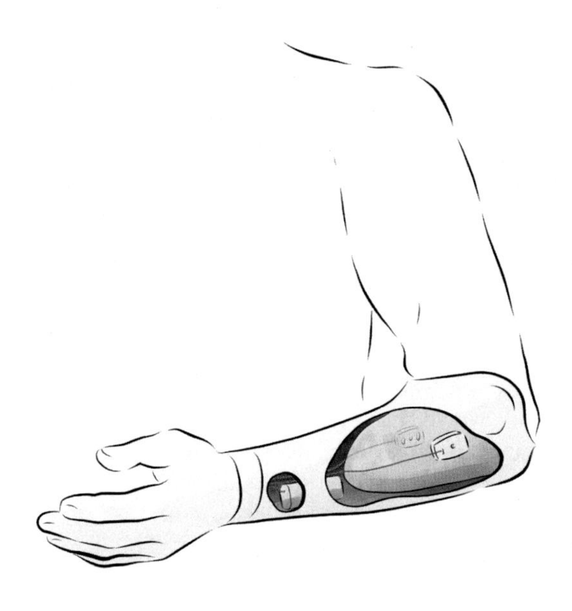

FIGURE 2.1

A prosthetic socket for a mid forearm loss. The socket fits over the remnant limb and may be held on by the elbow (olecranon) it also positions electrodes over the muscles in myoelectrically controlled arms. (Copyright: D8 Stavdahl.)

The majority of the entire human population can only be described as 'normal' within a most limited set of definitions. Historically, there were very rigid definitions of gender, intelligence, ability, race and physical attributes. All were arbitrary and are no longer subscribed to by the majority of the population. They existed to categorise and disadvantage different parts of humanity.

Until recently, to be called 'normal' (or to be more precise, not be referred to as 'abnormal') was also a judgement on the person's moral and/or spiritual standing. So to be physically different reflected something bad about a person's character. Even today this impression persists. Authors of popular fiction have used a physical impairment as a shorthand for moral impairment of their villains, from Captain Hook to the Joker in Batman. In some societies, such as Haiti, this judgement still lingers among some of its inhabitants. If someone had lost a leg or an arm they were seen as having transgressed in some way. One consequence of the Haitian earthquakes of January 2010 is that with so many people now carrying these marks, this belief has lost some of its power. I can only hope this stigmatising of the impaired continues to wither.

In the affluent world there persists a feeling of moral or ethical judgement, especially when the word normal is used. In this book it is hard to simply make a distinction between those who can use a prosthesis and those others who do not have an impairment that might require a prosthesis. In truth, the majority of the entire human population can only be described as normal within a most limited set of definitions, or if the definitions are drawn broadly enough then they will include the majority of the limb absent as well. So the term 'normal' is neither particularly accurate nor kind. As indicated, to get round this problem requires some form of circumlocution that would extend the text of this book considerably, without adding to the content. While I considered any number of words that might work (such as 'average' or 'the majority') I found none that did not have the same potential to be seen as judgemental in some way. In the absence of any inspiration, for this book I will use the words 'unimpaired', 'sound', 'typical' or 'natural', with the understanding I simply mean the part of the population who have a shorter limb, nothing else is implied.

A second area for careful use of language is how we refer to the part of the affected limb that does exist. It is not considered acceptable to refer to the limb as the 'stump' (even though the cloth interface that some users wear are referred to as 'stump socks'). The remaining limb is referred to as a 'residual limb' or the 'residuum'.

Another area for thoughtful language is how to refer to people who might use a prosthesis? As usual, the marking of a group of individuals with a single word or term is inadequate and cannot cover the diversity of experience and expression that the one word implies. I will use some single words implying only those to whom it applies. For example, it is common to refer to users of prostheses 'amputees'. This will not be used here. It is a classification of a group of people distinguished only by their clinical sign, objectifying them. Only some of the people it refers to have suffered any amputation. Other people are born entirely healthy but without parts of the body the majority have; as a result, they are complete, just different, with no amputation. The term is therefore inaccurate and excludes part of the population we are interested in. In this book I will use the terms 'user' and 'wearer'. While this becomes technically incorrect the moment they take off their prosthesis, or decide not to wear one, I believe it is neutral enough not to imply negative judgements. It can be assumed it is shorthand for 'nonuser'or 'potential wearer'. The only time I will use the word 'amputee' is specifically to reference a group of users who have suffered amputations. One exception to this is the term used for persons whose foetal development was impacted by the drug Thalidomide (4.4). The individuals themselves refers to themselves as 'Thalidomiders' and they allow those of us from outside to use it as well. It is shorter and pithier than any long winded neutral description and because they project it, the word has a positive connotation for a group of people who chose not to be limited by the mistakes of others.

The hand's importance in a person's life can be gauged by many things, not least the many expressions, similes and metaphors in most languages that use hand-based imagery, from 'on the other hand' to 'palm you off' to 'get your finger out'. The reader can probably think of a dozen more with little effort. These expressions are part of the language and I will not resist the temptation to use them, though most often it will not be an ironic or pointed use, but simply the way it would be used in usual speech.

One term I sometimes employ is 'Golden User'. This applies to the people with a limb difference, who dedicate their time to helping in studies of new devices or techniques. It is through their participation and patience in multiple sessions, waiting while we fix something technical that is broken, or suggest things that illuminate our ignorance. They teach us new ways to make our devices. Those of us who do prosthetics research learn what is right and what is fundamentally wrong with our work. It is these people who teach us what we should be doing and the very best experimental users (such as Jessie Sullivan in the USA and Campbell Aird in Scotland) become partners in the endeavour. These are truly Golden Users, for whatever we do, it is through their efforts that we succeed, and they make us look better.

One final area of language is that of anatomy. At first the different names of the muscles and bones seems daunting. The special language to describe the body makes sure that we are referring to the same thing. It is not uncommon to ask someone who says 'go left' to reply 'your left or mine?'. Thus the

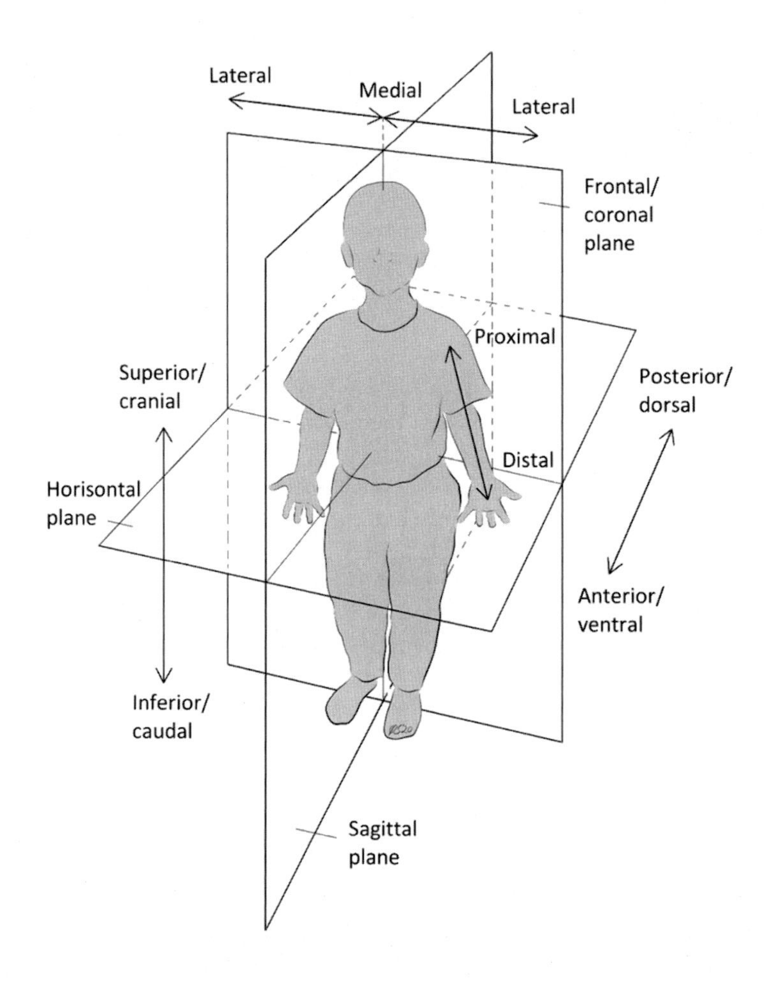

FIGURE 2.2

Conventions for the axes and orientations of the body. All directions are defined relative to the body in a pose described as 'neutral'. (Copyright: D8 Stavdahl.)

basic anatomical definitions are all relative to the same thing: The person in question. A body is defined as a figure upright and with the hands by their sides, palms facing forwards. Posture and movements are relative to this neutral position (Fig. 2.2).

Arms or legs *flex* their ends towards the body (Fig. 2.3). So hands move forward and up while feet move backwards and up, but it is the same name. To return to neutral means that the joint *extends*. If the joints move the hand or foot sideways, away from the body, it *abducts*. If towards, it *adducts* (i.e. it adds to the body). These terms are so similar in sound that in a medical environment there is a tendency to say 'ay-bee-duction' and 'ay-dee-duction' for clarity. While there are other joint rotations, the pair needed for upper limb are *pronation* and *supination*. It is often described as a function of the wrist. More accurately, this describes the two bones of the forearm (radius and ulnar) crossing and

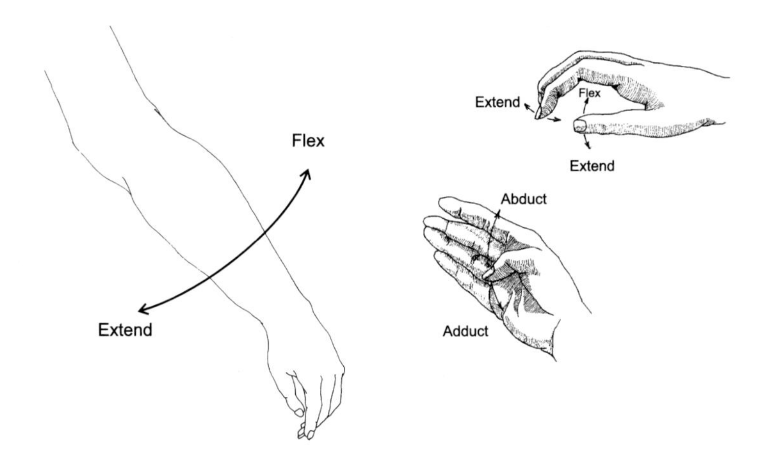

FIGURE 2.3

Conventions for the movements of the body. Directions are defined relative to the body *towards* the body is flexion and *away* is extension. (Copyright: W.P. Kyberd.)

uncrossing. In prosthetics the artificial joint is usually at the wrist level, so for this book that is how joint motion is defined. Telling one from the other is tricky, except that you pronate away from neutral so the hands point behind the body. Supination is required to bring the hands back to neutral. It is easier to remember if one considers that when the elbows are flexed to bring the hands to a horizontal position, the hands are 'soup-inated', they are in a position to hold two soup bowls.

Directions on the body are also carefully labelled relative to the centre of the body. Towards the middle is *medial*, away is *lateral*, so the outside edges of both arms are their *lateral borders*. The front is *anterior* and the back is *posterior*. The most important definition for the prosthesis reader is that the shoulder is *proximal* to the body (ap-prox-imately there), while the hands and feet are the most *distal* (distant) parts of the limbs. Other definitions are shown, but are of less use in this book.

2.1.1 The role of a prosthesis

The role of a prosthesis is very wide and dependent on the needs of the user at that time. A prosthesis can exist to make the wearer stand out less, to fill a sleeve or to provide a shoulder for the shirt to hang on. It can also be used to provide a limited number of holding functions, such as clamping, gripping or pinning to a desk. It might be used to perform actions similar to those the missing hand might perform, for instance picking up, holding, transporting and gesturing. To be able to do so the hand has to be placed in space in front of the person (positioning to the side is less common), and orientated to be able to grasp an object, so it must be possible to rotate it in three axes, as well as position it in three dimensions. This positioning can be achieved through using the rest of the arm and body, or the rest of the prosthesis (depending on the absence).

2.2 Types of prosthesis

2.2.1 Anthropomorphic prostheses

One important aspect of the design of a prosthesis is its appearance. The attitudes of society and the wearers are not trivial. A nonactuated, but anthropomorphically shaped hand/arm prosthesis, which might be a very highly detailed analogue of a natural arm, is generally known as a 'cosmetic' arm. These may be dismissed, especially by funding agencies who seem to see them as pointless or a fatuous sop to the person's vanity, and not worthy of funding. This is unkind and false. There are many reasons to believe that these are useful devices for many people. Firstly, they do create the look of a natural limb. This is a function, and for some people, an essential one, especially during the early days after amputation when the person might feel incomplete. That is enough justification to provide cosmetic devices. Moreover, those of us with two hands should not underestimate the influence our appearance has for some people. When I conducted a survey of users in the Oxfordshire region I found a few wearers of 'cosmetic' limbs who stated that they wore them for the comfort of others seeing them and not for their own satisfaction [16]. This bias reflects a prejudice that exists including in the research literature. One author wrote 'unsuccessful' users were: *"users (who) did not use a prosthesis, or wore a prosthesis for cosmesis without using it in a functional manner"* [17]. I disagree with this definition as the statement implies that a nonactuated prosthesis is not useful to the individual wearer. It also suggests that the device used in an entirely passive role, which is unlikely. I believe this attitude is diminishing with time.

The role of expression should not be neglected. An interesting experiment was conducted with a single person with a limb absence [18]. She was a total nonuser of any prostheses in her daily life. She had a series of very unlifelike prosthetic 'hands' made for her and her behaviour was observed. This user showed a strong tendency to use a more hand-like object in general gesturing. In the experiment, she started by wearing something on the end of her arm that was frankly nonanthropomorphic (a white rectangle). With this she left her arm down and the 'hand' in her lap. Then the rectangle had its corners cut off so the white card looked slightly more like a hand, and a 'thumb' was added (a strip of white card with one fold to 'oppose' the 'fingers'). This is the most rudimentary representation of a hand, but the user took the hand out of her lap and started gesticulating with it. This hints at appearance being important, and also shows that what is 'hand-like' is fundamental.

The additional overwhelming indication of passive functional use arises when standard 'nonfunctional', 'cosmetic' prosthetic hands are returned to the clinic. The fingers are often broken showing that the person has used the hand for passive prehension, i.e. jamming objects in between the fingers or leaning on the hands to rise from the table or chair. These prostheses are very simple structures with a skeleton of lightweight copper wires inside foam; the foam fills out the fingers of a high definition silicone glove to create the natural appearance. In everyday use, the internal copper wires are bent by the user to adapt the hand to the shape of an object or task. After a number of repeated bends the copper work-hardens which makes the copper easier to break. This is a strong argument for a new design of cosmetic hand that incorporates more adaptable fingers. One team has responded to the challenge (see 5.4.1). To improve designs we need to observe users more closely to see what they really do with their prostheses. This is beginning to be explored, using new technology (see 3.4.5).

The fact that the market has not found a better solution to making passive prostheses is based on simple economics. One of the advantages of cosmetic arms is that they are very light (and so easy to wear) and cheaper than any other design to produce. Improving their functionality means incorporating

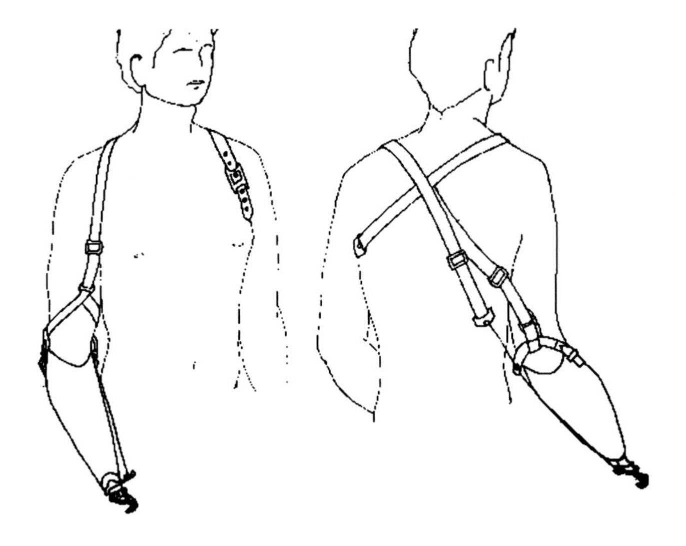

FIGURE 2.4

Basic body power set up for a person with a below elbow absence. The cable is fastened around the opposite arm and the relative movement pulls on the cable to open or close the terminal device. (Copyright: W.P. Kyberd.)

stronger fingers, incurring greater cost and making the hands heavier. This will take more money to build and will give the hands more mass. Neither of these factors can be easily circumvented. Like body powered terminal devices (see below), there is unlikely to be money to pay for novel development of these hands. Certainly, it would be hard for an independent company to justify the expense of developing a new semi-active cosmetic hand and then recoup its expenses by selling it at a profit, given the prices of the passive hands already on the market. The alternative route is to gain funding from a local or national government to support the research and design within the university, health or nonprofit sectors.

2.2.2 Body powered devices

Body powered prosthetic arms use the relative motion of one part of the body to another to pull on a harness to move the hand or joint. It would seem to be a simple idea to capture the relative motion to do some form of work. This is the principle of the trigger: The fingers move relative to the palm pushing on the stock of the gun/drill. The first guns with a trigger were produced long before 1800, so perhaps a cable should have been seen as a simple way to make a prosthetic hand move. However, there is no evidence that it was.

The most common form of body power (BP) for upper limb prosthesis is a harness that runs around the opposite shoulder, and pulls on a cable attached to the hand (Fig. 2.4). Ballif created the first such device in 1818. These prostheses are still very popular since they are simple, low cost, lightweight and can be used in any circumstance. Rapid advances in many areas of technology in the last half century, have not had that much impact on the designs of BP arms. Market forces tend to limit companies'

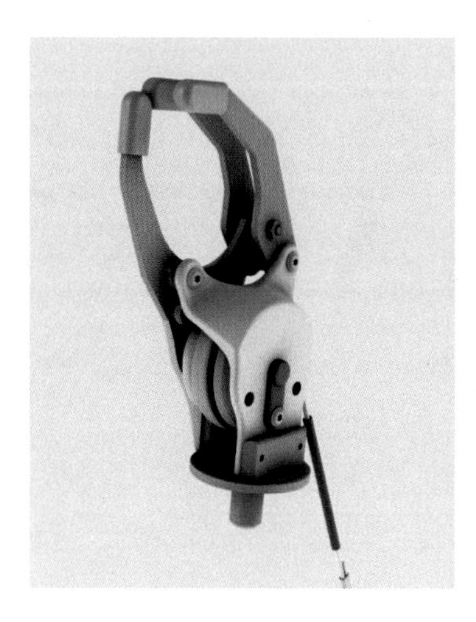

FIGURE 2.5

A body powered hand produced by Otto Bock. The hand is the same geometry as their popular single axis adult electric hand so that it can use the same external gloves. (Copyright: Ottobock SE & Co. KGaA, Duderstadt.)

ability to innovate (see 5.2.3). However, a small number of enthusiasts have continued to work in this area and Chapter 5 introduces us to two of the current pioneers.

Body power is generally associated with nonanthropomorphic, starkly functional terminal devices. These TDs have generally better manipulation capabilities, and attract those who value manipulation over anthropomorphic appearance. The appearance of the TD is of still important, but they do not need to be anthropomorphic in shape.

The range of commercial body powered devices is considerable. The traditional 'hook' form is common. The design can be a simple hook for grabbing and pushing items, or it can have two tines to create a tool that can hold as well. This is the 'split hook'. It has fine tips for precision grips and a broader base for more forceful grasping. The shape is varied to include the ability to hold or manoeuvre objects specific to the user's habitual requirements. Nonanthropomorphic designs are more varied than the powered devices with fingers or thumb (or both) driven (actuated). Prosthetic companies; Steeper and Otto Bock have created body powered hands that can use the same gloves as their powered hands (Fig. 2.5), but as will be described, the losses in the cables can limit the grasping abilities of these hands (5.4).

BP wrists can be passive or active, and many allow the hand to be moved with a limited amount of flexion to bring the hand more closely to the mid line [19].

2.2.3 Specialist devices

Some activities require specialist tools or devices. Most examples are in the kitchen, workshop and sports arena. Each area has special uses that are more likely to need a device that is designed to work there only. There are spade holders, ball throwers and knife attachments. The single biggest disadvantage of such devices is that it stops the user being spontaneous in their choice of activity. This is especially true with children who will freely change their activity without thought or preparation. The specialist prosthesis often uses a connector at the wrist so that the device can be changed, but the socket can remain the same. This is less popular with sporting prostheses since the increased sweating and force through the arm demands a special socket.

2.2.4 Externally powered devices

The final category of prosthesis employs some form of external power source. This source is usually electric, because it is easy to recharge such devices. Designs using electric power go back a century, but the first vaguely practical device came in the 1940s. Electric motor designs have continued to advance with increased performance and reduced mass. However, what all externally powered devices depend on is the power supply. Important questions about the supply are: How much power is stored and how long will it last? For electric motors this storage is batteries. For any other technology to compete with electrical actuation it must keep up with innovation in batteries and motors.

Advances in battery technology are still progressing at a considerable pace. There is pressure for improved function and endurance in consumer electronics so there is on-going investment aiming to improve battery performance. Prosthetics can take advantage of this progress. The advances mean that the energy density of batteries today is considerably greater than twenty years ago. Today's batteries are lighter, smaller, last longer and quicker to recharge.

Hydraulics is a motive power that has some promise, but limitations on size and power have always made this technology harder to use only recently have any commercial hydraulic prostheses appeared. For externally powered hydraulic hands the underlying power source remains an electric battery driving a motor used to compress the hydraulic fluid (see 14.1.5).

Despite all the advantages, electric motors are not the best means to create the movements that prosthetic arms must perform. By virtue of their construction and use, natural limbs move back and forward over short ranges. The elbow bends to pick up objects to the mouth and flex the arm back down to the table. Electric motors are generally rotary in motion. When we need such motion to be driven by a human we build machines to convert this cyclic action into circular motion (such as the cranks on a bicycle). Motors on the other hand spin continuously in one direction or the other. Electric motors are most efficient running at high speed in one direction. Changes in direction for a rotating motor are slow and cause inefficiencies as the system must slow down and start to move in the opposite direction. Motors usually have to be slowed down through gearboxes. This adds mass, and delays in changes of direction. Electric motors driving limbs are never efficient. Muscles, on the other hand, work best when moving slowly, so when driving a waving arm or striding leg they are efficient. However, the advantage that motors have over other actuators is they are compact and with a readily rechargable power source mean that they remain the principle power source commercially.

Like many in the industry, Bill Hanson of the prosthetics company Liberating Technologies Incorporated, acknowledged the role that battery technology, in particular lithium ion batteries, has in upper limb prosthetic products. *"They have twice the capacity and half the weight of nickel-cadmium batteries"* Hanson noted. *"They charge faster, don't lose capacity, seem to last as long, and don't have environmental problems"* [20] (at least not during usage). The batteries have allowed designers to use motors where they would have been impractical to use before.

Externally powered prostheses began with just a single motion, to open and close the hand (4.2.1.1). The partial success of the so-called 'Russian Hand' led to innovations in multiple countries, through modifications of the Soviet era technology or the development of new hands. Notably, Otto Bock began with motorising a wooden, body powered, Hüfner Hand (9), and then developed the mechanism that dominated the market for nearly fifty years. Other designs came and went between the 1960s and 1990s. Only with the innovation of David Gow's ProDigit design in 1993 for multifunction hands was this dominance challenged (8.2).

The next significant powered innovation was an elbow (Chapter 7). Two emerged from the USA before any practical commercial devices were developed in Europe (9.2). The control and lift capabilities have limited their creation and adoption, with body powered variations of two of the products being part of the evolution.

Wrist rotators have been developed, but these have tended to add to the length and mass of the prosthesis, so their adoption has been limited often to persons with a bilateral absence. Other axes, such as wrist flexion and humeral rotation, have tended to be in research devices only. Each and every joint has the same limitation of adding mass and power drain with insufficient *functional* gain.

2.2.4.1 Control inputs

One of the fundamental truths of prosthetics is that every user is an individual. This reflects not only their choice of device and what they want to do with it, but the choices the clinical team have when creating the prosthesis. The user's musculature and joints may allow the control of many motions or few. The user's temperament or cognitive capabilities may limit their ability to adapt to the prosthesis. All these factors play into the design. The team's role is to make a prosthesis that works for the user and allows them to get on with their life. In a body powered arm, pulling the cable not only moves the hand but controls how fast it moves or how wide it opens, and for some devices, how much grip force is employed.

Once the actuation and control are separated (as in external power), the way the person instructs the arm can change and the instructing action can be made remote from the device. External power sources change the way an arm can be controlled to match the user. The level of effort (e.g. a push or pull onto sensor) need not determine how wide the hand opens, but how fast it moves. If it is useful to a user, a hand can be made to work opposite to what might seem sensible; 'in' can be made to mean 'out' and 'down' to represent 'up'. What is important is that the clinician can choose a way that makes the control of the prosthesis easier or more intuitive for that user.

Over the years many different forms of switches, joysticks and levers have been employed to instruct externally powered prostheses, depending on the needs and capabilities of the user. One popular means to instruct powered arms is via myoelectricity. Myoelectric signals (EMGs) are the small electrical signals associated with muscular contraction. Most people are familiar with the ECG (electrocardiogram); this is the signal associated with the heart, this simply the myoelectric signal from the big heart muscles as it beats. Any other muscle can furnish similar signals.

From the earliest days of myoelectric control, the idea was to use muscles already associated with the motion of the hand or arm to perform the task (6.1.3). Signals associated with extension of the arm or opening the hand caused the prosthesis to open and flexion or closure was mapped to closing the hand. A myoelectric signal is tricky to obtain and is prone to many sources of interference. Over half a century of work in this field has now gone into making myoelectric signals simpler and more reliable to use.

2.2.5 No prosthesis

The limb absent person always has the option to not wear a prosthesis. This is quite reasonable, especially if the person was born without an arm. The end of their residual limb may be as sensitive as any finger and so covering it with a plastic shell would be like spending one's entire life wearing boxing gloves. Prostheses are heavy and hot and many people can be highly dextrous with their remnant limbs. Indeed, the majority of persons with one hand missing below the elbow are less 'handicapped' than 'inconvenienced'. Of course, many tasks are two handed and some require the arm to reach the middle of the person's body (such as feeding or using the toilet). However, the solution might be a much simpler device that is worn only during this task, such as a spoon with a velcro cuff that attaches to the far end of the arm during dinner and is not worn any other time. Or it may include a specially shaped rod that can reach the backside and hold toilet paper securely. Determining what tools are needed and how to make them is the role of the occupational therapist and prosthetist.

In the past the profession and the public have generally aimed to supply function with an anthropomorphic limb and not to create tools to make the job easier. More recently attitudes have changed. Occupational therapist, Sheila Hubbard of Toronto, refers to this tendency when speaking about treatment of the children who suffered the consequence of the drug Thalidomide, they were fitted anthropomorphic limbs that reduced their natural functional range [21], she believes this would not happen now.

2.2.6 Paediatric prosthetics

There is little historical evidence that children were catered for in terms of upper limb prostheses. No children's arm prostheses from the distant past have been found and examples in the literature are scant.

The needs of children are distinct, they are not just small adults. The difference is not just what the child can carry on their remnant limb, but also that their cognitive development limits what they can do with the prosthesis. The fitting of a child has to take into account what that child is physically and mentally capable of using, but may also address the need to foster good habits as a foundation for a lifetime with prosthetic interventions. The process is fitting the family as well as the child, which can bring in its own challenges.

The considerations of development make the discussion as to when, or if, a child is fitted with a prosthetic limb even more complex than adult provision. As observed, for independence, the most important consideration for any prosthesis, is the need to reach the mid-line of the body. While a child might be able to reach their mouth with their remnant limb, the levels of contortion needed to achieve this can be substantial. Some individuals can perform a wide range of tasks with their feet and their flexibility lasts well into adulthood. I have seen an inspirational speaker without either arm, who even in later age could hold a knife with his feet, and slice potatoes with ease. Unfortunately, like all other talents, this flexibility is not distributed uniformly throughout the population. While some people may remain able and flexible, others will always need other means to reach their mouths effectively.

These considerations also include the development of tactile experience and staging of fittings. If a child is effective without prostheses, covering up sensitive skin on the ends of their residua with shells will reduce their ability to explore the world with touch. Secondly, as children develop as individuals, self determination and a strong will make daily battles over wearing clothes, hats or gloves a feature of parenting. The influential Swedish occupational therapist, Lottie Hermansson of Örebro, does not want the prosthesis to become part of this battle of wills. Thus she prefers to wait to fit the first prosthesis until after this phase has passed [22]. Additionally, any device for a child needs to be small enough and light enough to be worn by the child, so some milestones are size based. Specific prosthetic technology can be fitted as the child becomes large enough to use a particular device. However, Sheila Hubbard of the Toronto Centre is equally certain that the child who is fitted younger, integrates their device into their body image, while a child fitted later always uses it as a tool [21]. This view has fallen out of favour in the UK.

One practice to help with the child's development is to fit a child with a flexible mitt-like hand (a 'crawling hand') at six months. It balances the child, giving them arms of a more similar length. The soft nature of the hand protects the child and its carers from accidental collisions. The child is fitted with a powered hand by eighteen months. The child is encouraged to use bimanual actions with the hand by offering it objects on their prosthetic side. An electric hand is opened by contracting one muscle in the forearm, something that needs practice. Initial control is often discovered by the child, and when the hand opens the use is encouraged by the therapist. This first device is often a one electrode controller. The muscle signal opens the hand, no signal closes it (Voluntary Opening Involuntary Closing - VOIC or 'cookie crusher'). An external switch on the outside of the socket allows the parent to reach over and override the child if the hand becomes stuck or the child is recalcitrant ('parental intervention' or 'mommy' switch). As the child grows an additional electrode can be added that allows them to open and close the hand separately (Voluntary Opening Voluntary Closing - VOVC), so the hand remains open when they relax. The child receives additional training as they grow and develop.

Of course the parents and guardians of the child play a vital role in the development and normalisation of the individual. Their response to what might have been a shock to them when they are told their child is to be atypical in some way often influences how they and the children progress. If they can find a way to approach the situation with a positive attitude and help the child develop as they would any other, the individual can grow up accepting of their situation. It is also not uncommon for a teenager who has obediently worn their arm since they were small to decide not to as they mature. This a common element of growing up and they may later return to some sort of prosthesis, when they find a use for one.

2.3 Problems

The choice of no prosthesis is always the option of the individual. However, if a person with some part of their arm missing chooses not to use any sort of prosthesis they risk curvature of the spine (scoliosis) [23]. The spine is a pile of vertebrae without a rigid joint between them that works a little like a tower which is held up through the tensions of the wires running from it to the ground. If a wire is out of tension the tower will bend. In the case of a short arm the mass of the arm itself and the muscles needed to support it help to maintain the spine's direction. Lose some mass or strength, and the spine may bend. An actuated prosthesis tends to reduce this effect. As with all matters connected with the body, this does not always happen, or develop to the same extent in everyone with a short arm who opts for no prosthesis.

2.3.1 Overuse

A second consequence of not using a prosthesis is overuse of the other limb. Overuse is when you employ a joint more often, over a greater range or with greater than typical forces. Tennis elbow is the result of too much (or unaccustomed) use of the arm, whether it is playing tennis or painting the fence. Wrist overuse injuries can result from using a computer keyboard too much with the hand is unsupported, or with poor posture. The result of this injury is pain, often intense, every time the joint is used for a task.

If someone has one hand, the overuse problem becomes more acute. The leverage required to perform tasks becomes greater. The shorter the remnant limb, the greater range is needed, compared to a full-length arm. A prosthesis extends the arm but the short limb inside a prosthesis must apply greater forces to move the limb around in space. Overuse injuries can occur with the use of a prosthesis or without, but the prosthesis user does not have to reach as far if the device is used actively. Users can pick up or hold objects on the prosthesis side easily and not reach across their body from their other side.

Even with a prosthesis, overuse can occur when the wearer of a prosthesis with many powered joints chooses to take short cuts. For example, a person often needs to rotate their wrist to position the hand to pick up an object. Even with a powered wrist, it is still often faster and requires less mental effort to use their shoulder to raise the elbow sideways away from the body to rotate the hand, rather than selecting to rotate the wrist and driving the hand to the target orientation. While this approach may be very much more convenient, it has tremendous potential for overuse injuries. Alternatively, the person may reach across their body and use their natural hand. This again uses more motions and more effort, which has potential for injury. If the amputation is on their dominant side, the user will become considerably more impaired. The loss may prevent them from performing the simplest actions of self care or daily living, with many concomitant limitations to their abilities and life. This consequence is something the prosthetics profession must guard against. Whether a person chooses the most advanced prosthetic limb or none, the clinical team need to help the users, and guide them to methods of action that reduce the chances of injury.

Side dominance impacts on how a recent amputee faces their loss. No prosthetic limb is yet as effective as the natural one, irrespective of which side was dominant before any loss. The natural hand will become the new dominant. Learning new skills becomes more involved when the dominant hand is lost. A natural consequence of the causes of loss that many lost hands were the dominant ones. Again, it is testament to the adaptability of people that they will persevere and become entirely capable with their other hand.

When studying persons using a prosthesis to perform routine daily acts, we found the changes in posture affected every part of the body, including the opposite (intact) limb, spine, legs and head [24]. This means that any part of the body could have overuse injuries, not just the short arm. Also someone who has not used a prosthesis when younger, no matter how capable they are in their youth, may begin to have joint problems in later life. Indeed the most well known members of the limb absent population, those affected by Thalidomide suffer badly from this problem, as they moved into their 60s [25].

2.4 Causes of absence

The standard distinction in the causes of absence of a limb is between those with a congenital absence and those with an *acquired* loss, a neutral and impersonal expression for a life changing event. This second group will have either suffered a traumatic amputation due to an accident or injury or have the arm (or leg) removed in response to infection or cancer. Accidental loss may not be a clean severing of the limb. If the limb is intact, surgeons will attempt to salvage and reattach it. This is met with has mixed success. Damage from crushing or blasts makes reconstruction difficult. Surgeons go to great lengths to try and salvage as much of the limb as they can (see 2.6), but there may come a time when it is more traumatic or dangerous to the health of the patient to keep the hand than to lose it. Of course, sometimes it is necessary to quickly and efficiently remove the remains of traumatic damage to stabilise the patient and save their life. When this happens, the surgeon may have to operate again later to improve the end of the bone and skin to make the stump suitable for use in daily life. Follow up operations should not to be considered lightly. Multiple procedures can be stressful or dispiriting. For this reason, advanced treatments for loss (grafts, direct attachment, or neural interfaces) may not be desired by an individual. When an infection or cancerous growth needs removing, then the opportunities to carefully plan where to cut and how to go about the operation are far greater. Even then the prosthetic choices are limited by the state of the remnant.

In the past, before the knowledge of infection and antisepsis, simple cuts and splinters could cause infection that would spread up the limb and eventually kill the patient, if the limb was not removed through fast and radical surgery [26]. In the modern era the more common reason for gangrenous limb loss is a systemic infection, such as meningococcal septicaemia. Here the infection causes the blood vessels to lose their ability to retain the blood and so blood seeps into the skin, pooling in the ends of limbs. The blood is not returned to the lungs to be re-oxygenated. The cells at the ends of the limbs are starved of oxygen and begin to die. Previously, the person would simply have died as the infection spread throughout the body, but modern techniques can keep the person alive until the infection is beaten. However, the consequence is that the peripheries may have already died. This leads to a small but increasing number of persons who have lost parts of two, three, or four limbs, creating greater challenge for the profession. This would have been very rare is it now slightly less rare. One hard learned lesson is that surgical teams have determined that they should leave the person's body to 'self amputate'. This means the gangrene will cause some parts to die and simply fall off, but it is surprising how much of the peripheral parts (especially digits) may recover. In general, the longer the limbs the better prosthetic functional outcomes. If any functioning digits remain the person will be more able than with a prosthesis. Surgeons now admit they were surprised by how much the body can recover. If they had operated early to remove affected areas, they would have taken away parts that ultimately proved to be viable and useful. This is an example of the two approaches in medicine of 'watch and wait' or 'heroic intervention' [26]. This dichotomy is explored more, in Section 2.6.

The other category of absence is persons who were born without a limb, or part thereof. Typically this is due to spontaneous interruptions in growth in the womb. The absences are generally without any cause or action on the part of the mother, even if she feels responsible. There is a slight bias towards the left side and to females. The bias in acquired loss is far more pronounced, with a greater number of males suffering amputations on their right side. Most acquired loss occurs between late teens to thirties, in a group with less regard for personal safety. The important distinction between acquired losses and congenital absences is that while someone who has had an amputation is missing something, someone born without a hand is not incomplete, they generally do not feel incomplete and only some will wish for a prosthetic replacement.

A person with a congenital difference may have some of the nerve endings in the end of the shorter limb which would have gone to the fingers in a fully developed limb. This gives the person touch sensitivity that a prosthesis would hide. Small growths (called nubbins) may occur that may or may not correspond to undeveloped fingers, but again they add sensitivity which the person can use. In the past these nubbins have been removed early by well meaning surgeons, unaware of the role they can play in fine manipulation and sensory exploration of the world.

The remnant limb of a person with congenital absence is not like a person's whose limb has been amputated. Many physicians and engineers assume the arm is merely short. This is generally untrue. The underdeveloped limb has been interrupted in its growth, so the bones may have a different shape and the muscles may not have grown or developed in the same way. In the past, efforts, to explore these limbs have been treated with scepticism. The growth may have in fact been disrupted for some direct cause, so the nerves, muscles and bones may be different to usual growth patterns. Someone with an absence below the elbow may not have the same range of motion as someone without a limb absence, or they may be able to bend their elbow beyond straight down but not flex to their humerus (upper arm). The result is the prosthetist may have to adjust the line of the prosthesis to give them a more conventional range, providing the hand an angular offset so that it can reach the mid line.

When muscles are used for prosthesis control the assumption is that the signals from all users are the same. I found that this is not true [27]. The muscles in a limb that has not grown or developed, and which are not involved in moving a joint above the muscle, do not have the ordered structure of other muscles and do not generate exactly the same electrical signals.

2.4.1 Extent of the loss

The smaller the loss, the greater the chance of occurrence (very roughly). Finger tip losses are more common than hand losses and far more common than a loss from the shoulder. In general the more of the arm that is lost, the greater inconvenience to the person and the harder it is to provide a prosthetic replacement, as it needs more separate motions to replace the missing joints. Increasing size or complexity of prosthesis will result in greater weight and expense.

Someone with one natural arm is far more effective at performing everyday tasks than if they have lost both. With a single loss the person can use the prosthesis to hold an object or pin it down to a surface, while all the complex motions are done with their living hand. With a bilateral loss, a person must be able to perform all tasks with their prostheses. They must be able to put on and take off both arms without the help of another person, or the user can never be truly independent. For someone with this amount of loss it is even more important that the prostheses can bring objects to both the mouth and the anus, fasten or unfasten buttons, pass objects to others, reach the top shelf and tie shoe laces. A prosthesis for a person with a single sided loss is different to devices someone with a bilateral absence [28] (see 5.3.1).

Fig. 2.6 summarises the range of prosthetic fittings available for persons with an arm loss. Left to right they also show the options from unpowered, to cable operated (body powered) to the most sophisticated devices to be obtained today on the market.

Much of the rest of the book will look at externally powered prosthetics technology since this is where most of the innovation has happened in the last seventy years. This does not mean that, world wide, external power is the most popular or effective way to activate prosthetic arms. Even in the poorest countries, the wish to appear like others means that many people with limb losses desire anthropomorphic looking arms, even when these prostheses are less effective than starkly functional manipulators. The problem of providing prosthetic solutions for a thousandth of the cost of devices available in wealthier countries is occupying the minds of people on both sides of the financial divide.

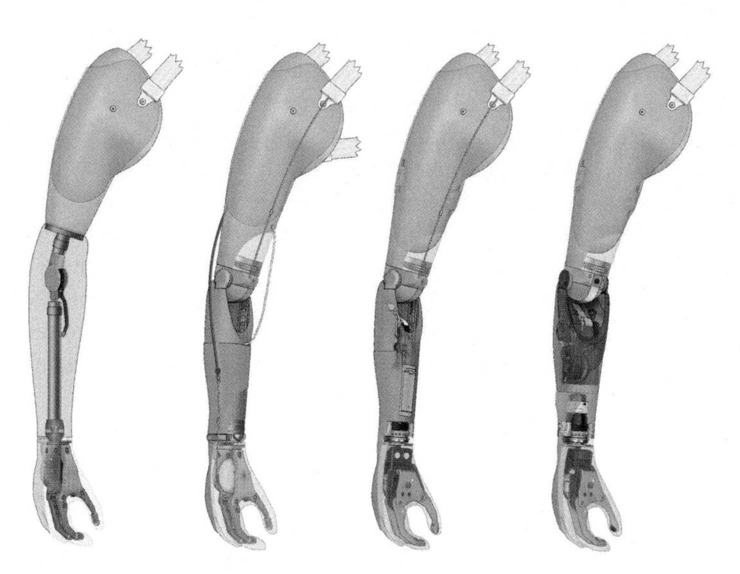

(a) Options for losses above the elbow. Second from the left is the ErgoArm.

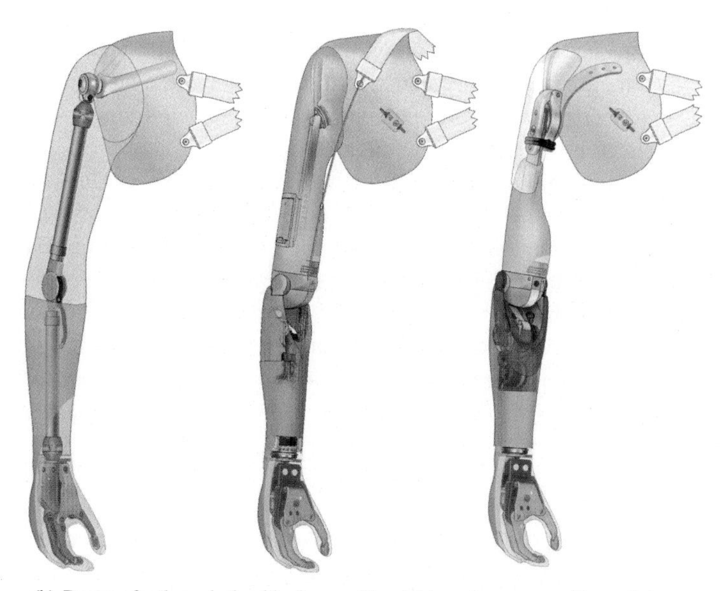

(b) Devices for through shoulder losses. The right most arm uses a *DynamicArm*.

FIGURE 2.6

Prosthetic options from Otto Bock from entirely unpowered, through body powered to electric arms, all electric arms are hybrid needing some nonmyoelectric inputs. (Copyright: Ottobock SE & Co. KGaA, Duderstadt.)

2.5 Choice of prosthetic limb

The choice of prosthetic limb depends on a number of factors. Demographics play an important part in shaping the market. Mercifully, the market for upper limb prostheses is small. In the USA there are about 6000 upper extremity amputations a year. The US Veterans Administration has 40,000 users currently receiving services in its healthcare system [29]. The majority of upper limb amputations occur as a result of trauma and involve young people [30], and this has been decreasing compared to disease-related limb loss. Injuries to servicemen in the recent conflicts in the Middle East are not numerically large, but their presence introduced a population with multiple limb losses, that have a greater public profile than the typical potential user. This has changed the focus of the US market and the public's perceptions of injury, loss and prosthetic replacement. In the past half century in the United Kingdom, most traumatic amputations occurred as a result of industrial accidents [31]. Major upper limb amputations (excluding thumb and fingers) were about 15% of all acquired major limb amputations in the US. In other countries the figures were 3% in Denmark, 17% in the UK, and 5%-6% in Australia [32]. Roughly ten times more prosthetic legs are fitted than arms.

A number of factors influence the kinds of upper limb prostheses that gain acceptance by users. These include the amputation level, what muscles are available in the remnant limb and how easily the prospective user can control them, user needs (functional, recreational and vocational), availability of health care resources, and insurance coverage [33]. Cultural factors are important. For instance, roughly three quarters of upper limb prosthesis users in the United States preferred hooks as a terminal device; whereas in three European countries this percentage was lower, varying between an eighth and a third of the user population. This is confirmed by the buying preferences observed by US prosthetics manufacturers, TRS [28] and Motion Control [34].

More body powered upper limb prostheses are fitted than powered ones [35]. Advantages of body powered prostheses are, in general, they are lighter, more durable during heavy work, and less susceptible to damage by environmental exposure (mechanical, electrical, chemical). They cause users fewer skin problems, have less heat build-up in the socket, suffer fewer technical failures, have a quicker response time, offer more choice of interchangeable terminal devices, are easier to learn to use, may be fitted by a greater number of clinics and experts, and require no battery charging [36]. However, they may also be seen as ugly or old fashioned. Significantly, the cables can interfere with or damage the cuffs of shirts and are generally mechanically inefficient, requiring large forces at the arm to convert into small forces at the finger tips [3]. While body powered prostheses have lower up front and maintenance costs, surveys of users found that myoelectric hands are preferred for light manual tasks, office work, activities of daily living and social occasion. Body powered prostheses are preferred for heavier jobs and farm work [29]. For more than two generations, powered prostheses have enjoyed far greater interest by companies and researchers [37].

2.6 Surgery

Some surgeons see heroic intervention as an important part of their response to injury and disease [26]. They feel they must employ some sort of surgery to reconstruct the limb, lengthen it, take parts from another area of the body and make something that the person can use. Surgeons may do as much as they can to save the limb, but if some of it is lost, they may adapt or alter the residual limb surgically to try and help. There have been a number of surgical solutions proposed to firmly mount a socket on a shortened humerus. In the Marquardt procedure (named after the surgeon who perfected the approach) a piece of bone is attached to the end of the humerus at an angle to the main bone, creating an extension off to the side of the humerus. This gives the bone a nonuniform, nontapered shape, that provides a surface to fix the socket stopping unwanted rotation [38]. Similarly, a second technique inserts a metal implant into the end of the bone (in a similar fashion to an artificial hip prosthesis) which creates artificial condyles for prosthetic stabilisation [39].

An obvious solution is to anchor the prosthesis directly the bone. To do this the surgeon must insert a peg into the bone, which then passes out through the skin, so that the prosthesis can hang off it. This was the ideal sought for many years before Per-Ingvar Branemark found a solution to the problem (see Section 11.3). Unfortunately by passing through skin, the implant creates a path for infection to travel into the bone. This must be prevented, if further amputation is to be avoided. Hence perfection and adoption of the technique has taken a long time.

When a person is injured sufficiently to warrant amputation, a surgeon is needed. While the early days of surgery, especially orthopaedics, were dominated by amputation, today surgeons would rather find a way to rescue the limb, or restore as much of the function as possible, and will go to enormous lengths to try to save the limb. Indeed, I suspect (with some anecdotal evidence) that surgeons do not care for prosthetic limbs, because it reminds them of their failures, when they had to give up and amputate rather than save. Amputation also passes the control to others (prosthetists and engineers).

The second medical tactic of 'watch and wait' [26] has a long pedigree. Here the doctor will do what they can to stabilise the patient and wait for natural processes to take effect (such as post meningitis, Section 2.4).

2.6.1 Reconstruction surgery

While the majority of surgical interventions are restricted to amputate and then 'tidy up', every so often a surgeon comes up with an idea to help the population through more adventurous surgical interventions. The Kruckenberg cineplasty is an extreme intervention [40]. It creates a sort of pincer from the remnant limb. The radius and ulna bones in the fore arm are split apart and shaped like pincer tips, then sensitive skin from the arms and hands that has survived the trauma is put over the areas towards the tips and inner surfaces of the bones. What results are two very large 'fingers'. This is a substantially disfiguring intervention, but persons with it can achieve a high level of dexterity. It is considered a particularly useful procedure for someone with a bilateral loss and impaired vision. While rare, enthusiasts still promote it for lower income countries. Intellectually it can be seen to be a pragmatic response to a difficult situation, however, emotionally the surgery still looks like the most outrageous form of mutilation and the poverty of the user is not an excuse. Given its small population it is unlikely to have much attention from research. Conceivably, the right design of socket could be designed to hide the arm while using the sensation from the joints at the base of the bones and on the inner surfaces of the bones to help drive and control a custom prosthesis. This in turn would create a novel and effective way to control a prosthesis.

Another surgical intervention is tunnel cineplasty [41]. In this procedure, a loop is made in the end of the muscle following amputation. This loop can be used to pull on a cable and actuate a body powered prosthesis. This is very effective since all the subtle internal feedback paths that exist within the muscle are retained. A person can feel when the prosthesis is opening and how much force they are using. If a pair of opposing muscles are used, the person can drive the hand open and closed themselves without relying on a spring for one direction as they would in a conventional body powered device. This makes for very good and fast control. Persons with this cineplasty can operate the prosthesis with fewer attendant cables required for a typical body powered hand. As the person is less encumbered by the socket, they can operate the hand behind their back or above their heads which is difficult or impossible with a conventional cable prosthesis. Unfortunately it is also disfiguring. It has been suggested that the procedure was performed in the 1950s, as part of the initial amputation by surgeons who did not have the express, prior permission of the patient. If so, then it is hardly surprising that the patients did not care for the new body part and rejected its employment. Without care, these unnatural pockets can also become dirty or infected so a rejection of the procedure might lead to problems. Whatever the reason for its rejection, tunnel cineplasties are rarely performed. Dudley Childress tried to resurrect the idea, proposing much smaller loops in the skin to pull on cables and provide feedback for externally powered limbs.

More recently, Todd Kuiken of the Rehabilitation Institute of Chicago (RIC, now the Shirley Ryan AbilityLab), pioneered Targeted Muscle Reinnervation (TMR). This involves taking nerves that were used for controlling the missing parts of the body and placing the nerves in muscles that survive but now have nothing to move. This procedure provides more relevant muscular signals to control externally powered prostheses (see Section 12.2) and much more besides. This is a modern 'heroic' intervention that is gaining ground and is part of a larger trend towards body adaptation. Fundamentally, nerves are programmed to sit in muscles and tissue and transfer signals to and from the brain [42]. When the nerves are left without a home, any signal generated can be interpreted as a pain signal by the brain. Removing pain associated with the missing parts is a difficult and often an intractable problem. When cut nerves are left in the remnant limb, they form little balls or neuromas. If these connect with the skin tissue, they can then detect pressures on the skin, or other bodily changes, and send random signals to the brain, which can be considerable and disabling. To reduce this effect, the nerves are often cut further up to reduce or remove the pain. This does not always work. Kuiken maintains that once the cut nerves have been attached to a muscle they will settle down and work more predictably. Thus it is the *"healthy way"* to handle such nerves [42]. This is a most powerful argument for using TMR.

In the last few years different variations in the treatment of nerves and the remnant limb have reinvigorated surgeons to want to take a much greater part in the post amputation treatment of limb loss. Regenerative Peripheral Nerve Interfaces (RPNI) are a technique where small sections of muscles are wrapped around the nerve to create small muscular elements similar to TMR. While TMR requires the target muscle to be denervated before connecting the nerve. RPNI takes small sections of muscle away from its original nerves and moves the section of muscle to the nerve. The result can be the same, with a nerve settled in a muscle not creating painful signals, and providing new EMG signals for prosthesis control.

One of the most high-profile amputees of recent years is the engineer Hugh Herr. He was already at MIT when he lost both feet in a climbing accident. As a result he changed his focus to prosthetics. His first design was a knee with adaptive damping to compete with the existing commercial knees. Only later has he moved to feet and ankles where he personally would benefit. His most significant insight was about postamputation surgery, here the ends of the remaining muscles are typically fixed to the bones so the muscles cannot shorten when they contract. For a foot amputation, the muscles that flex and extend the foot (tibialis anterior and soleus) are thus fixed; they cannot shorten when contracted. This feels to the amputee as if their foot is fixed in space, much as in a ski boot that cannot

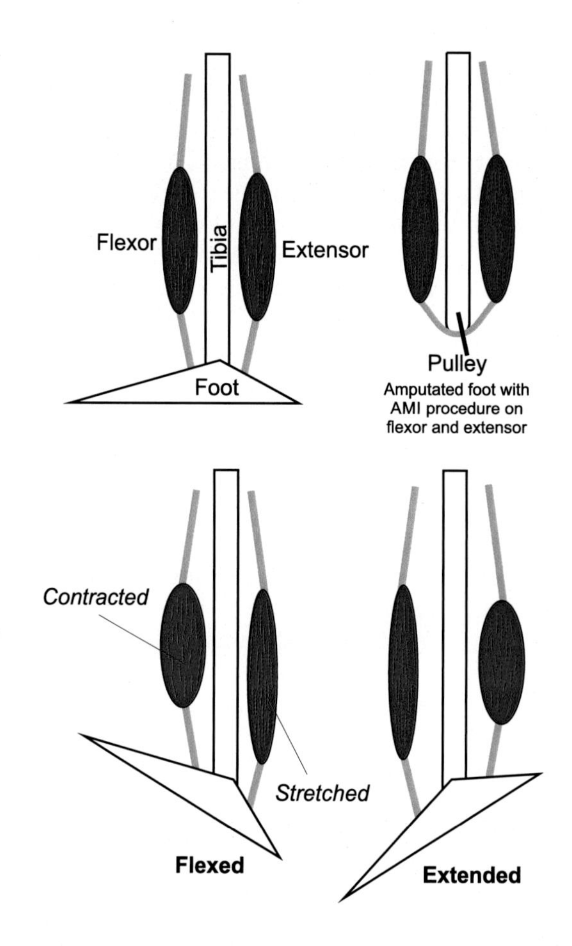

FIGURE 2.7

Most movement in the body uses opposing pairs of muscles (flexors and extensors). Following an amputation, instead of fixing flexors and extensors that controlled the next limb segment, (such as the foot) to the remnant bone, the Agonist/Antagonist Myoneural Interface (AMI) attaches the muscles across the end of the remnant limb so that they retain in an arrangement that stretches one when the other contracts. Restoring a more natural length relationship between the muscles.

flex or extend. This can cause pain and cramping. In a typically developed leg, the two muscles are linked to each other via the foot, so if one shortens to raise the foot the other lengthens and vice versa. Herr's insight was to ask the surgeon to use a tendon removed from the damaged ankle and reuse it as a pulley, linking the two muscles together (Myodesis) so that tension in one, causes lengthening in the other. Not only does the Agonist/Antagonist Myoneural Interface (AMI) free the foot from its virtual boot, but this also creates the most natural control input for a powered ankle [43] (Fig. 2.7). The first recipient of this procedure was able to control a prosthetic ankle more naturally than with a conventional myoelectric controller and was recorded in an idle moment bouncing his prosthetic toes up and down, much like a natural foot. It is early days and more subjects and more tests are needed, but its appeal to intuition is persuasive.

We must be cautious of the aims of a surgeon. Bob Radocy (a prosthesis user) reports that early in his prosthetic career he witnessed a surgeon seriously suggesting that he should take off an extra few inches of arm to make space for a fitting a battery *"It made me shudder... I knew how valuable ever inch of your forearm was"* for function. *"[It was] everything I could do to stay seated and not jump up and start screaming"* Today: *"I would stand up and say something now I might regret"* [44]. He hoped no surgeon would suggest that today.

2.6.2 Arm transplants

Probably the most heroic of the surgical interventions is to attach a different living arm to a person. While it seems to be a completely wild and unexpected idea, there is actually a logical progression from reattaching the severed limb (autografts) to hand/arm transplants (allografts) [45].

All of this is reconstructive surgery, when the surgeon attempts to overcome the injury by reconstructing the person with parts taken from other areas on their body. The earliest such procedures were skin grafts, using skin from one part of the body to repair burns and other injuries at other parts of the body. Often there was not enough skin to cover the space, so a balloon is placed under the donor skin and inflated, causing the skin to grow; the balloon is inflated repeatedly until enough skin has been grown. Early pioneers working on badly burned pilots in the second world war attached skin from one part of the torso to the badly burned face, keeping the old blood supply. The skin could grow on its new area and a new blood supply was established. Then the old links were severed and the surgeon could continue the long process of reconstructing the face.

For the hand, grasp is most important, so a digit to oppose the fingers is needed when the thumb is lost. The large toe can be used to replace the thumb, or the index finger can be moved to oppose the middle finger. If the damage is not too bad and if the surgeon is skilled, the resulting hand is extremely functional and almost completely invisible to the general public. Intuitively, we know this to be true, since many cartoonists make this same compromise, giving Mickey Mouse and Homer Simpson only three fingers and a thumb, so most people thus repaired will generally go about their business unnoticed.

The best clinical outcomes for reattachment come from a cleanly severed limb that is kept cold for long enough to allow the surgeon to reattach it. Structurally, there is no difference between this, and taking a healthy hand from a deceased donor and transferring it to the injured site, thus autografts (moving and reattaching a person's own limbs) become allografts (attaching parts from a donor). The one significant difference is the need for immunosuppression to stop the host from rejecting the new hand. Rejection is where the body recognises the alien tissue and destroys it. This is part of the way we defend ourselves from invaders such as bacteria and viruses. Immunosuppression means that the body's reactions are damped down chemically. This is the most difficult medical aspect of this procedure, because the other defensive measures are muted too, meaning that the body will not defend itself from infection and perhaps even cancer. Anyone on immunosuppressants is likely to be sicker more often.

It is more possible to understand and control the body's reaction to a single type of transplanted cell (such as blood), which is one of the reasons why blood donation is so well established as a routine technique: get the right blood type and everything works well. Organs like hearts and hands have many different types of cells and therefore many different possible reactions to the stranger's presence. Other sorts of complex allografts have been used for decades, such as kidney, heart and liver transplants. However, these organs still have only a few types of different cells. Arms have a wide range of tissue types, from skin to nerve to bone and nail. The other difference from internal organ donations is that these are the last option for a person before complete failure of the organ will kill them, the risks of immunosuppression are easier to justify. A great body of knowledge about the effect of immunosuppression on the host in kidney and other internal organ transplants exist. Even so, little data is available on what impact immunosuppression has on the long term life expectancy of a kidney recipient. Life expectancy in these patients remains regrettably quite short and no change in life expectancy has so far been observed. Since all kidney recipients are very sick when they get a transplant, it is hard to disentangle cause from effect. Does the new kidney extend life, while the drugs shorten it again? Or were all of their other systems already damaged by the poorly functioning kidney not extracting toxins from the blood? Are they closer to dying from organ failure than a person with a missing limb? For transplanting fingers or hands, few recipients will be anything like as ill. How much shorter (if any) will their lifespan be once they get their new hand? We still do not know, and will not for some time as the data needs to be collected as people live with their new hands.

We do have many years of information on what happens when the host and donor cells combine [46]. The immune system is complex and one possible side effect is not the host rejecting the implant, but in the spirit of 'man bites dog', the graft may reject the host. Lessons learned from experiments and experience with earlier grafts taught us that if bone marrow from the donor is transferred along with the hand, this form of rejection is very much reduced. Even if it happens, compared with heart or kidney rejection, life should be easier. In an ideal world, the use of drugs would be avoided or kept to as small a dose as possible. The drugs would be administered only when rejection is taking place. Once the body has settled down to the graft, the drugs would only be required to start again if the rejection returned, keeping doses to a minimum. With a heart or kidney allograft, the organ is buried deep in the body and rejection is only visible when the donor organ fails altogether, which may be far too late. To protect kidney patients, physicians tend to prescribe large does of suppressants for a long time. For a hand recipient the skin of the hand is an indicator of the success or failure of the suppression. The team can keep lowering the amount of drugs taken until any rejection symptoms show up on the skin, or only reapply the drugs when rejection is happening. This should reduce doses considerably and if immunosuppressants are a factor in shortening life, it should lengthen the recipient's life expectancy.

The number of people who have had an external transplant is still quite small. A person with a single side loss is not particularly disabled it, and in the past professional opinion was that the risks associated with immunosuppressants were too great to ethically justify a single side replacement. The first recipient of a replacement arm was Clint Hallam in 1998, an Australian who had lost his hand in 1984 [47]. Hallam was also an interesting choice as a patient, as it was reported he was in prison serving a sentence for fraud when he lost his hand [47,48]. The operation did not work out well for Hallam. According to the Austrian allograft surgeon Milomir Nincovic, Hallam returned home to Australia after the transplant. As he had undergone the procedure abroad, he was unable get his immunosuppressants on the state health system, and could not afford the considerable costs himself. The rejection process started and in the end he persuaded a surgeon to amputate the new hand [49]. Subsequent grafts on other individuals have been much more successful.

An important aspect of a transplant is psychological. The hands are one of the most personal parts of the body. While the face is important to the rest of the world, it is the hands the person themselves sees all the time. For decades, popular culture has fixated on what a transplanted hand might have done in its previous existence. Lurid stories abound of murderers' hands continuing their grizzly ways after transplantation [50]. While there is no evidence for this in the real recipients, Hallam reported that he had problems getting used to a stranger's hand at the end of one of his arms. At least two responses can be made to this concern. The first is the one adopted by Nincovic's team: They perform only double amputations and the hands must come from the same donor [46,51]. The result is that *the hands match each other*. A second approach is exemplified by the team at St James' Hospital in Leeds, UK. They have a very comprehensive set of protocols. The medical team will attempt to match skin tone and appearance of the donor hand to the other side (the contralateral), so the hands look as if they belong together. This means that this team can perform unilateral grafts.

Whichever approach is taken, the follow up physio and occupational therapy is extensive [51]. Many months are required to recover adequate function when a severed hand is reattached. The nerves must regrow within the hand and allow the control and feeling to realign. Even in a reattachment of a person's own hand, the nerves are not restored to the same routing as before. It is not uncommon for the nerves to go to a different part of the hand, so initially the patient may feel the bottom of their hand when the top is stimulated. The mind remains plastic, so once they have consistent stimulations (vision and nerves), the recipient quickly adapts to the new sensation and soon the feelings come from the correct places.

Nerves require time to regrow, and as they do more control and more feeling returns. The process can take several years. The nerves have to grow from the reattachment site down to the finger tips. Given how slowly the nerves regrow (roughly a millimetre a day), it is easy to see why, so far, allografts have generally been at the wrist level.

When recipients' reactions are reported, they are generally very pleased with their new hands and do not want to go back to the prosthetic solution. However, not being able to use their prostheses in the first place is one of the indicators for performing an allograft.

The entire process of an allograft takes many years and a large budget. The commitment from the patient is considerable, with many years of exercises and occupational therapy. It also requires a determined attitude from the recipient. There are many problems associated with the technique. For example, the team may never find the appropriate donor. Even with improved techniques, the surgery requires at least two full surgical teams working at the same time, and there is no way to encourage nerves to grow any faster. This procedure, like other more radical interventions, will only be one of a number of possible solutions for persons who have suffered an amputation, and is unlikely to become the routine solution.

When the user of a double allograft was presented at a conference alongside the user of a pair of multifunction hands, it was clear that neither had a noticeable functional advantage. It was also clear that they both had extensive therapy and put in a considerable amount of time in practice. When asked, neither would swap with the other.

2.6.3 Amputation surgery

The ideal relationship between prosthetist and surgeon is one where the surgeon takes the prosthetist's suggestions into consideration when planning the amputation, if circumstances allow. This permits the remnant limb to be set up for as easy and comfortable socket design as possible. Of course the injury itself can easily make many options impossible. Residuum length is important. A long stump makes for shorter prostheses and lower forces on the skin. If too long, the stump will intrude into the area needed for the next prosthetic joint, making the prosthetic design more complex. In this case the prosthesis will either have the joint in a nonanatomical place (for instance, an elbow below where the elbow is on the other side), or no joint at all. If the radius and ulna are removed, but the humerus remains intact then there is no space for the elbow. One solution is to place external hinges around the elbow area and hang the prosthesis off the humerus; this uses the increased width of the end caused by the condyles (the joint surfaces on either end of the bone) to hold the prosthesis on. An alternative solution is to shorten the humerus: A surgeon can take a few centimetres from the middle of the bone and place the condyles further up the arm, making the upper arm shorter, but keeping the nontapered shape to fix the prosthesis to. This sounds quite extreme, but given the right circumstances this can be a straightforward procedure.

The surgeon may have to shape the end of the bone so that it does not stick into the flesh that wraps around the end of the limb. Also the bone must support the weight placed upon it without the skin breaking down. For prosthetic legs, if the loss is through the knee, the surgeon may take a few centimetres out of the femur (the bone above the joint) and then put the condyles back. The resulting prosthesis has space for a prosthetic knee and the patient stands on their condyles (as they did before the amputation) and not on bone ends.

Following amputation surgery, the patient will need time to recover and build up their strength. Also the wounds will need to heal and swellings to come down before the person can wear a definitive prosthesis. Once they can, it is time for the physiotherapist to help the recovery of mobility and the occupational therapist to help them learn how to use their prosthetic limb.

One effect of blasts and other blunt trauma to muscles is that they can build up calcium within their structure. This is known as Heterotopic Ossification (HO) [52], and it results in sharp spikes of bone forming within the muscle that can cause pain if pressed. While this phenomenon can happen any time there is trauma to muscle (such as a blow on the forearm from a bat). OH is more common when an amputation is caused by a bomb blast. In the conflicts in Iraq and Afghanistan the use of land mines by local forces has been common. Improvements in military body armour mean that many more soldiers than previously have survived the mine blasts [53], suffering extensive limb injuries. With this has come considerable amounts of HO that grows in the patients' muscles after they have partly recovered from the trauma and are beginning to undergo rehabilitation with their first prostheses. At this point the surgeon may have to return to the wound and remove the HO, which slows down the recovery considerably.

2.7 Supply and provision

Successful prosthetic limbs require the production, distribution and fitting of the prosthetic components to the user population. Manufacturers create the basic components and these are combined in clinics of prosthetists and therapists to produce an appropriate assistive device for the individual. The clinic can be part of a state health system or a company that might be contracted by the state (nationally or locally) to run the clinic. The company that holds the contract may be part of a prosthetic manufacturing company. This in turn creates challenges to avoid conflicts of interest between the supplier and the clinician making decisions on which components to use. Small local clinics comprised of a small number of clinicians can specialise in types of prosthesis (upper limb or sports) or patient groups (paediatrics).

2.7.1 The origins of clinical prosthetics

As a profession, the title of 'prosthetist' emerges after the second world war. Before this time prosthetists were referred to as 'limb fitters'. This form evolved after the First World War as standardised components allowed a more routine fitting process. Limb making was a craft based occupation working in local independent centres. It was only later was the prosthetics profession formalised [54]. Fabrication centres made the socket to match the residual limb shape, often made from leather moulded to a plaster cast taken by the limb fitter. The arm was assembled by a team with individual skills, building the structure to match the person and integrating standardised elements such as wrists, elbow units, harnesses and operating systems. For upper limbs, a range of devices for occupational activity and perhaps a hand for 'best' together with a pair of leather gloves were supplied. Usually the hand was passive and made from a lightweight linen and doped laminate, sometimes with a spring loaded or cable controlled thumb. For function, a gripping split hook with one fixed and one moving jaw was supplied. Users would be trained to be proficient in specific equipment for their occupations and recreation and for potential employment by an experienced user. Later this role was taken by occupational therapists in a specialist rehabilitation unit within a hospital. On discharge a user would often be provided with a small case containing a selection of interchangeable tools [55].

2.7.2 The prosthetic clinic

When a potential wearer attends a modern prosthetic clinic they are assessed by the clinical team, who will identify and manufacture the prosthesis that most closely matches the person's needs. The form of the clinic has changed over the years and varies in different countries. From the 1950s onwards there has been more of a tendency to house the limb centres on the site of hospitals, especially when the provision is paid by health insurance or state healthcare and requires a physician to prescribe the work.

In these forms of healthcare, the device is fitted within the medical system and it is therefore prescribed by the physician and fitted by the prosthetist, with final approval from the medic. The tendency has been to move towards a multidisciplinary approach (proposed as far back as 1954 by Klopsteg and Wilson [56]). The team provides a much richer range of knowledge including the physician, the prosthetist and the therapists. The clinic should have access to a psychologist to help people who have undergone an amputation recover from the trauma. There are other members of the team who can provide assistance where appropriate, engineers, and other professionals who are concerned with the person's welfare. For the person with an upper limb absence, the key partnership is between them the prosthetist and the occupational therapist (OT).

The design of the socket matches the prosthesis to the wearer. The socket must stay on and not move about the remnant limb when it is being used, but should not be too tight or too hot to wear. This is a difficult task. Historically sockets were made of wood and were heavy. Later they were fashioned of metal. Most recently plastics and rubber sleeves have taken over. However, as I can attest, as late as the 1990s the dominant sound emerging from the Oxford prosthetics workshop was of metal sockets being adjusted. This was because the World War Two veterans retained their choices for the rest of their lives. This brings up one of the important factors in the progress of new technology, especially technology that is to be used by people for a long time. Once a person has worked out how to cope with their difference, they will adopt strategies that can allow them to function well. The strategies often require some forethought and practice. Once users have found a good solution, they become disinclined to give up this hard won knowledge and adopt a new device or find a new approach. This makes newer devices less readily adopted by established users than, say, in the smart phone market, where the novelty of the system and the requirement for learning to adapt to continual changes in operation seems to be welcomed. The difference perhaps is that a person with an amputation, struggling with a new prosthesis, is reminded of their 'handicap'. While the early adopter of new consumer technology might see the handicap of learning another new operating system as a reminder to all how cutting edge they are. When World War Two veterans stuck with their tin legs and did not to want to change, they were not being especially difficult, or unusual, *simply pragmatic*.

Sockets may fit the person closely, but many materials will not allow the skin underneath to breathe efficiently. Humans lose heat and moisture across all their skin, and the loss of a limb means a loss of some of that surface. Adding a prosthetic socket over more of the skin substantially reduces the surface area still further. The result is overheating, sweating within the socket, discomfort and sores. The practicalities of keeping the socket clean when the user might sweat in it is also a constant problem. New materials that can conduct heat away more readily are being developed to reduce these problems [57].

2.7.2.1 Prosthetist

Becoming a prosthetist requires training and practice. The usual approach is degree program of several years of lectures interspaced with practical training and internships in working clinics. Following this period, there is additional practice in a clinic or clinics to gain experience under the guidance of an experienced professional before the person is fully qualified. There is an internationally recognised form of program [58], but not every school, even in the affluent world, subscribes to this set of standards.

The requirements to be a prosthetist are craft based, in the sense that the practitioner needs skill to make moulds of the remnant limb, then manually adapt the mould to the correct shape and then make the socket from one of several different materials. Next the prosthetist builds up the socket to mount the commercial mechanical components on it and shape the outer dimensions to position the terminal device in the right place for functional use. The prosthetist incorporates controls, such as cables or myoelectric amplifiers (Fig. 2.8). Finally, the device is made to look pleasing to the user. Fitting a limb is a highly skilled and painstaking process. Additionally, prosthetists must have a good three-dimensional sense and be technically inclined, so as to adapt the commercial components to the particular user. Every user is different, and the challenge of the job is to find the right solution for that person. This means identifying ways to operate the prosthesis that is not tiring and ensuring that if there are two controls (operating a hand and an elbow), one action will not interfere with another, such as, the hand not opening as the elbow moves. This clearly comes with knowledge and experience. Historically, prosthetists needed to be skilled in working with leather, metal and wood. These days many plastics, rubbers composite materials and computer controllers need to be mastered.

Like other aspects of the field, socket manufacture is being impacted by advanced manufacturing techniques. It is possible to scan the residuum and use direct manufacturing techniques, like 3D printing, to create the mould or the socket itself. There are some limitations in that the flesh on the end of the limb compresses or shifts when it is stood on or leant on by the user. If provision for shape changes is not built into the socket the prosthesis can be unwearable. This is part of the 'feel' of being an experienced prosthetist. It is much harder to incorporate into a computer program, although progress in this direction is being made.

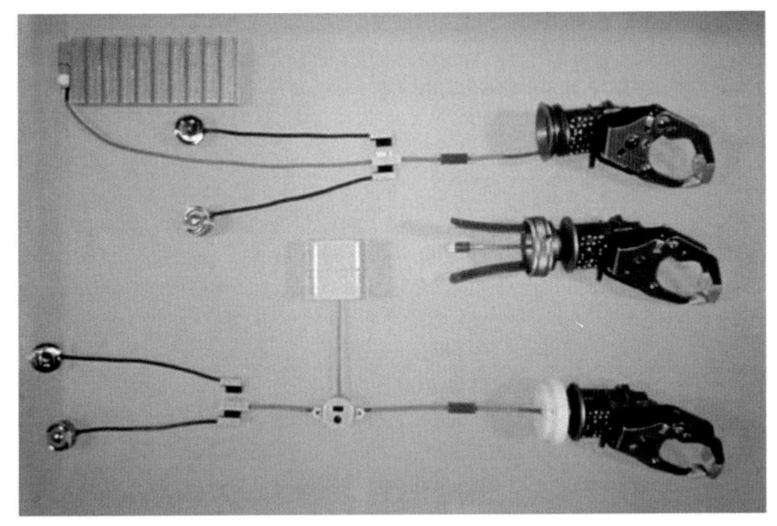

(a) Otto Bock electric components in 1972

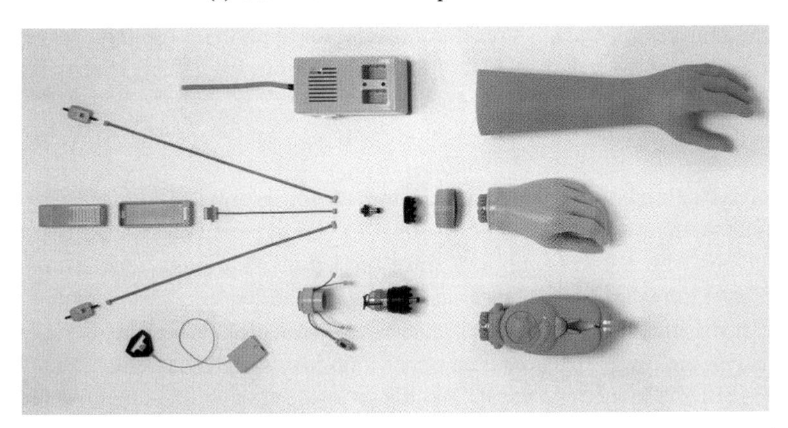

(b) Otto Bock electric components in 1990, now include Quick Disconnect Wrist, powered wrist and other terminal devices

FIGURE 2.8

The prosthetist has to integrate into the socket the wiring for the electrodes, batteries and the terminal device. Over the years the job as become more complex as the range and sophistication of the devices has increased. (Copyright: Ottobock SE & Co. KGaA, Duderstadt.)

The usual process to obtain a prosthesis is to go to a clinic and see the prosthetist. In many countries users do not typically buy direct from product manufacturers. Selection and fitting is led by the prosthetist at the clinic, so prosthetists have an important influence on device sales. One way a company can increase their revenue is to operate the service at hospitals. Many companies have a clinical service, with contracts to supply and fit prostheses in the clinic, which means the in-house prosthetists at a hospital are under contract to the company. From this position the company can gain an idea of the needs of the user population and a ready supply of interested users to trial their latest devices, whilst carefully maintaining a distance from any conflict of interest in which manufactures' products they fit.

2.7.2.2 Occupational therapist

In the past, occupational therapy taught skills to enable a patient to function in society through the use of basic tasks. Occupational therapy has moved on from a time when the therapy for those injured or impaired was to teach simple occupations so they could be 'useful to society'. Now the role is to help them function and be independent individuals. Any part of the person's life can be studied and they can be guided to improve their methods. The prosthetic OT helps the user get the most out of their device, and to learn to work with and without their prosthesis.

An OT will work with the user to help them find solutions for day-to-day living. Such things are not obvious and do not come naturally to every person. Solutions may be similar to the way a person uses their natural hand, but sometimes are not. A different approach might be more efficient and involve actions less likely to cause overuse injuries. Alone, a person might take a long time to find the best ways to achieve a task. The therapist passes on knowledge, skills and experience. The OT will also make sure that the person does not develop habits that will introduce problems later on, such as overuse injuries, by bending in an unusual way, always reaching too far, or applying large forces with the rest of the body. If a person does not work with an OT, or is unable to learn from them, they will have to learn by themselves and may suffer injuries as a result.

2.7.2.3 Physiotherapist

A physiotherapist (PT) might work with a person with an upper limb loss after surgery to restore range of motion, muscle strength, etcetera. After this time the OT typically takes over therapy. PTs remain essential for persons with a lower limb loss since the leg user's routine activity is very much more physical and relates to endurance, balance and strength. The two professions should work together seamlessly, but management decisions often make them rivals for funding and employment. For example, there many more lower absences then upper limb losses. In a busy clinic with limited funding, an OT might be seen as a luxury and the PT's skills are close enough for some managers to justify the reduction in staff by one. This economy has been suggested as an effective compromise on numerous occasions, but it clearly could adversely affect the service. The proposed cut is invariably in favour of the PT, making (at times) occupational therapy the more beleaguered profession, and more defensive of their position. The financial saving may be at the cost of injuries later on.

2.7.3 Prosthetic manufactures

The other part of the provision of prosthetic limbs is the manufacturers. Prosthetics is a small market, which gives anyone who would be interested in creating a company in the field to pause. Some are companies which were founded by an amputee with an idea, or created to exploit the result of a research and development project, or to fill a perceived gap in the market. Whichever the reason it is clear that there will never be a huge market in artificial hands alone. This tends to make a friendly atmosphere at conferences and shows. There are times when the different players have to behave in what they think is the best interests of their company, but they will generally respect the work of their rivals.

The history of prosthetic manufacturers reflects the history of manufacturing in general. Prior to the Second World War, small workshops of 'limb fitters' provided prosthetics services to the local population. This would have been without support from any government or charity, thus the markets were small. Even after amputation became a less traumatic operation the market remained limited to the richer part of the population. The price could be reduced and volume of limbs could increase once techniques of mass production made the basic components standard (even though the final device is custom). A change occurred during the First World War, a large number of soldiers were injured and the companies that existed could not respond easily to the sudden demand. For example, in the UK many of the workers for the companies had volunteered to fight, although prosthetics was recognised as an important service and it became a 'reserved' occupation, i.e. they would not be conscripted to fight. Companies in the United States flourished at this time as their market and workforce was less effected by the war [54]. In the UK market, staff were reduced by being on active duty, and equipment reduced by shortages. The designs of the limbs had to change as a result of a wider range of persons requiring limbs being able to afford them. Previously, poorer labourers, would have not been able to afford limbs and so the requirement to create prostheses to undertake manual labour was not great. Following the mass injuries from the war there were many more men who needed devices to help them return to work. After the war, the British government provided limbs for their injured, but even there the class system prevailed with one device; the Carnes arm, being provided only for the higher ranks and thus became known as the 'Officers' arm' [59].

The impact of total war in the Second World War resulted in many shortages. The UK government took over the control of the production of essential supplies. This meant that in the UK, this control was needed for every aspect of life, from production of food to the supply of entertainment. Prosthetics was included, and the various companies that existed in Great Britain before the war, were put under state control for the duration. After cessation of hostilities the companies reasserted their independence. Hugh Steeper emerged as the supplier of arms and Chas Blatchfords focused on prosthetic legs (although before the war they had supplied both).

State provision of healthcare has been a political selling point originally used by Bismarck, Chancellor in Germany, *before* the First World War [26]. The impact of supplying prostheses for no cost to the user had a significant impact on the market in the UK and elsewhere. Not only did the proportion of men wearing prosthetic limbs grow, but the market grew specifically to now include women and children in a way that had not happened before.

Japan remains an interesting example of a different cultural response. There is no evidence for prostheses in Japan ahead of contact with Europe and much of the surgical and amputation information came from Japanese translations of books by Europeans such as Ambroise Paré [60]. According to Takechi, most prostheses seen historically in Japan have been based on European designs. In the modern era Japan took up myoelectric prostheses much later than other wealthy nations, although a small number of campaigners have been working hard to rectify this situation.

In 2015 there were six established manufacturers of upper limb myoelectric control components and a number of smaller companies, some of which had recently been formed. Within three years this landscape had changed. Two had been absorbed by other prosthetic manufacturers. This is not a particularly unique phase in the history of prosthetic manufacturers, this pattern has occurred many times.

- **Otto Bock** of Germany, has designed myoelectric components and hands since the 1960s and has the largest market share of any of the companies (see Chapter 9). Its upper limb myoelectric group located in Vienna.
- **Hugh Steeper** of Leeds, UK, like Otto Bock, is a long-standing manufacturer of prosthetic limbs and other rehabilitation products. The company launched the bebionic multifunction hand, which was the second entrant into the recent market pioneered by the Touch Bionics. The bebionic hand (10.1.2) was sold to Otto Bock, and Steeper bought the Espire elbow from College Park (7.1.3.1), but still make simpler upper limb prosthetic components (see 10.1).
- **Liberating Technologies Incorporated** of Hopkinson, Massachusetts, USA, was a spin-off from Liberty Mutual Insurance. It was sold to College Park (at the time a lower limb specialist company). Their elbow was totally re-engineered and lunched in spring 2019. LTI only made products for the upper limb market, but was a third party seller of other manufacturers' products in North America (see 7.1.2). In the summer of 2021 Coapt acquired LTI.
- **Touch Bionics** of Livingstone, Scotland, was formed by David Gow to produce his novel design of powered fingers (ProDigits) in the form of the first commercial multifunction hand to be produced in numbers since the ES Hand (successor of the Sven Hand 11.2) in the 1980s (8.2). The company was bought by the Icelandic company Össur in 2016, (see 8.2.2), who later acquired College Park in 2020.
- **Fillauer LLC** based in Chattanooga, Tennessee owns Motion Control and Therapeutic Recreational Systems (TRS). *Motion Control* is based in Salt Lake City, Utah. It began development of its Utah Arm in the early 1970s and launched the system in the early 1980s (7.2). Motion Control began with one product, the second commercial powered elbow, but has since added hands prehensors and places an emphasis on the robustness of its products, having the first waterproof arm components. *TRS* is the creation of Bob Radocy and specialises in devices to assist in activities, especially sports. It generally finds a different way to solve the problems to other manufacturers.
- **Centri** of Sollentuna, Sweden was a company that manufactured rubber products. They were brought into prosthetics by the Sven hand project (11.2) to produce gloves for the multifunction hand. While the Sven hand was never a product, the ES hand was the first commercial multifunction hand. Centri went on to produce prosthetic mechanisms for adults and children.

Changes in the prosthetics industry since 2014 have included the appearance of a number of new companies and some mergers and acquisitions. In 2015 College Park expanded from being exclusively the provider of prosthetic legs by buying LTI. The new company produced a completely new design of elbow to replace the Boston Arm, before themselves being bought out in 2020 by Össur. Össur had also started by making lower limb prostheses only, they too moved into arm prosthetics by buying and upper limb company, TouchBionics in 2016. During the later purchase of College Park, the US Federal Trade Commission ruled that the resulting company could not own both the Espire Arm (7.1.3.1) and the elbow that TouchBionics had been developing, so Össur sold the Espire to Steeper. Steeper already marketed the Espire's predecessor, the Boston Armin the UK, and many other College Park products.

The following now sell upper limb products and/or services.

- **Vincent systems** produce the smallest multifunction hand on the market, and have the only child's hand with that description. Based on work by engineers from Karlesruhe in Germany it started as a response to the Touch Bionics' ProDigit fingers which were too long to replace fingers lost at the proximal joint.

- **Coapt** is a spin off from the Rehabilitation Institute of Chicago (12.1). It produced the first commercially sold EMG pattern recognition system since the ES Hand from Sweden (11.2). It specialises in myoelectric controllers. Its products can control a range of the major manufacturers' components (see 14.1.2). In 2021 it acquired LTI from College Park.
- **Hy5** is a Norwegian company. Their hand has two motions and is hydraulically driven. It offers an intermediate step between the conventional single degree of freedom myoelectric hands and the newer multifunction hands (see 14.1.5).
- **Open Bionics** is based in Bristol, UK. This company produces hands for children using Additive Manufacture techniques. This allows Open Bionics to take the customisation of a prosthesis much further than conventional devices. They also aim to produce more affordable prostheses (see 14.1.4).
- **Taska Prosthetics** is a New Zealand based company. Started by Mat Jury in his garage, the Taska hand is a multifunction hand that is more water resistant than a conventional prosthesis (see 14.1.3).
- **COVVI** were formed by members of the team in Steeper who worked on the bebionic hand. They have produced a multifunction hand that exploits the experience gained from this process.

In North America a number of myoelectric component manufacturers have come and gone. Fidelity Electronics, Leaf Electronics, Veterans Administration-New York University, Variety Abilities Systems Incorporated and UNB all used to manufacture upper limb myoelectric products and components, but are no longer in the business.

2.8 The literature, media and prosthetics

An uneasy relationship between the media and prosthetics goes back many decades. It is not uncommon for the evil adversary in a fiction to be disfigured or disabled in such a way that their external impairment seems to reflect an internal malaise. This runs through popular literature, from Frankenstein's monster and Captain Hook to Batman's many adversaries and on to Darth Vader in Star Wars. While some aspects of this have run contra in the last four decades, with technology creating prosthetic supermen (Bionic Man and Iron Man), this feeling of moral insufficiency continues, although a study of the literature and films suggest there are a second set of stereotypes; the injured hero struggling against the adversity of injury or disfigurement [61]. The experience of people with limb loss is neither of these stereotypes. In the study above the authors suggest there a similar number of stories of crippled villains, and those of the 'supercrips' and 'heroes' living through inspirational adversity. Sometimes the authors observed, that a character was none of these stereotypes, just a person getting on with their life.

The media like to dwell on the injured as villain. In the UK, the radical Islamic cleric, Abu Hamza, was convicted of promoting interracial and religious hatred [62]. However, the British press dwelt on the fact that he had a bilateral absence and prosthetic replacements. Their rhetoric focused on his choice of hooks and that they were something that seemed to reflect his aggressive and militant aims, rather than that he chose the practicality of body powered devices. Perhaps they were the only devices that he could afford.

For those growing up with an atypical limb, the lack of genuine positive role models in the media can have an impact on their self belief and confidence. It is only in recent years that individuals such as

Alison Lapper, the artist with significant congenital limb absences, have begun to change the way that the public thinks of persons not within the range of what might be considered 'normal'.

The changes in attitudes in the general population to prosthetics and other aids following the high profiles the Paralympics have given disability might increase the public knowledge and acceptance of the disabled as a group, but it is a distorted view. This publicity emphasises those who can excel and the athletes are seen as heroic and worthy of praise. These champions are as rare and different to the general population as an able-bodied Olympian. Those who make it to any international competition are physically and mentally exceptional, whether they have four limbs or none.

All announcements of new medical technology need to be couched in the sort of circumspect language of academics that the press loathes. The press wants a pithy sound bite with no qualifiers and no mentions of the limitations, and we as researchers and innovators need to strenuously avoid such things. The term 'bionic' has been used freely in the press for decades to refer to basic myoelectric devices. The hype has increased expectations and subsequent disappointments from potential users. Contributors to this book have commented on the negative impact of media hype on the progress of medical technology. At the same time the wish for something better has at times been a spur to companies and governments to provide better devices.

All of this can be seen against the background of increasing awareness of disability and apparent tolerance of difference. However, progress is slow. Some physically atypical commentators dislike being held up as 'inspirational' simply for getting on with their lives [63].

Even in the most popular culture there is a gap. The Star Wars movie released in 2017 'The Last Jedi' featured the much-vaunted return of the character Luke Skywalker. In the series he had a sophisticated prosthesis fitted at the end of *The Empire Strikes Back*. The instalment prior to *The Last Jedi* had hints that Luke would return, with shots of an ungloved prosthetic hand flashed to the audience as clues for the cognoscenti to relish. However, when the character returned at the start of the new film, the mechanism of the hand was briefly animated on the actor (something very simple to do with modern digital techniques), and for the rest of the film the hand was either hidden from the camera or the character was wearing a black glove. In the final film of the series, released in 2019, again the character appears, but this time his right arm never appears out from his flowing cloak. It would have been simple and cheap to animate the prosthesis in many shots, without making a feature of the device. This would normalise its use, moving it from a feature, to something that was just part of the character. The signal the Disney corporation sent (intentional or otherwise) was that even with the most famous prosthetic hero of the modern age, it was still unacceptable to have their bionics on show. We still have some way to go.

CHAPTER

The hand

3

3.1 The anatomy of the hand

The structure of the human hand evolved from the earliest fish [11]. A fossil of one of these fish was discovered in July 2004, on Ellesmere Island (80 degrees north), in Canada. It was 375 million years old, and named *Tiktaalik* meaning 'large freshwater fish' in Inuktitut, the local aboriginal language. It was an intermediate stage between fish and tetrapod. That is, it sort of swam and sort of walked in the margins of the tide. Tiktaalik had a wrist structure which effectively allowed it do press-ups in the shallows. There it found a safe niche that allowed it to get away from the very many predators that likely abounded in the seas. Evolution had already laid down the pattern that almost all limbs follow today. As Neil Shubin puts it; *"if you follow the bones going down the arm from the shoulder to the fingers they all have the same pattern: one bone, two bones, lotsa blobs, fingers/toes"* [11]. These are the bones that became the humerus (upper arm - one bone), radius and ulnar (forearm - two bones), wrist bones (blobs) and fingers. The blobs allow the arm to bend at the wrist and let Tiktaalik do its push ups. By 365 million years ago another fish/tetrapod, Acanthostega, had rudimentary fingers, and by 275 million years the full arm/leg was present [11]. One consequence of this evolutionary path is that the motors for the natural hand (the muscles) are mostly in the forearm, not in the hand. This structure has always led to greater complexity for designers of prosthetic hands.

The flexibility that this evolution has given us with our hands is considerable. The evolution by hominids of the combination of forward-facing binocular vision combined with an opposable thumb define the transition to homo sapiens. The problem with this argument is that there are other animals with either or both of these traits; they can sometimes make tools and some have considerable manipulative skills, but they are not the same as us. The history of modern humans is full of unsuccessful attempts to define what is unique about homo sapiens. For instance, we aren't the only talkers. There are other creatures that can make and appreciate art. Some animals can tell lies and clearly know what is in another creature's mind, and some can recognise themselves in the mirror. All of these traits have at one time or another been ascribed as defining what makes humans unique. What separates humans from other species is not one trait, but that we combine all of these qualities, and we do all of them so much more than any other animal. So we handle more objects and make more tools than any other animal, and there are some things we can handle that an ape would find hard to hold.

Our complex hands allow us to manipulate with flexibility. When we find a limit using just our hands, we create tools to extend the capability further. This flexibility makes the replacement of all functions with a single prosthesis very hard. In any specific case, it is more effective to have a range of different prosthetic terminal devices matched to different tasks. Taken to the limit, this would require a separate specialist prosthesis for every task, which is clearly not practical. Instead, it is better to have a generalist manipulator, as the natural hand is.

Making Hands. https://doi.org/10.1016/B978-0-12-820544-0.00012-3

One important consideration when designing any machine is to know what that machine is to be used for. For the prosthesis designer, this is a problem because the hand is so capable; it is hard to know what it can't do. It is equally hard to know what a prosthesis will be expected to do. With so much capability the subtleties between one grip and another become a matter that people in the field discuss endlessly. The problems start when we simply try and define what grips the hand can form in order to recreate them in a prosthesis.

Napier [64] was not the first to try to systematically define the grips of the hand, but his classification was the most effective. He divided the grips into two grip forms and related those to the tasks the person was trying to complete. They were: *Precision* and *Power*. *Precision* means the grips where the tip of the thumb meets (opposes) the tips of the fingers. We cannot impart large forces on an object held like this, but we can make it move in amazingly precise ways: We write, paint and draw with the precision grip. *Power* grips are made when the fingers wrap round an object. The thumb may also wrap round the back of the fingers so that greater force can be brought to bear on the object. This grip also has a greater contact area between the hand and the object, which means that the hand can transmit the force more effectively. We hold a bat, a hammer or any handle this way.

Napier's classification works well as long as the focus is exclusively the task. However, if it is the shape of the object that is of interest, then Napier fails to capture the grip between the thumb tip and the side of the fingers (lateral grip). A number of other researchers have wanted instead to capture the full extent of the range of grasps the hand can produce. Cutkosky [65] was the most comprehensive. He starts with Napier's pair and then breaks down every form with a description of each of their shapes. This is its flaw; it catalogues every grip and small variation of the grip. It is like a Victorian naturalist, cataloguing a multitude of insects without any idea of their genetics or the niche they fill. The end result is a dry list of every possible species without knowing its environment. This approach appeals to the engineer, as it is comprehensive and clear, but it does not help the prosthesis designer as much as other ways to look at grasp classification. More recently, another comprehensive classification system has come out of studies conducted at Yale University in the USA [66] (see 3.4.5).

Thea Iberall [67] used the concept of 'opposition space' taken from Arbib [68]. She began with the same basic focus as Napier, but extended it. Her idea was that in any grasp there is always a 'virtual finger'. This can be one finger or a collection of digits or even a hard surface. The fingers apply forces opposite each other through the object to be gripped; holding and manipulating an object (prehension) requires at least two opposing forces, and friction on the gripping surfaces. If the fingers and object were perfectly slippery a minimum of seven forces in different directions would be needed to prevent the object from slipping out of the grasp. This is why holding a wet bar of soap can be hard if we grip it from opposite sides only.

Iberall models the hand as a palm area (rectangle) with simple fingers linked along one edge (Fig. 3.1). The fingers oppose forces from the fingers or the palm to grasp the object. She describes three basic primitives: *Pad Opposition, Palm Opposition, and Side Opposition,* which encompass Napier's grasps. Pad opposition is the pads of two or more fingers come towards each other on opposite sides of the object. *Palm opposition* occurs when the fingers heading towards the palm as in a fist or power grip and Side opposition is the tip of the thumb heading towards the side of the index finger. She also added the idea of an additional finger, (termed *Virtual Finger 3* or VF3) which is any combination of gripping surfaces, such as all the fingers together, the palm, or a desk the object is held against. All prehensile postures captured by Cutkosky are combinations of these primitives. Iberall's summary of

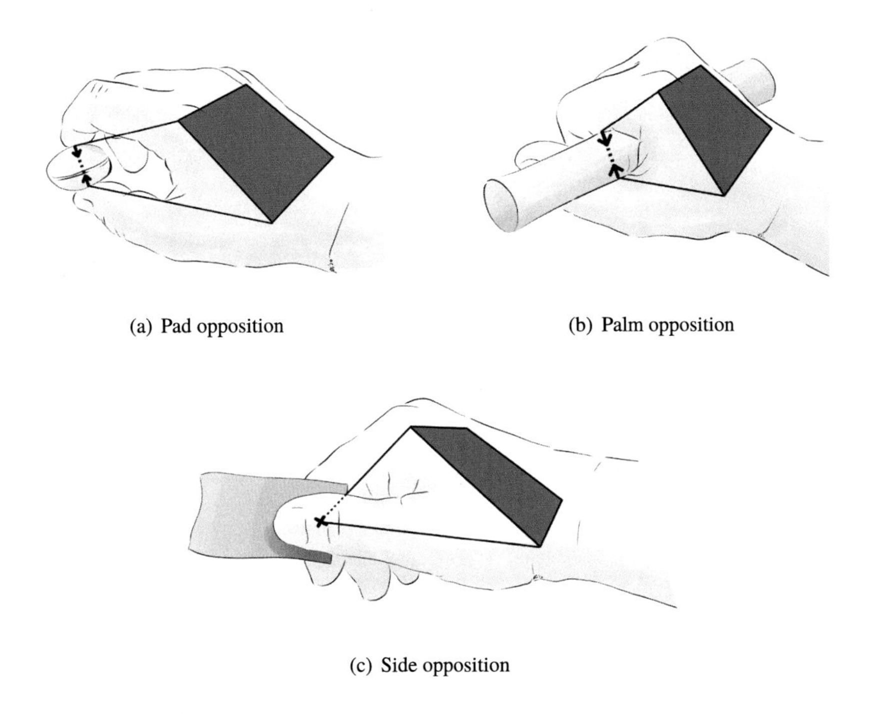

FIGURE 3.1

Iberall's 'opposition spaces'. (Copyright: D8 Stavdahl.)

postures identified by other researchers is shown in Fig. 3.2. Examples are the hook grasp, where the fingers act as a VF3 in opposition to gravity; and holding a tray, where the whole hand acts as a VF3.

This classification by Iberall is *practical*. It covers what the hand does and how it does it. Using the same analogy as before, it is a modern ecologist's description of the insect within its entire ecosystem: From it we could know how to manage an environment to help or wipe out an insect population; Iberall's classification can help us design hands.

Modern prosthetic manufacturers classify the grip patterns of their prostheses using an extension of Napier. Thus for tips opposition there can be one, two or more fingers involved. For the palmar opposition, the thumb wraps around a sphere or a bar.

In the race to show advantage for each new product, the manufacturers have attempted to show their hand is capable of more grips than their rivals. The implication that *more grips equals more function* has not been proven. Lottie Hermansson, one of the most influential occupational therapists of her generation, regards the menagerie of grips as distracting [22]. No single user can switch to every grip easily, or even remember how they can get to a particular grip, so that any individual uses a small number of the available grips. This is not really a surprise or too much of a drawback, as it is very much how the majority of people use the features on their car or computer or personal electronic device: They select only those features that they can use and ignore the functions that they cannot. The Otto Bock Michelangelo Hand has really only two grip forms: basic tips and lateral opposition. This does not stop

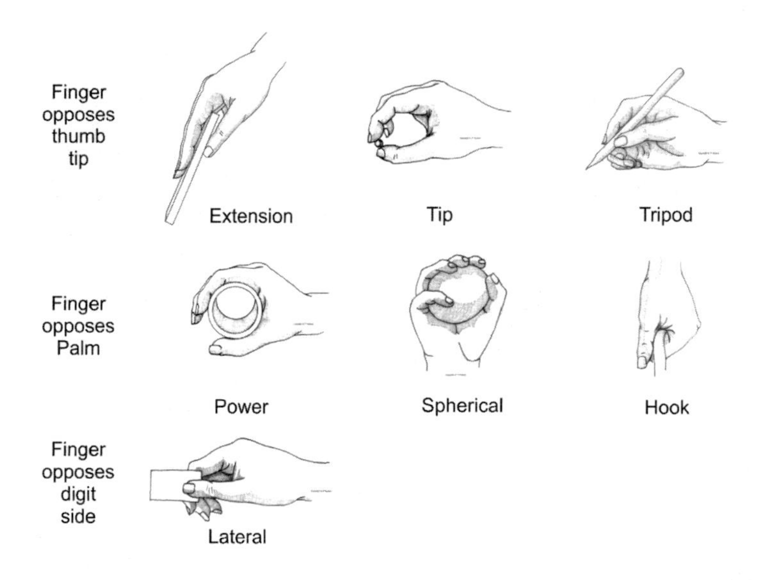

FIGURE 3.2

Hand grip patterns used by prosthetics manufacturers, mapped against Iberall's opposition spaces. (Copyright: W.P. Kyberd.)

Otto Bock claiming many more grip forms, nor does it stop it being a very capable terminal device. One variation they introduced is effectively Iberall's VP3. Never before had a manufacturer made a virtue of a flat hand being usable for holding trays, Otto Bock was the first to do so because the palmer surface of the Michelangelo is more anthropomorphic than other prostheses so it can hold a tray or cradle a bowl successfully.

One unexpected result of multifunction hands is that the lateral (side or key) grip is used a great deal more often with prosthetic hands than it is with natural hands. According to studies [69], in an average day, a person will use the lateral grip about 20% of the time. Reports from clinicians who have fitted multifunction hands suggest it is used much more often than this. This makes some sense. A thumb opposing the side of a finger is a single point bearing down on a flat surface. As such any larger object that is mostly flat, such as a wallet or a credit card, can be pinned against the flat area more securely than if it is held between the tips of the thumb and multiple fingers, which form an uneven platform. While it is not the way the majority hold that object it is a simple and easy way to do the job, and the ever practical prosthesis user will adopt it to good effect.

3.2 The control of the human arm

While there are similar numbers of joints and muscles in the arm and leg, they are used for very different purposes. Our nearest cousins, the great apes, can use their feet for manipulation, usually we do not. Walking is a complex task, but it is not as involved as manipulation. We can tell this by

FIGURE 3.3

Penfield's sensory homunculus each body part is proportional to the amount of the sensory cortex devoted to the processing of sensory information from the part. So the hands are far larger than the arms. (Creative commons licence.)

looking at the amount of the brain (cortex) given over to controlling the arms compared to the rest of the body. The best representation of this is Penfield's homunculi (Fig. 3.3). They are like the electoral maps that change the size of the land in proportion to the number of votes cast. The first homunculus is for motor control, and one can easily see that the hands have much more processing power devoted to their control than the legs and torso combined. The second map is of the sensory processing. It maps how many sensors are located in an area and how much processing is used to handle the data. The back is so insensitive that two points three centimetres apart feel like a single point, while in the finger tip two points three millimetres apart are still felt separately.

The most striking part of both homunculi is the size of lips and tongue. They are huge. This may seem disproportionate until one thinks about the way a baby explores the world. They grab an object with their hands, turn it over feeling its shape and surface, then it goes straight into the mouth. Not only will they feel with great acuity the shape and texture of the object, but they then get an extra dimension: Taste. This can tell them something about the object beyond their sight and touch.

The control of the hands and arms has evolved over the many different uses, from swimming to walking to climbing to holding to shaping. Different parts of the cortex perform different roles in the control of the hands. The results are the sorts of oddities that create 'illusions' and 'paradoxes' that psychologists and magicians like to exploit.

We have evolved to perform dexterous tasks with our hands. The complexity of controlling as many degrees of freedom as the hand and arm have, meant that some of the actions are normally not under our conscious control. If they were, we would not be able to do two operations at once (see 3.4.4), but would be stuck focused on one operation at a time. We would be much like a beginner at a new task (like skiing) who cannot see the world around them, and has no time to react to the changes in circumstance that the experienced person would see and avoid easily.

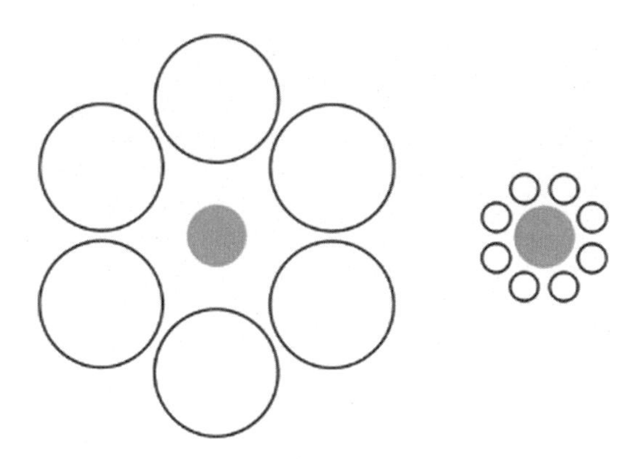

FIGURE 3.4

The Ebbinghaus illusion. The central discs appear to different sizes depending on the size of the discs around it, even though they are actually the same size.

One of the complex operations we perform effortlessly is reaching out for an object. The multiple digits form into the shape associated with the task we have in mind and open just enough to admit the object and then close on it. This happens very much in the same pattern every time. The margin the hand opens relative to the size of the object is roughly the same each time, irrespective of the object size. It is just enough to clear the object and a little more (roughly ten millimetres) [70]. If we are perturbed so our elbow is pulled back and then we reach forward again, our hand closes and then opens again as it if were mechanically linked to the distance the hand was from the object. Based on the maximum size of the aperture, we can measure how big our subconscious thinks the object is.

One example of the number of different cortical areas that are responsible for the control of the hand is the *Ebbinghaus illusion* (Fig. 3.4). The illusion occurs if two plain coins of the same diameter are presented to a person, with one surrounded by small discs and the other by large discs. If asked, most people will feel that the one surrounded by the smaller coins *is larger*. However, if they are asked to pick up the coins they will open their hands the same amount for both coins, meaning that a part of the cortex subconsciously recognises they are the same size. This is the result of there being two different routes for the data for vision and the limb control (dorsal and ventral streams) and each having a different role in the processing of that information [71].

When the brain has to control a prosthetic hand, things are different. The level of feedback from a prosthesis is small. For a conventional device the user can see the amount it is open. If the hand is body powered (see 2.2.2), they can feel the amount the harness is being pulled and the force they are imparting on the cable. If the hand is motorised (see 2.2.4), they can feel how much they are tensioning their muscles; also they can often feel or hear the amount the motor vibrates as it opens or closes and so they may have additional information when the motor stalls and stops turning. This is useful but does not mean the control of the arm is simple or effective. Attempts to supply additional feedback have generally been unsuccessful. While in the laboratory, some users may be able to discriminate between different objects/textures/forces, using different sorts of tactile feedback from plungers to vibrating discs, when it comes to measuring how effective the person is with the device, no study has shown an advantage [72]. Jon Sensinger of the University of New Brunswick in Canada (13.2) suggests this is because the quality of feedback from any of the sensors is less than that from the eyes [73]. He comes to the reasonable conclusion that, given limited information from the artificial feedback or the wealth of high quality data from the eyes, one is going to believe one's eyes.

For a long-time the study of the control of prosthetic limbs was left to engineers [74]. A significant study made by Alan Wing and Carole Fraser of the Applied Psychology Unit in Cambridge University looked at the different ways we use our fingers and thumb during a reaching and grasping exercise. They noted how we open and close our hand as we reach [75]. They found that we do not open our thumb and fingers the same amount. The thumb comes closer to an object first and then the fingers close onto the object. It seems likely that the thumb is usually more visible than the fingers when we reach for an object, so we can use it as a reference point for targeting our hands. The team studied one young lady with a body powered anthropomorphic hand [76]. The mechanism of the prosthesis moved differently to the natural hand. The prosthetic fingers and thumb were geared together so they flex the same amount as the operator closed it. Despite this, the motion relative to the object was still much more similar to the natural motion than would be expected. Wing and Fraser hypothesised that the subject was raising and lowering her elbow to stop the thumb from closing on the object until later. They noted this observation (using very limited technology) and then moved on. When I read this some years later it seemed to him that the results called out loudly for greater study with more modern higher quality measurements. He could see that the user was employing unusual motions, not to perform a task, but to perform it in a way that mimicked the body's natural control. It was possible that this design of the prosthesis was making the hand harder to use. Leading to the question: *"How should one design a prosthesis to make it easier to control?"*

This experiment also shows that by observing people with a limb absence using prostheses we can gain an insight into the way that we all use our bodies under routine circumstances. The differences between the two groups can show us the way that the body is controlled or used. Experimental medicine and psychology are full of tragic stories of people with injuries and impairments that have given scientists clues as to how we work. By observing what is affected by specific damage it gives us a clue to how it works in the unimpaired.

3.2.1 Body representation in the brain

Neuroscience is the study of the nervous system and it has learned much about the way the brain works through removing or disabling parts of the brain or body and seeing what effect this has. Thus far, any results relating to limb control have not have much direct impact on prosthetic design. One question that neuroscientists are interested in is how the body is represented in the motor cortex and how that differs when a body part is removed. Amputation has been used to observe this process. In a series of studies in 1980s and 1990s, monkeys either had the nerve connections to their arms destroyed (deafferation) or underwent full amputation. The experiments studied the area in the brain used to control the hand. They observed that this area became responsive when other body parts were stimulated. From this, neuroscientists developed the idea that if someone is not using an area of the brain, then the brain will find a different use for it. It supported the idea that the brain is *plastic* and adapts. In the case of upper limb amputation, the area of the cortex adjacent to that responsible for the hand maps to the face. It was found that following deafferation or amputation, the map of the face grew to include the hand space.

These results have several limitations. At the time of the studies cortical maps were measured by placing individual electrodes in the brains of monkeys and recording the electrical activity. The electrodes record activity in a very small area of the brain close to the electrode. To get the bigger picture across all of the brain requires laborious sampling many individual points less and one millimetre apart. This creates snapshots of the changes. Clearly, it was not something that could normally be done in human volunteers, while in monkeys it was impossible to demonstrate that the changes in structure of the brain had any functional benefits for the monkey.

This all changed in the 1990s when functional Magnetic Resonance Imaging (fMRI) became possible. Magnetic Resonance Imaging uses very intense magnetic fields to detect the presence of hydrogen atoms in living tissue, by causing the spins of the hydrogen atoms to align and then flip to a different orientation giving off detectable and characteristic energy signatures. This innovation opened up an enormous world of possibilities for the clinician and scientist. With the parallel development of high-powered computing, very highly detail three dimensional images of living human beings became possible by making many individual scans and adding them together. The next innovation was to be able to increase the speed of a measurement so that it was fast enough to see things in transition. It was now possible to observe the increased flow of oxygenated blood to specific areas of the brain as it processed information, i.e. watch the brain in the act of thinking. This is *functional* MRI. Using this, it became possible to do neural imaging in human amputees to see what differences the amputation made to their cortex.

Earlier studies found a strong correlation among people who underwent a shift in the brain hemisphere that represented the face and hand, to how much they felt phantom pain. This finding appeared to support the idea that the shift (observed in monkeys) creates a mismatch between the signals and their apparent source and this confusion is interpreted by the brain as pain. This fits within a class of disturbances that result from such a mismatch. For example, motion sickness is the result of the signals from the inner ear not matching the visual signals, such as a rushing scene in a virtual reality display not matching the stationary signals from the inner ear. The small problem with this explanation for phantom pain is that the area under study *is not responsible for pain.*

This is part of the odd world of the phantom. If any part of the body is missing, such as a hand, leg, or breast, the person may continue to feel its presence. Some people can move the phantom, whereas others feel only pain or cramp. Some find the ghostly feeling a comfort that the part isn't entirely gone. It is a shame that it was given the name 'phantom' as this can be a barrier to understanding. Some people feel judged by the name, they feel that others think the feeling isn't real, just imaginary. It is worth stressing it *is* real, as we will see, there are still structures in the brain that correspond to the missing part. The problem is that it is such a weird and highly personal experience that some amputees may never even tell their spouses of the phantoms, for fear of incomprehension, or of ridicule [77]. Cambridge occupational therapist, Carol Fraser, discovered this when she interviewed couples about the limb loss. Wives announced that after many years of marriage this was the first time they had heard of this phenomenon in their husbands.

When the phantom causes pain, drugs seldom reduce it. One treatment that is claimed to work is *Mirror Therapy*. It is supposed to reduce or remove pain and perhaps remobilise the paralysed phantom [78]. To perform mirror therapy, a mirror is arranged between the arms with the reflective side pointing away from the injured limb so that the subject can see the reflection of their intact hand where their missing hand should be. However, for some it has no effect at all. It is hard to predict who will benefit from the treatment or why it does not work in others.

Tamar Makin, an Israeli neuroscientist who was interested in the field of body representations (how the body is represented in the brain), came to a similar conclusion as I did, that prosthesis use is a good way to explore human control through the study of the differences. She wondered how an amputation changes the representation of the body in the brain and how the world is represented relative to the altered body [79]. At her disposal was a powerful fMRI scanner. She was able to map the places that become active ('light up') when people move their limbs. This is called premotor activity. It indicates what parts of the brain are put to work when the body part is moved. What she found has turned the entire field on its head [80].

Her team's work repeatedly and comprehensively showed that there is *no reorganisation* as previously reported. Instead, when they observed the activity of the brain during manipulations, the cortex actually lights up when the *other* hand is used. This finding does not fit the theory for the neighbouring areas of the cortex taking over the control of the area for the amputated limb, they are in different hemispheres of the brain. What it shows is the changes are not the result of cortical plasticity, but something else. It is a clue to the real reason for the original results. The monkey was performing manipulations with its mouth that it would previously have performed with its hands. To perform these new movements, it used control circuitry already in place to perform the task. It needed that part of the cortex trained to perform the manipulations. The brain is not changing its role, simply re-employing it to control the mouth when performing that task. Similarly, writing uses the same circuitry no matter how you hold the pen; Dudley Childress was fond of showing that signing one's name does not change if one uses ones fingers in the hand, or the forearm, or holding the arm stiff and using the shoulder. The signature is more similar than one might imagine, because the control 'program' comes from the same part of the central nervous system, it is just used by different muscles. It is roughly analogous to a single website driving a smart phone, tablet or desktop display.

As the monkeys tended to use their mouths to do some manipulations, the motor cortex was recruited to assist in the control. This would happen if the feet were being used, or the other hand. The scientists were not looking for this, so misinterpreted the results.

In every one of the people with amputations Tamar and her team scanned, they saw that the original representation remained intact, even in those who no longer feel their phantom. The fMRI technology is now at a high enough resolution to enable them to delineate individual fingers. This is the brain organisation that is observed when they try and move their missing fingers, even if they could no longer feel them. There may be a little blurring around the edges of some of the areas, but the patterns are very stable [1,81] (Fig. 3.5). When Tamar gives talks presenting these results, she shows a scan/map for a person with an amputation beside one of an typical limb, and challenges the audience to say which is which. Rarely is any response better than a random distribution of answers. So little in the brain has actually changed that it is impossible to tell them apart. Of this result Makin says:

> "This is absolutely a canonical organising principle. We find it in people who have been amputated for three decades, [and] we find it in people who have been amputated for three months, it doesn't matter, everyone has it" [79].

When this is understood, the variable results of Mirror Therapy become clear. Mirror Therapy is supposed to work by tricking the brain through illusory feedback into reversing the maladaptive plasticity, so the structure returns to before, removing the conflict and therefore the pain. However, as there is no remapping in the first place, this is not happening. Instead a very powerful placebo effect is triggered. This varies considerably between people, hence the results of the mirror box are unpredictable.

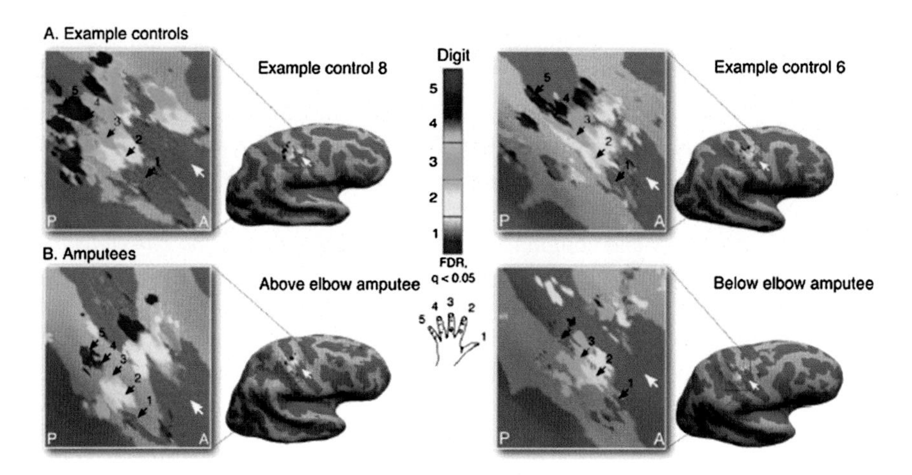

FIGURE 3.5

Maps of the cortexes of two persons with arm amputations decades in the past compared with those of two persons without a loss. What is clear that despite a long established belief in the brain remapping following amputation, what is shown is that the areas for the fingers seem very much more fixed an unaffected by the loss of the body part. Adapted from [1]. (Creative commons licence.)

Placebo is not a small effect. The belief in a treatment or physician can make a considerable difference to the outcomes. Makin points out that psychology students asked to train two sorts of mice ('smart' ones and 'dumb' ones) to perform some sort of task (such as finding food in a maze), found that the 'smart' mice performed better, *"even though they were genetically identical mice"* [79]. The students' faith in what they had been told impacted on the way they worked with the mice. Equally significant changes in outcomes can result when the patient believes in their physician. Considerable effort in the design of an experiment is required to remove any chance of the influence of a placebo in an experiment. The ideal experimental format is randomised, fully blinded, and placebo controlled. So a cohort of test subjects is divided into two groups and the groups are given two things that appear to be the same. One is real and the other is a sham, a potential placebo. Those receiving the treatment and those administering the treatment cannot know which group any patient is in, otherwise subtle cues from the administrator could easily make a difference. The assignment to the groups must be truly random. At this time no randomised, controlled trials have been performed with the Mirror Box to unequivocally demonstrate its effectiveness.

Makin believes phantom limb pain is based on the nerves being disconnected. The hypothesis is that some nerves are firing and as the pain sensors are most easily triggered, the person will feel pain. The problem is finding a treatment that is effective without causing damage elsewhere. Targeted Muscle Reinnervation offers one potential treatment (12.2).

Makin sees the results of the brain scans as very positive. *"It is exciting for prostheses because it means that you do have an existing functional architecture that knows how to represent hands, that knows how to control hands, that is just sitting there waiting. The question is how to harness these brain resources to work for the prosthesis"*. To control the prosthesis well *"You want to take advantage of the amazing architecture that exists in our bodies"*. Rickard Brånemark (11.3) the Swedish implant surgeon agrees: *"The brain is fantastic at sensory and control systems... the better the signals we can supply to the brain, the more the brain can do"* [82].

The way Tamar believes prosthetic control should develop is to employ the system as it is. We should use the existing structures in our peripheral and subcortical systems in order to filter and integrate the signals so when they reach the cortex it makes sense to the brain and all the neural processing can be used. The motor cortex alone is not designed to control the arm, it requires all the input and output circuitry of the peripheral nervous system to convert the signals into meaningful signals. This is why Tamar finds pattern recognition of muscular signals from the forearm exciting (13.2.5). The input to the prosthesis is already organised in the right format to control a limb. *"The false approach is the direct brain machine interfaces"* [79]. Control of an arm starts as a thought of an action in the brain. The thought is converted into a series of neural instructions to the muscles in the forearm and when the muscles contract, they control the movement of the arm. Pattern recognition observes the patterns of the contraction of the muscles and relates it to the resulting motion. The neural signals are complex and very different to the resulting motions. If we wanted to use the neural signals we would also need to design electronic controllers that take the small signals from the brain and convert those into instructions for a mechanical arm. It is a very much more complex task. A similar challenge would be to convert thought into text as a brain computer interface. Instead of tackling this task, we allow thoughts to be converted into the actions of the mouth and lips to create sounds and then do speech recognition on the sounds. This is far less difficult.

When a baby is born it has only vague control over its body. It is driven to acquire control and knowledge. Often this improvement in control is by dint of very many repetitions. This is much to the frustration and fatigue of its parents or family as the child wishes to repeat operations endlessly. Through thousands of examples of this sort of activity, we all hone our skills to control our bodies, so that by the time we are sixteen or so, we have matured and can perform complex activities.

Ironically, it is this facility that creates the barrier to using a prosthetic limb. Good prosthesis use is not an accident. It requires good preparation by occupational therapists and dedication to practice by the user. Without the effort they will not use the prosthesis and it will hang lifeless by their side, or immobile, left behind in a closet. However, no user is going to put in the same level of effort they originally put into their natural limbs. A sixteen year training program for each new amputee will not be tolerated. Any prosthesis has to be usable very quickly, even if the true refined control takes a little longer. What is 'easily usable' is, of course, another complex question. We might feel we already know how to use our hands, so an anthropomorphic hand might seem the obvious choice to replace the missing function. The truth is we have forgotten how much effort we needed to control the hand the first time, and so a complex hand is not easy to use. A simple hand and skilful guidance can be more effective. The clinical team help to bring this about.

An unexpected consequence of Makin's fMRI studies relates to the observation that hand and tools are represented differently in the brain. What became clear is that the prostheses are neither represented like hands or tools. They are not tools nor are they hand replacements. They have their own place in the cortex [83]. At this time, it is not clear if better signals from the prosthesis might change this representation. This finding is in conflict with the impressions of some occupational therapists, that early adopters become integrated with their prosthesis, while later adopters have them only as tools [21].

3.3 The design of upper limb prostheses

The requirements for the design of prosthetic arms and prosthetic legs are different. While arms are used to push or pull things, hold items close to the mouth or clamp them so we can manipulate them with other parts of the body or the other hand, prosthetic legs have to hold people up, without fail. They should also be roughly the same length as the other leg. Thus if a person is missing part of one leg, basic needs for the prosthetic leg are easy to see. It is hard for a person with a leg absence to get away without using any aid, be it a crutch or a prosthesis, whereas a person missing a single hand or arm can often get along very well without a mechanical replacement (and many do). It also means for any individual, for simplicity, the different uses that *their* hand/arm are put to must be expressed in their particular prosthesis, without adding other unwanted functions. If one person needs to look 'normal', then a lightweight arm with the right shape and colour is what they require. If a different person wants to be able to work with machinery and tools and does not care about appearance, then a purely mechanical device created to hold or control their tools, which is tough and not particularly attractive, may suffice. There is less need to make one device that covers both situations. If someone wants both at different times, then it is possible to exchange one model for another in the same way we change our clothes and shoes for dirty work or social scenes. However, most people would not want to be constantly swapping hands or arms, so the designer has a series of compromises to make.

In the 1970s the US Congress Office of Technology Assessment (OTA) study [84] observed that the objectives of prosthetic technology are not obvious. Little has changed. The report found that the powered elbow (the Boston Arm, see 7.1) embodied only a subset of the original functions of the human arm. Choosing among them was a complex process, involving disputed judgements depending on the priorities of the users. One of the participants interviewed for the study said: *"It would be difficult to overstate the divergences of viewpoints expressed by the prosthetics experts interviewed for this study. What was a critical feature for one was a red herring for another, and, although some disagreements rested on data, others concerned ideas about what an artificial arm should be expected to do"*. Half a century later, this remains at the heart of the difference between projects to improve prosthetics. Over the intervening decades, some have chosen to focus on one area, others on another. Regrettably, many have ignored the input from the users themselves.

3.3.1 Prosthetic mechanisms

The choices in the design of a prosthetic limb have an impact on its use and range. It is a compromise between many factors. One of the questions designers need to consider is: *"What is more critical, that the arm move like a human arm, or be easy to use?"* Over the years, the form and design of the movement (actuation) of prosthetic hands has occupied the minds of many ingenious designers. Generally, every designer wishes to produce a device that mimics the action of a human hand/arm, but is as simple and light and easy to make as possible. Where each designer draws the line depends on their own choices and probably prejudices. We will see the differences in two designs of an elbow that came from the same starting point (Chapter 7).

Additionally, the more joints that move and move separately on a prosthesis, the greater the complexity, but potentially the more functions the device can perform. However, with complexity comes weight and also expense. The choices made in the design phase have a big impact on the eventual device. There are a number of choices of features that are common and will impact on the eventual design, some of which are described below:

Number of working fingers Not all fingers in the prosthetic hand may move. Some may be driven by motors or pulled by cables from the user's harness. Other digits may be dummies formed from the shape of the glove and may be fixed to follow along with the driven fingers. In the most common conventional electric hands, two fingers are formed by the glove and two are driven directly.

Number of joints per finger The simplest prosthetic fingers only flex at their base, moving as a single piece. This makes them simpler to build and maintain, but they look less natural. If there are more joints along the fingers, then they can wrap around an object, look more anthropomorphic and spread the grip force more evenly across the object. Fingers such as these are more complex and generally more expensive to make, and likely to be more easily broken. An effective compromise is to have the distal (furthest out) finger joint fixed at a slightly flexed angle and actuate the next joint down as well as the one at the base of the finger. This creates a reasonable shape for grasping objects.

The drawback is that with an exactly anthropomorphic finger with a rigid joint, the tips of the fingers will not flex fully to get to the base of the fingers. A hand like this would not be able to grasp small objects to the palm, hence would suffer some loss in function. A hand produced in Canada in the 1990s known as the *Montreal Hand* had the rigid distal joint, and the designers altered the lengths of the joints so the tip of the finger could flex to the palm with only one intermediate joint [85]. This is an acceptable visual compromise as with other short cuts if it is hand-like enough we will not notice it, even for those of us with long experience of hand design. When I first saw the Montreal hand on a person in a lift in Chicago, we they rode over 20 floors before I could work out how the designers had created the hand.

Thumb moves separately to fingers This can allow precision or power grips, with the tip of the thumb opposing the tips of the fingers, or closing later than the fingers and wrapping round the back. If, in addition, the thumb can move off to the side then the lateral or side grip can also be achieved.

Cables or links This determines how the drive from the motor is transferred to the motion of the prosthetic fingers. Cables are a straightforward way for the actuation mechanism to pull on the digits, they do not need the same precise alignment as less flexible links. They can go round corners via pulleys or run down the fingers, pulling the tips in, causing them to curl. They tend to require fewer bearings and fixtures in the design, compared with rigid links, which cannot go round corners as easily or go across complicated routings. However, cables are far harder to attach to the places the designer wants to actuate. They may also stretch or break more easily. Early designs of the Utah Hand in the 1970s suffered from the latter problem [86]. When the cables run over joints some resistance is introduced into the cable and some commercial body powered hands require tremendous forces to close with quite a light grasp [87]. Despite this, numerous designers have preferred the advantages and continued to use cables.

Number of separate drives Every finger can have separate drives, this adds extra complexity of the mechanism, and possible additional mass but makes controlling them separately or in combination an option. As the fingers not only flex to the palm, they also move side to side (ab/adduct). Few prosthetic (or robotic) hands have axes that are not in a line, so that the grip is not fully anthropomorphic. The Michaelangelo Hand from Otto Bock is one of the few hands with this feature to make it into the field. Its passive appearance is exceptionally life-like. The way that the fingers come together as the hand closes makes for a very firm and reliable grip.

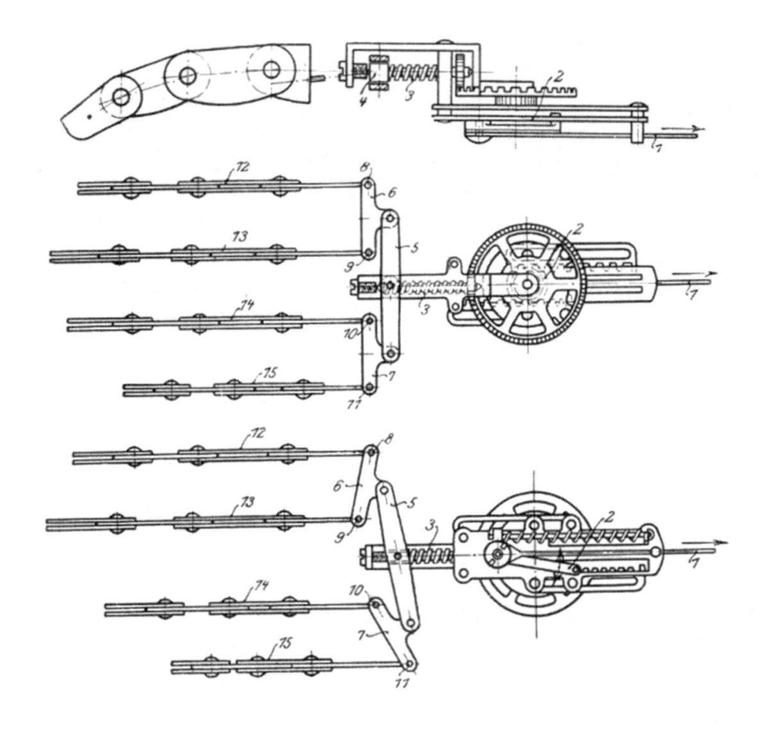

FIGURE 3.6

An example of a pair of two branch whiffle trees used to allow the fingers to adapt to the shape of the held object. (Taken from [2].)

An alternative is to have one cable or driven link pull more than one finger. If they are rigidly linked together then the hand will not adapt to held objects. Most everyday objects are not uniformly shaped, or grasped each time in exactly the same way, yet a prosthesis with a fixed shape needs to grasp every object the same way, meaning the hand must approach the object from the same direction with the same orientation. This restricts how the user grasps the object and limits what they can do with both the hand and the object.

It is possible to link the separate drives to each finger across an equalising mechanism, in the same way that two horses pull one carriage evenly through a series of links (or the single hand brake lever of a car pulls two brake shoes evenly). When horses are rigidly linked together, then the lead horse takes all the strain. Instead, if they are linked through an additional lever, it allows one horse to move forward ahead of the other. As the horse moves forward, the other end of the lever moves backward, taking up the slack in the harness, until the two are pulling evenly. This device has various names, one is *whiffle tree*. It is a very common mechanism, in robotics and prosthetics design it allows the fingers to be *underactuated*, which means there are more outputs despite fewer motors (actuators) than one actuator per finger (output), Fig. 3.6.

This mechanism is also part of what are arguably some of most successful machines that mankind has ever built: It is how the wheels of the Mars rovers *Spirit, Opportunity, Curiosity* and *Persever-*

ance linked to the chassis of the vehicle. Each rover had three wheels on either side of the body linked together through a whiffle tree. Because of this, the wheels tended to stay in contact with the Martian surface despite how much it undulates. Similarly it is possible to make a hand so that three fingers remain in contact with an object, despite its variable shape (see 10.2.1).

Grip force versus prehension speed There is a trade-off between the force required from the hand and how fast it can work. Electric motors spin fast and require gearing to bring the speed down to a manageable level. If the speed is kept fast the resulting grip forces will be smaller. Conversely, slower fingers can be much more powerful, but the hand may be too slow to be tolerated.

A stable grasp is only possible if sufficient force is transmitted over sufficient area. This means that if the grip does not conform to the shape of the object through some mechanism (see above), then there are likely to be a small number of contact points between the hand and the object. The natural hand conforms around an object and we generally use as little effort as possible to hold an object. For a conventional single motion hand there are few contact points between the hand and the object. To hold an object stably, the grip force must be high in these small points. Thus the Otto Bock single motor hand (9.1.1), is rated at a 120 N pinch. This is very large (most grasps of a natural hand are below 12 N [88]), and yet it still is considered 'weak' as objects still slide from its grasp as its inflexible fingers grip at only two or three points. The flexibility of the natural hand spreads the load more evenly.

To achieve a high grip force requires a great deal of gearing. This means the hand must move slowly. It is clear that people enjoy the use of a quicker hand. In a heavily geared mechanism, all the gears (in the chain to the fingers) add delay to changing speed or direction. It takes some time and effort to start the hand or to stop it once it is moving. One solution is to use two different drives: A fast and weak one for the big movements, such as opening and closing the hand, and a slower, more powerful one to grip and squeeze an object. Another solution is to have a gearbox that changes ratio when the hand is closed on an object. The Otto Bock single axis hand has a very effective gearbox that automatically changes gear when driven to squeeze down on an object, but allows the hand to open and close quickly too. However, complexity always tends to introduce cost or reduce reliability, or both.

Size constraint Every person is a different size and shape. Each individual's residuum is a different length, each residuum a different shape. The natural human hand has its drives (muscles) all the way up the forearm. If a prosthesis is to be fitted to the widest range of people (one person with most of their arm intact, and another missing from above the elbow), a prosthesis should have all its mechanisms within the fingers and palm.

The second constraint is making the size of the hand proportionate to the wearer. The smaller the person, the smaller the prosthesis. In order for a prosthetic hand to be usable by more than a few large persons, it needs to be smaller than the biggest hands, or come in a range of sizes. Additionally, more women and children are clearly smaller and cannot carry larger heavier hands comfortably. However, the smaller designs are harder to make. As a result, the first designs of prostheses have generally been used by larger males, before smaller hands appropriate for women were devised. Children have often been served very much last with prosthetic innovation, as the mechanisms need to be even smaller. The first children fitted with powered hands received adult hands (see 11.1).

Location of motors The size constraint also has an effect on where the motors of the hand reside. Evolution has placed most of the muscles of the hand in the forearm. After amputation the muscles remain available for use to give instructions to the hand. If designers choose to place the motors in the forearm, this will further limit the number of people who can use the arm to those who have lost a significant proportion of their forearm. Sadly, the first attempts at building new advanced arms for American soldiers injured in Iraq and Afghanistan were driven this way, despite years of experience which advised against it as someone with an amputation or loss just at the wrist would not be able to wear or use such devices. The idea of reducing the forearm surgically to make room for mechanisms has been suggested, but this is unacceptable as it is a one-way decision and new, more compact, technology might well make it unnecessary in future (see 2.6.1).

Locking of the grasp If a hand is to maintain a grip on an object then the fingers have to be held in a grasping position. Either the drive has to continue to be applied or a brake must be applied. The first will require effort to be maintained all the time the grip is applied, which is inefficient. If the fingers are to be braked then it requires a large braking force to stop the fingers opening. This force either must come from the mechanism in a powered TD or the operator in a BP device. For body powered devices the grip force and the resistance to being opened can be the same thing (see below). Naturally, if the effort needed to hold an object is great it will be used less often.

For externally powered devices the amount a motor and gearbox train resists being opened (back driven) comes from the design of the drive train. For example; a gearbox can be back driven open if it is greater than 50% efficient. If it is less efficient then it will passively resist opening forces. Most hands and elbows work this way. There is a greater loss of energy in the drive train, but the benefit is there is no need for additional brakes.

Brakes in line with the drive train have been used in some applications. An early electric hand (the Viennatone 9.1.1) had a brake that prevented back driving. It employed a mechanism that used the rotation of the shaft to drive wedges outwards to jam the motion. It was complex and relied on very precise manufacturing to allow this to work reliably and repeatedly.

Mass distribution Every part of the prosthesis must be carried by the user. It must be retained using the socket. Since the prosthesis will always be at the end of the arm, every gram of the prosthesis is felt by the wearer. When the hand is raised (perhaps with an object in it), it will be resting on the skin of the remaining arm. This means that the mass of the hand and arm need to not only be as low as possible, but placed as close to the body as possible, as the leverage of a distant weight will be felt when the wearer is at full stretch.

If all the motor/drives have to rest within the hand, then this is an important consideration when choosing the number to be used. Multifunction hands show differences in design when addressing these considerations. For instance, the i-Limb and Vincent hands are built up from individual powered fingers, with all the motors very far down the hand. By contrast, the Michelangelo and bebionic hands have their motors in the palm area. Similarly, as each motor adds considerable mass to the arm, it will also affect where the mass is on the arm. Natural selection has spread the weight of the arm along its length, with the forearm and upper arm containing most of the muscles. Likewise, the muscles to raise the shoulder are distributed across the shoulders and chest. Because the drives are in the prosthesis, this takes them further down the arm and makes the arm feel heavier.

Side It is obvious, but needs to be considered. Arms come in two forms: Left and right. To manufacture two complete sets of components, one for each side, increases the cost of any prosthesis as the manufacturer is effectively producing two devices for every product. Therefore manufacturers will attempt to make much of their devices easily reversed, adding a few specific components to determine the side.

Size In a similar way, the fewer changes to be made between the smallest and the largest, the more affordable the product should become. The size range from child to large adult is quite considerable. Children's hands are generally not the same design as the adult hands. Stefan Schultz of Vincent Systems took the bold step to make a simpler and smaller multifunction hand for children from a subset of the same components as the adult hand. The Vincent adult hand was also the first multifunction hand to be small enough at the initial launch for the majority of the female users, rather than most of the male range. This was a breakthrough in the production of these advanced hands as the three previous multifunction hands were aimed at the middle of the male range and had no easy way to be made any smaller. Prior to multifunction hands, manufacturers produced up to four sizes for adults. This size was determined by the distance in inches around the palm area. The newer hands are simply designated *small, medium* and *large*.

3.3.2 Power sources

The choice of where the power comes from in externally powered limbs is an important one. Electric power is the common choice. Other options have been explored, below.

Electric motors are inefficient at low speeds. The cyclic motion of the arm is not the most efficient use of a rotary motor. To work, motors tend to be spun quickly but prosthesis motion is slower, so that they have to be geared down with gearboxes, which are often heavy. High speed motors and gearboxes can also be noisy. Electric batteries are easy to recharge, requiring a small and portable charger. They can be recharged almost anywhere, so a user travelling away from home can still keep their arm working. Any historical drawback concerning the need to recharge electric batteries is removed entirely by the ubiquitous nature of mobile electronics today. Recharging is simply a part of everyday life.

Pneumatics The differences in the sizes of the input and output pistons in pneumatic and hydraulic systems can change the output. Greater force can be traded for increased movement, or vice versa. Both can work with or without an external power source. Additionally, both forms work much better at low speeds than electric motors and they do not need gearboxes, making the actuators much lighter in mass. In addition, the piping can go round corners more readily than fixed mechanisms. In pneumatics, the gas is compressible, making the output flexible and compliant. Uncontrolled the output will tend to bounce. A major drawback is that it is far harder to replenish a gas bottle than an electric battery. The bottles are heavy, while the piping is light. The amount of force developed is proportional to the gas pressure and higher pressures require far more robust and heavy pipe work. They also tend to hiss when activated.

Hydraulic systems have the similar advantages as pneumatics (without the output compliance), but they operate at a higher operating pressures, and require more robust pipe work. The actuation itself is silent, but it if it uses a pump, this will make some noise instead. When hydraulics leak it is generally messy. The work of the group in Delft has been to reduce the inefficiencies in body powered devices using passive hydraulics [89]. The Norwegian company Hy5, has created an hydraulic hand that does not have any pipework, increasing reliability significantly (see 14.1.5).

Alternative actuators There are some other novel forms of actuation. Many are based on a physical property that causes some materials to change shape or length with heat or electric power. This makes them attractive as they look a lot like muscles. Their problem is that they tend to require high temperatures or voltages, which makes them impractical for a portable prosthesis. The high temperatures and voltages mean the user needs protection. The addition and removal of heat is slower than is useful for prosthetic limbs [90]. An example is a nickel titanium alloy (Nitinol), which when cool is soft and easily stretched, but when it heats up it returns to a set length. Hence is it known as *Shape Memory Alloy* (SMA). Because it can be stretched and then pulls back like one muscle in a pair (see 6.1.3.2) it is also referred to as *'muscle wire'*. However, if it is overstretched its qualities are destroyed. Since it needs to heat up and cool down quickly it must be arranged in fine wires which cannot pull much and are easily over stretched. It is less like a full muscle and more like a muscle fibre. It is about as useful. In the 1990s both David Gow and I had students investigate different aspects of the use of SMA in prosthetics and found it impractical [90,91]. In the end, I predicted that it needed an application that was not fast moving and had a great supply of cooling air or water to make it work. The idea of a robot fish sprang to mind and indeed since then at least one such robot fish has been made [92]. However, despite some extravagant claims of force generated, none have been used to create a practical prosthesis.

3.3.3 Control

How the hand is actuated often dictates how it is commanded (or controlled) by the operator. If the device is body powered, then the control and the power source are the same thing; the pull on a cable provides the power to open the hand. If the power source is electric then there are numerous ways to control the hand, from switches to taking body signals and using them to instruct the electronic controller. As electronics have become more sophisticated and the number of motions of the hands have increased, the types of commands needed have also become more complex, perhaps at times too complex. The control has to match the device, but also the desires and abilities of the user. Children's hands tend to be less sophisticated, partly because of the users' cognitive development. They have yet to fully comprehend how to control the hand. There is an argument that if a child was fitted with a more sophisticated hand and controller which could expand as they grew cognitively, it could grow with the child. A great deal of the rest of the book details ways that people have investigated prosthetic control.

3.4 Assessment

In any development process it is important to be able to measure the progress towards a goal. This is as true in prosthetics as it is in any other activity. The problem facing a designer, clinician, user, or provider of financial support, is that this is a difficult question to answer. Prosthetics has many stakeholders, from the users to the clinicians to the those who pay for the fitting. Each of the stakeholders would like to know if a change of the design of the hand, arm, socket, or the way they go about the task will benefit the user. To justify any investment in a change its effect really needs to be measured. The simplest example of this is if one can measure some aspect of the user's life (such as the effort they need to perform a task) and see if it is altered with the proposed change. However, this is quite clearly not simple. For a start, every person in the process above is interested in a different aspect of that story and so needs a different way to measure the change.

For example, the designer of a hand mechanism needs to know: What items can the hand pick up? Does the hand open wide enough to admit the objects they want to hold? Is the grip stable enough to retain the object? The designer of the control system needs to know if the person can perform the actions to control the hand. Also, can they operate the hand when it is over their head, or when the arm is holding a heavy object. A prosthetist will be curious if the socket is comfortable, and when it might not be. Also they will want to know if the person can place the hand in the right orientation relative to the object to easily pick it up. An occupational therapist will want to find out how they orientate the hand and their body when they perform a common action, and how often they really only use their sound arm (if they have one). The funder will wish to determine when the injured can return to work and pay taxes back into the system, and how long the new device might last. The user may just want to be able to put a worm on a hook while they are fishing with their grandfather. Everyone, all the way back to the designer, needs to know what people really want out of their hands so they can try and match it. If they do not match the needs then the device will not get used. A neat solution to a problem no one needs to solve is nice, but pointless.

From the above examples one can see that the needs of each group are different and so the way to test for the outcomes will have to be different too. There is no single measure of the performance of a prosthesis so different tests must be developed to measure the different factors. A further problem is that when the different groups talk to each other they may well mean something very different by the same words. For instance, unless they have already come to an understanding about terminology, *function* is likely to be interpreted very differently by an occupational therapist and an engineer. The move towards the goal of a common standard began nearly two decades ago. Around the turn of the 21st century, it was becoming obvious to many workers in the field that there was a need for evidence on which to base their clinical and scientific decisions. Prior to this different assessment tools had been developed and used, but they tended to be employed by small groups. Often a tool developed in one place was modified by others to fit the local circumstance [93]. This is a huge error as the results cannot then be compared.

Measurement is only simple when the tool and the measurand are simple, such as when measuring length or mass. Even then, to get a consistent result the measurer must know how to use the tape measure and be sure to use the same set of weighing scales in the same gravitational field as the people they are talking to, otherwise the measurements cannot be understood by someone else. For example, on one side of the Atlantic length is measured in feet while on the other it is metres. Even so, standards and conversions are well understood and established. It is easy to measure these properties and convert them. Usually, an item made in one part of the world will be the right size and shape when delivered to another.

Prosthetic function is a much vaguer concept. What is chosen to be measured (which aspect of function) has to be tested to ensure it does measure what we think it measures, and does so reliably. The tests are designed to make sure that if the same person measures the same user/prosthesis on two different days, the same (or similar) result is obtained. This is called *repeatability*. It also should be the same result if two different persons make the measurement on the same prosthesis, this is *interrater* reliability. Clearly if this was not true and results depended on the person or the time of day the measurement was made, it would be a fairly useless measure. Consider the idea of measuring the length of a room for a carpet if it depended on who measured it and when. It would be impossible to get a carpet fitted satisfactorily.

There are many other factors that need to be taken into account including the following:

Does it measure what we think it measures? For example, height and weight are closely related, but are not uniquely tied together. If we had been using a person's height as a measure of their weight based on stocky people, it would work quite well until we chose to measure a skinny person. Suddenly the weight estimate would be way too high.

Are we really measuring the attitude of the subject? If a person does not care about the score, they may not try too hard. Conversely, someone with less talent may get a good score if they are keen.

How sensitive to a change is it? If we measured the time of a hundred metre race with an egg timer, it would be hard to tell between the best and the worst runner. Similarly, if the best and worst prostheses have almost the same score, what does the test tell us?

Does it measure all the important factors? This depends on what is considered important and will depend on the audience.

Can we use measurement to predict performance? What will happen if a particular person is fitted with a prosthesis? When someone is prescribed with a new prosthesis, it is nice to be sure that it will make a difference to them. Can we say a person will be more effective with this prosthesis than the other one? And how much more effective are they?

These (and other factors) are referred to as the test's *psychometric properties*. They can only be determined by repeatedly making measurements with many people. If they are made and compared against other tests that have known properties, we can ensure the test is behaving as it should. If there are no existing tests for comparison this becomes more complex as someone needs to make many measurements under many circumstances to get an idea if any of the above questions are answered. If they do check out then the *validity* of the test is demonstrated and as long as people follow the same processes and procedures they can be confident that they are measuring the same thing as yesterday or tomorrow and as someone on the other side of the world. They will also know that when they have measured something, it means what they think it means.

The crucial part of the design of a test is the first question: Does it measure what we think it does? It is much like formal written examinations. Do they actually test how good someone is at the subject, or just how good they are at taking exams?

Any test must be treated with caution. The user must understand the limits of a test. A set of weighing scales can only work reliably if used on a planet with the same gravity as the Earth's; if it is taken to the Moon or Mars, then the results must be corrected for the new planet. If a test is carefully calibrated and someone changes the way it is used, e.g. they take only half of the test to save time, then the results are likely to be meaningless.

In 2005 a group of therapists, engineers and prosthetists got together to address this problem; the Upper Limb Prosthesis Outcome Measures group (ULPOM). The group determined which tests existed. This included tests for natural hand function and any form of hand impairment. They then worked out if the tests had been tested sufficiently to determine the psychometric properties and so be able to say they were *valid* tests of *prosthetic* function. If the test was not yet valid, then they also tried to determine how much extra effort was required to make them valid [94]. Since ULPOM first published their work, the American Academy of Orthotists and Prosthetists has collaborated with the group to extend the range and scope of the work [95]. Encouragingly, others from outside the original team have also set about working to bring the unvalidated tools to a point where they can be used with prostheses. Others are using this model to develop tests for aspects of use or function previously unaddressed [96].

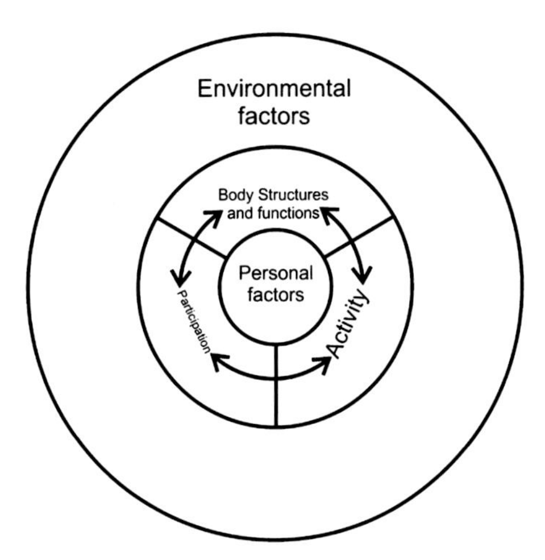

FIGURE 3.7

World Health Organisation's International Classification of Functioning, Disability and Health (WHO-ICF) - Describes the domains of knowledge within the area of disability. When used in connection with prosthetics it can divide up the areas for assessment so that it is clear that no one tool can measure all three domains successfully.

3.4.1 The World Health Organisation's International Classification of Functioning, Disability and Health

One other contribution of the ULPOM was to promote the idea of which tools are useful in which area. For this they took the World Health Organisation's International Classification of Functioning, Disability and Health (WHO-ICF) as a model. The ICF was developed so that a disease or disorder could be objectively categorised, such as what sort of impairment a person had, and so tie it in with what would be needed to alleviate these problems [97,98]. The ULPOM used this to get a practical insight into what was being measured when we try to measure a prosthetic hand (Fig. 3.7). The model divides the application of any medical or rehabilitation tool into three domains:

Body Structures and Functions are the technical aspects of the device or the physical attributes of the person. For a prosthesis: How big is the grasp? How fast does it open? Which objects can it grasp? These factors can often be measured with standard engineering tools such as rulers or timers.

Activity concerns the operation of the device and what it *can* be used for. Measures are based on activities, often with domestic equipment like cups or other items that simulate Activities of Daily Living (ADLs). These are tasks that we would perform routinely, generally abstracted to be performed in the lab.

Participation is what the users *actually* do with their prostheses. To measure this one can ask users what they do using questionnaires, which have to be carefully designed to ensure the respondent understands what the questioner means. Alternatively, one can log how the device is being used, but this is quite invasive and raises many concerns over privacy, although with sporting activity loggers and small head mounted cameras this previously unexplored area is opening up to measurement (see 3.4.5).

It should be clear that these are three quite separate sorts of question. One might realise therefore, that there cannot be a single test that measures all of the above. There is no one measurement tool that covers all three domains; it would be like expecting one test to measure the speed of service of a garage and the pleasure the owner had in driving the car. The different aspects are so dissimilar that it should take separate tests to measure all aspects.

To design and validate a new tool takes a long time; Virginia Wright estimated may take as long as a decade for all the steps to be taken [99]. After a test has been devised and tested on a number of subjects and the measurements repeated with different testers, it should then be compared with other known tests to ensure that it is measuring what it says it is. Each stage needs to be scrutinised by the profession through peer review publication before it can be accepted as demonstrated.

The first tests of prosthesis function were simple and based around straightforward activities. Later tests became more sophisticated and subtle. The Assessment for the Capacity of Myoelectric Control (ACMC) was developed by Liselotte Hermansson in Örebro, Sweden (see 11.1.1). It aims very specifically to measure the way a person uses myoelectric inputs for prosthetic control. It cannot, and should not, be used for other sorts of prosthetic control because it has yet to be validated for such an application. Its aims and measurement techniques are very specific. The test's format is to ask a person to perform a familiar task. A trained observer scores how well they do it using specific criteria. By using a familiar activity, it avoids one of the big problems with many tests: How novel is the activity to the subjects? If the activity is something they have never or rarely done, then the subjects may not perform them as well as they might a different task. Hence they will get a lower score than they would with a different, more familiar, test then the result does not truly reflect the capability of the subject.

Ideally, a test should measure the level of whatever it is we are interested in and nothing else. As the subject becomes familiar with a test they should not become better at it, which is why Hermansson chose a task the person could already do. An analogy would be if one was measuring length: Improving scores would be like measuring the same piece of wood several times and getting increasing lengths. What we want to measure is the fundamental and true length, not some related quality. We should want to do the same for prosthesis function, but it is rarely straightforward.

When my student Arthur Zinck was designing a test to measure the way a person with a prosthesis uses their prosthetic wrist, or compensates with other motions when they don't, we encountered a small problem: To make such measurements Arthur chose a series of two handed domestic tasks that the team thought everyone did regularly. The idea was that subjects would immediately get close to their best score. It turned out that Canadian teenagers in the new century do not hang out laundry or sweep the floor, so that they were unfamiliar with clothes pegs and handling a broom, and had to learn what to do as they were tested, which impacted their assessments.

3.4.2 Tools for measurement

One important consideration for how a prosthesis is used is to understand the way people move when they perform the task, and perhaps what sort of compensations they might need to make to achieve it. The most straightforward method is observe the way a person moves. This can be done by an expert who has studied movement and can judge subtle differences. This is difficult and hard to achieve objectively. A more quantitative method is to use tools to record the movement and analyse them with a computer. This is referred to as *motion tracking*. Conventionally, infra-red cameras track the movement of the limbs by placing small reflective markers on the subject and shining an infra-red light on them. In this way the movement of what is of interest to the engineer is made clear and it does not greatly interfere with the motion of the person. This is a popular way to record motions for the movie and computer game industries and in this form is known as *performance capture*. Motion capture was originally created to measure the walking movements of humans in the laboratory. As technology has grown it has been extended not only to the movies, but also to many other animals from sloths to elephants. It has a long history and the first such motion analysis was by Eadweard Muybridge in the nineteenth century [100]. He took a series of photographs of horses walking, trotting and galloping to investigate how horses ran. This serial photographic method was employed in motion studies for many years, until the 1970s and 1980s when computer-based systems became first experimental units, and then commercial products.

This form of measurement can have many applications. If there is damage or malformation of bones or muscles or their control, it is possible for surgeons to detect the differences and to plan surgery to modify the skeleton to change the motion to be easier, quicker or with less long term-damage to the joints.

Walking (or gait) was the first movement to be analysed because it is cyclic, that is, it repeats regularly and predictably. One stride can serve as a representative measure of the motion. Once this is recorded, deviations from the typical are easily observed. Gait analysis then looks at the differences to predict what is wrong, suggests treatment and, if used after recovery, can measure improvements/changes created by the treatment. Its hallmark is that it can show that a problem with one joint (such as the ankle) will result in limited motion of joints higher up (a limp where the hip flexes little). Assessors who understand the motion, can separate out the clinical problem from the effect and treat the problem.

Other bodily motions, which do not repeat, need more information to be recorded before an understanding of what is typical motion can be achieved. Once the average is established, differences from normal motions can be observed. This understanding can then be used to decide what might be done to correct these differences. In gait, varying factors like walking speed and stride length can be removed if complete strides are stretched or compressed in time so that they lie over each other on a graphical plot. If this is done, everyone without an injury or impairment fits within fairly narrow ranges, while anything outside that indicates some sort of problem or impairment that can be detected, and perhaps fixed.

The movements of the arms are much more variable. We can perform the same task in very many ways. As movements are influenced by complex physiological, social and cultural factors, as well as individual responses to impairment, it is challenging to compare the way people use their arms. Researchers in the field aimed to develop tests of upper limb motion that created a quantitative measure of the person's function, compared with the typical population. Simple measures, such as the ranges of the joints were used, but they were unable capture all of the different movements people use to perform routine tasks. Instead they chose manual tasks that people should know how to perform, and then compared them with the actions made by prosthesis users.

3.4.3 Creating a measure

To illustrate the process of creating a measurement tool, I will describe some of the choices made when a specific test was devised by my student Arthur Zinck. The test was part of a program to develop a new wrist prosthesis. The aim was to see if changes made to the device or its control could have a positive effect on the way hands were used [24].

The first part of the process was to decide on the activities that were used. This is complex. The task must be easy enough to perform without being too easy, hard enough to stretch the users without them having to give up. Once our occupational therapist had chosen those tasks they considered appropriate, we set out to see how prosthesis users and physically tytical subjects performed these tasks. We aimed to control the tasks sufficiently closely that most subjects performed the tasks the same way and any changes observed with a new prosthesis were due to the changes created by the prosthesis. The change we wanted to measure might be in the design of the prosthesis or the controller or the way the person was trained.

This proved to be far more complex than we hoped. Different people perform very routine tasks in very different ways. The question became: *Is this difference physical, or due to personality or social conventions?* For example: One prosthesis user was asked to cut a piece of bread and bring it to their mouth as if they were feeding. They held the knife in their good hand and the fork in the prosthesis. Once cut, they brought the bread to their mouth by swapping hands and using their natural hand to hold the fork. This looked very much like an adaptation to their prosthesis: They always used their sound hand for the most complex part of the task, when they could. A perfectly reasonable adaptation. Except at least one of the typical subjects did the same, swapping into their dominant hand to feed. Culture or convenience? We could not say.

One of the other tests we chose was to cut a circle in a piece of paper with a pair of scissors. The most usual way this task is performed is to start at roughly four o'clock and cut over the top until they reached about ten o'clock. At this point, they would withdraw the scissors and rotate the paper until the start of the cut was at about four o'clock, when they would repeat the process two or more times until the circle was complete.

One of the prosthesis users chose to stop at ten o'clock and then, not moving the paper so not needing to open her prosthesis, she cut again from four o'clock in the *opposite* direction until she could not get her hand round (at about seven thirty). Only then did she rotate the paper. By this method she continued to cut the paper with the paper close to her body. This is a magnificent adaptation. It cut down on the number of times she had to reposition the paper in her prosthesis and kept the cutting area closer to her body, making it easier to control. This makes the comparison with any other subjects less straightforward. What aspects do we compare? We can measure ranges of motion and speeds of action, but it does not fully capture the ingenuity and skill this person showed.

In the end we need to find a measure that reliably separates one subject from another and depends as little on their underlying skills or abilities, otherwise the most talented kinesthetically will get a good score irrespective of the design of prosthesis. In this way the measure will tell us that and not what we want to learn. The best measure is therefore one that the subject has no immediate control over, such as weight. No matter how light a person thinks the scales will give them an undeniable weight.

3.4.4 Cognitive load

Another aspect of the control of the upper limb that it makes sense to explore is how much cognitive effort is required to operate a prosthesis. Recalling Penfield's homunculi. We can see a great deal of the brain is dedicated to controlling the hands. We use our hands for many different purposes and our exquisite control is in part due to this provision. This is one of the reasons that prosthetic control is hard.

We cannot access all the facility to control so we have to make simpler devices which we must control more consciously. This means that while natural hand control is very much subconscious, prosthetic control requires cognitive effort. If the person perceives the effort is too much, they may well abandon the limb and opt for something simpler. In order to determine how much effort is required and if a new way to control the limb might prove to be easier, we need to be able to measure the cognitive load needed to perform a task. Measuring cognitive load, or even defining what we mean by it is also not at all simple.

3.4.4.1 The dual task paradigm

One method of measuring cognitive load is to challenge the subject to perform two tasks concurrently. This is the *dual task* paradigm. If you are concentrating hard on the first task, then adding a second will make life tougher. In its simplest form, a dual task requires a person to perform one task, the one we want to measure (the primary task), and perform it again while undertaking a challenge (the secondary task). We measure how much worse they are at both tasks when they are performed together compared with each separately [101].

The most archetypical test might be to see how easy it is to walk in high heeled shoes. To make the measurement, we could see how far subjects walked in a set time, for example, a minute. Then we would see how much of a secondary task they could do in a minute when standing still, and finally measure how well they did when both tasks were performed together.

The choice of the secondary task is important. It must be possible to do while walking and be *sufficiently* challenging mentally. It must also be possible to do without stopping the subject performing both at once, i.e. walk and not do the second task, or vice versa. Finally, it must not be so easy that they can do both without slowing them down at all. One such secondary task might be counting backwards in sevens from one hundred. The assumption is that both tasks need effort to perform and so will be slower when performed together.

The procedure would be to perform each task separately and then together. The result would be the combination of both scores, i.e. the drop in performance in the first task (the fewer metres walked) and in the number of subtractions performed. To test how much harder it was to walk in heels, two groups of walkers (habitual and novice heel users) would be tested walking with flat shoes and heels and performing the maths test separately. Then they would do the maths while walking in the flats followed by the same with heels. It is probable that the majority of men would be slower in the heels anyway, but the impact of the counting might be quite significant as the effort to concentrate on walking in the strange shoes would tax them too much to be able to count backwards easily. They may stop counting, or stop walking in order to complete the other task. If successful at both, the difference in the overall scores of counting backwards while walking with ordinary shoes and heels would show how much more difficult it was for the novices to walk in heels. Habitual heel walkers would be expected to have a similar score both times.

This form of testing relies on the fact that each time the performance is the same. The novices must not become better at walking in heels over time, and all the subjects must not get better at counting backwards in sevens with practice. If they did improve then this would change the last measurement relative to the first, and so affect the result. This is again like a changing a length measurement over time.

Tests like these are popular with psychologists. They use them to test aspects of human control or cognition. The groups they test are often uniform, being student volunteers, young with the same sort of mental agility. What is being tested by the psychologists can be something fundamental to humans (effect of a drug in learning or impact of an injury), so any findings can be generalised to the rest of humanity. Prosthesis users are much more varied in age and mental agility, meaning the results are not nearly as clear, or readily generalised.

However, the most fundamental flaw when using something like this in prosthetic testing is the population of users is very much *unlike* a population of undergraduate students. Measurements are based on the idea that your sample somehow represents the entire population and that a score your test subjects get is the same as if you had tested everyone in the world. Leaving out some groups of people can affect the results in important ways. Without being able to measure these other groups you can never be certain your experiment is not flawed and only applies to a single population.

Students volunteers can be assumed to have good mental agility and confidence, especially at maths, unfortunately some people are scared of maths. It is more likely that you can find a member of the general public who is scared of numbers, compared with a science undergraduate. So asking someone less confident in maths to perform a maths related task could be one of the worst things you might ask them to do. It is possible that they would not want to take part when they heard what was involved. If you excluded them, you would exclude a part of the population, which means you would not know what that meant for the results. It could mean anything; perhaps all the maths phobics were better users of the prostheses. The results of this test would be different than if one could measure it in some other way. This is one reason why other forms of test of cognitive load were developed.

3.4.4.2 Measuring evoked potentials

One possible tool is evoked potentials [102]. This looks at the cortical effort used while performing tasks. Electrodes are placed on the head and the changes in brain waves are measured. It is already being used in research looking at the way people use computers; human-machine interface research. It has the potential to provide an objective measure of cognitive load.

One example of its use is to present the subjects with two types of information and ask them to detect the difference between them. This would be the second in a dual task test. For example, when we hear sounds there are many layers of interpretation: Are the sounds sensible words? Are the words build into something that is a sentence structure? Does that sentence make sense? This last requires a large amount of processing to come to a conclusion. However, if the subject were asked to react by pressing a button or speaking a response, many other factors, such as their dexterity, would come into play, which would make the measurement hard to do reliably. What is more certain is the relative time it takes to process the information *before* the person reacts to reply, this is more consistent. In this sort of experiment the secondary task is to ask a person to listen to sentences (while performing manipulation tasks). The difference in the brain wave patterns is recorded and the difference is the key.

When a subject is asked to listen to a series of phrases, sensible and nonsense and to tell them apart, it takes a person a fraction of a second longer to think about a nonsense sentence. All phrases are grammatically correct, but some are rubbish. A sensible sentence might be: *The woman walks the dog*, a nonsense phrase is: *The man planes the cat*. With evoked potentials we can observe if the person is working to decode the sentence and this delay increases with cognitive effort.

Even this is quite an invasive method needing electrodes to be stuck on to the head. This might prove to be too much for an already overloaded subject who may perform badly, or refuse to take part.

3.4.4.3 Gaze tracking

Another cognitive load test is to use the person's visual attention. When performing a task which is made up of a series of actions, the majority of persons will tend to look ahead to the next phase of a task. If they are going to pick up an object they will be looking at the object. By the time their hand is on the object they will already be looking at where their hand is going to be next, such as where the object is going to be set down [103,104]. My students have tested some users of conventional prostheses, and their gaze tends to follow the prosthesis, especially when the user is new to the prosthesis [105]. If the task requires the hand to hold the object for some time while the task is being completed then the gaze must move about. For example; if the person is pouring from a carton into a jug, the typical subject will concentrate on the spout and the water to make sure they are not spilling the water. The prosthesis user may have two things on their mind: One is not spilling the water, but the other is not dropping the carton. Their gaze flits between the two sites, the carton lip and the fingers of their prosthesis. From this it is possible to see that the user of the prosthesis, without reliable feedback, will need to concentrate more on the task.

To the average person this additional effort might not seem much, but to someone with a physical limitation every task needs to be planned and executed to ensure the person does not drop objects. As some people will not tolerate this extra effort to control the hand, they will tend to not use the prosthesis for active manipulation, or not use it at all. Thus researchers are interested in what changes in prosthesis control could improve the experience for the users.

3.4.5 Assessment in the wild - the pioneers of modern assessment

In the immediate future, with the vast array of different ways we have to monitor ourselves, from body mounted cameras everywhere to activity monitors, the assessment of prosthetic function (and all other medical conditions and treatments) will see a big change in what is measured and how that information is used. This information should be studied to help improve our knowledge of how prostheses are used and how we may improve all aspects of their design.

One of the crucial failings in the range of tools that have been developed to assess outcomes from prosthetic fitting are the measures of the Participation domain. Until now, the only way to know how a person really used their device was to either ask them with a survey or observe them in the field. It is very difficult to design an assessment tool to do this that generates generally replicable data.

The questions a survey asks must be clear and unambiguous. There must be no chance that the subject will misinterpret what is meant by the question. A survey that I used in several clinics [106, 107] asked questions about the 'reliability' of the prosthesis. The aim was to know about the users' experience of the longevity of the devices in the field, how often did they break and so on. Despite this, it was clear that some of the answers were more a comment on the clinics' service record than how robust the prostheses were. These two aspects are important, but distinct. A further concern is if the replies are truly accurate and objective. People may exaggerate as they have a wish to help or to impress the surveyor, or their answers may be inaccurate because they simply cannot remember what happened precisely.

Similarly, if a survey is to be used in another culture, or country, the language has to be checked for meaning. Native speakers are rarely aware of the wide range of meaning and nuance any particular phrase has, which can easily lead to misunderstandings or misinterpretations. The best way to remove such potential flaws is to translate the survey into the new language and then back into the original language, using different translators. If the survey returns altered in some way, then some meaning has clearly been lost or changed in the process. For example, in the first international space mission between the American astronauts and Soviet cosmonauts in 1975, the crews spoke the opposite language. The Americans spoke Russian and vice versa [108]. This reduced the chance that someone speaking in their own language would assume a meaning of a word. Mistakes should be less likely to be made if the precise meaning of the word and no other was used. In this instance mistakes had the potential to be fatal.

The best way to find how someone uses their device is to study them actually performing their daily tasks in their own environment. The disadvantage is that this has been difficult to do without it being invasive. The presence of the observer making notes in the subject's home is likely to change how the subject behaves. The information does not reflect what they actually do with their prosthesis. What is required are tools to monitor behaviour unobtrusively.

The development of compact recording devices that people are prepared to wear to record their activities for personal monitoring has opened up the field in a way that was unpredictable a few years earlier. While the technology to make small digital cameras and compact computer memories was easily predictable, the idea that people would willingly burden themselves with the equipment to record their every movement was less credible. Even so, a market has developed that produces such devices in a cheap form. This has made the adoption of new ways to monitor activities for scientific or therapeutic purposes far easier and cheaper. The assessment field has changed dramatically. Two pioneers of the field are Alix Chadwell from Laurence Kenney's group in the University of Salford University (home of the English prosthetics and orthotics school) in the UK, and Adam Spiers who was working in Aaron Dollar's lab in Yale University, in the USA.

3.4.5.1 Activity monitoring

In the early years of the new century Laurence Kenney began considering what were the barriers to successful deployment of prostheses, especially myoelectrically controlled hands. Over the last fifty years the basic design of the socket has not changed significantly. Small enhancements in socket and electrode design have been made, but the prostheses are no more successful than in the past and are still often rejected. Unreliability is often cited as a reason for rejection [109]. Simply, if the user cannot be sure that the hand will open to pick up an object when requested, or it inadvertently opens when carrying the object, the user cannot rely on the hand. It follows that no matter how effective the hand is as a manipulator, it will not be used routinely.

Alix Chadwell was recruited to investigate the key factors that have stopped the technology from being used in the field. To study a topic, process or device, one always has to separate the various factors that influence the results and identify the impact of one of them on the rest of the process. In the case of prosthetic control, it is the result of a complex set of interactions between the prosthesis, the user and the environment. A significant problem when using a prosthesis is that there are times when hands open spontaneously and may drop whatever they are holding. There are a number of possible reasons why a hand might open on its own. When this happens it is difficult to know which is the cause. Without this knowledge it is difficult to improve the reliability. Alix needed to separate the users' ability to control their myoelectric signals from the reliability of the signal to be translated in to the device, and how long the prosthesis took to respond [110,111]. To simplify matters, Chadwell chose to look at standard two channel myoelectric control. This is the way that conventional single degree of freedom electric prosthetic hands are controlled, where the signal from one muscle is used to instruct the hand to open and from the other to close it (see 6.1.3.2).

The first question to be answered was: Has the person issued the right signal? Generally, this means one muscle contracting enough to be detected while the muscle under the other electrode stays quiet. There are times when the person will fire both oppositional muscles in their arm, perhaps to steady the arm when held out from the body. Someone with an amputation will be likely to use their muscles together as they had before the loss, hence both muscles will contract. If the controller is not designed to respond to this it is unlikely to issue the right command to the hand, i.e. it may cause it to open inadvertently.

Secondly, if the operator has generated a nice clean signal, has the signal been picked up by the electrode correctly? If the arm has changed orientation or the hand is now holding a load at the end of it, then the socket may shift. If the electrode shifts sideways across the muscle, the electrical environment under the electrode has changed and this will be amplified as if it is a real signal from the muscle. A second possibility is a shift is away from the arm: If one of the two contacts on an electrode loses connection to the skin, then the balance between the two inputs is broken and the difference is amplified. A small shift then looks like a big contraction (*electrode lift (off)*).

A third question that became more important as the project ran was simply knowing how was the arm being used day-to-day and how *much* it was being used.

This led to an ambitious program of experiments with a much larger cohort of subjects than are usually tested in clinical prosthetic experiments. It is fortunate that there are relatively few people with an upper limb absence, and fewer who use powered prosthetic arms. But this means that in any one centre there is only a small number of users who can be recruited to an experiment. Most tests have a handful of volunteers. Chadwell wanted a more representative cohort. She recruited an impressive twenty subjects. To do this she needed to travel from Manchester, where she was based. She got help from six clinics: Salford, Strathclyde, Wythenshawe, Roehampton, Nottingham and Sheffield. To undertake this work she had to travel to each clinic, set up and make measurements, before sending the volunteers out with activity monitors on both wrists (natural and prosthetic). For each subject there were a range of tests across a single day. They aimed to measure differences in performance, with and without the influence of the weight of the prosthesis and socket. Finally, the hand's use in manual tasks was observed in the lab, followed by one week of activity monitoring at home.

With the exception of the last activity, all of the above were repeated ten times in each setting, before repeating the tests with the different electrode/arm configurations. They performed them with their standard set up; again with a five hundred gram load on the hand, and finally with electrodes alone taped to the arm. This last set up was supposed to be the ideal location of the electrodes without the load of the arm to shift the socket. It would have been a more ideal test to have been able to disassemble the socket and make sure the pressure of the electrodes inside the socket was optimal. This would have required changing their routine prostheses, which was not possible. If anything had been disturbed during the process, it would have impacted on the subjects, something that is unacceptable ethically. So a compromise was chosen [112].

What Alix observed was a very great level of unpredictability in the response of the devices to the signal. If the myoelectric signal is observed in the laboratory, it is seen to be clean and separable. There is very good contact with the skin, which means it has high reliability. However, the nature of the way a prosthesis is used in the field is very different. The socket has to be donned and doffed easily. The electrodes cannot be pressed into the skin hard for many hours a day. This makes the signal far less predictable and prone to all of the interferences discussed above. This jeopardises any fitting. With any hand, if you can't get the signal reliably into the hand it is going to be unreliable. This conclusion creates a problem for pattern recognition based control (see 6.1.3.3). This technique is going to continue to be less reliable than it might be, as long as the acquisition of the signals is poor. Although Chadwell did not comment on this, it returns myoelectric research back to the idea of electrodes buried in the muscles as the way to improve performance. These results suggested that in the absence of in-dwelling electrodes, an adjustable housing for the electrodes is worth pursuit, and this was a follow-on project in Salford.

3.4.5.2 Alix Chadwell

Biomedical Engineering is one of the few parts of engineering that has a more even balance between the genders. As a teenager Alix was studying the subjects that would lead to engineering (maths and design), but her perception of engineers was the conventional one: Someone who builds engines. In the UK there is still a strong social bias against the technical and towards those activities that are seen as creative [113].

Growing up, a family friend who had a below knee amputation so Alix had many conversations with him about his devices. Later, when Chadwell saw Heather Mills (the model, campaigner and high-profile prosthesis user) on television, she realised this was an area she was attracted to. Around this time, TMR was receiving its first publicity in the UK. Alix found a direction to head, she did not initially associate working with patients and cutting edge technologies with engineering, but decided that working with prosthetics was the sort of career she wanted to pursue. She investigated training as a prosthetist, but this could only take place in the English or Scottish prosthetic schools in universities in Manchester or Glasgow. Meanwhile her school were guiding her towards a higher ranking university. Independently, Chadwell discovered the discipline of biomedical engineering and then a degree program within the mechanical engineering department at Bath University. The degree was then discontinued half way through her studies, so she was the only student to graduate in medical engineering that year. However, her final year project was in prosthetics. It shows that the end goal makes the difference to the interests of the engineer. The topic of the project was the delamination of carbon fibre in foot springs. These devices are the well known 'blades' of the para-Olympians. *"If it had been delamination in term of racing cars I probably wouldn't have cared, it was the application of it to prosthetics. It's been my driver all the way since teenage years... I want to work with prosthetics".*

She had no intention to do a PhD, but an opportunity to do prosthetics presented itself and she took a look. It was a hard decision. By the 2010s an undergraduate degree in the UK was expensive to gain. There was little direct support from the government any longer. It was mostly self-funded, leaving the majority of graduates with a significant debt. Her choices were a salary, a graduate engineering scheme and professional chartership in two years, and pay off the debt, or continue to be a student. She is glad she took the option into prosthetics.

For her, the ability to meet the recipients of the technology is important. During her degree, even in medical engineering, she did not talk to any patients. Its only during her PhD and after that she has seen the user needs and applications side, but this is an important aspect of the job.

> "A multidisciplinary team... you design something together. Once you start to work with clinicians, ... you start to meet patients in person, you start to hear different stories. Each new person you meet they have a different outlook on what their experience has been. If you meet someone with a different experience it might throw something different into your thought process" [112].

A major outcome from her work has been seeing the results of using activity monitors to objectively measure what is going on outside the clinic. It was necessary to work through the mass of information to get usable results. One important comparison made was between the use of the prosthesis and the contralateral hand. Prior to the work, the likely perception from the profession was that the prosthesis would be used intermittently but often, and in support of the other hand. This was not reflected in the data at all [114].

Perhaps the most important finding was that, based on the relative activity of each arm, there wasn't any correlation between wear and use, where use is measured by recorded activity of both arms at the same time [111]. This is perhaps the first time we have had objective confirmation of the suspicion that has existed in the profession for years; there was an important difference between wear and use [106]. This demonstrates what is possible when new measurements become practical.

The activity monitor measures when the hand is being moved about, and when it is not. The latter suggests when the device is sitting in a lap, or lying in a cupboard. There were periods of light activity of both limbs. This probably happens when the arm was being swung as the wearer walked. Although the users were asked to keep a diary of when they wore the arm, it is similar to the standard survey, the person may forget to fill it in or record it inaccurately. Thus the gap left is that we cannot tell if the prosthesis is being used actively (grasping) or passively (pushing).

It is not difficult to build monitors and recorders into commercial prostheses. In the past a mechanical counter could be added to record the number of openings and closings, but this could be the result of the user 'playing' with the hand, not its active use. By placing the recorder within the hand, it allows the controller to record the accelerations along with the EMG commands opening and closing the hand. Thus active use can be put in context and compared with the passive use of the hand. This cannot be added by researchers to an existing commercial hand without replacing the electronics, which is modifying the hand more than the manufacturers tolerate. Some manufacturers have recorded some use data, and have reported it a little at conferences, but when asked, none have so far been prepared to release this information to clinicians or users.

3.4.5.3 Task tracking

Chadwell's work identified when the prosthesis was being used, but it pointed to a gap in knowledge that would remain hard to measure without an instrumented hand; *how* was it being used? This can only be done with direct observations, and this too can now be achieved electronically.

The approach taken at Aaron Dollar's lab in the USA was to apply head mounted cameras to record people undertaking their daily jobs [115,116]. This experiment looked at the range of prehensile grasps performed by a small number of healthy able-bodied users. This was before the GoPro and other head mounted cameras had reached the market, so the team had to develop their own devices. They recorded the work activities of four individuals: Two mechanics in a workshop and two professional house cleaners. What they found was the subjects used a wide range of different ways to solve the same problem. When Adam Spiers took on a similar task for prosthesis users it was not a surprise that the results were also very variable.

3.4.5.4 Adam Spiers

Adam Spiers obtained a degree in Cybernetics from the University of Reading in the UK. He was interested in biomedical engineering, but the work in Reading was on neural interfaces which wasn't for him. He moved into robot manipulators for the nuclear industry, before going to the Bristol Robotics

Laboratory to do a PhD on manipulators. His aim was to apply some of the robotics techniques to prosthetics, but the more he read the more he realised it wasn't that straightforward. There were too many differences that made the application inappropriate, although he thought that many others did not come to this conclusion and continued to work in the wrong direction [115].

Adam's PhD aimed to look at the human control of robot arms. As the human arm has many joints it is possible to move the hand from one place to another in many different ways, bending the joints in the arm, shoulder and back to move the object though many different paths. Humans usually employ very similar solutions, unless they are injured or something is in the way. It feels natural to reach out with the elbow pointing downwards, even though moving it out to one side or the other by thirty degrees involves little additional effort. Despite this, our ability to measure where our arms are in space is not that precise. For example, if we reach for an object with our eyes closed, we rarely get our hand to exactly the same place twice. This rarely matters because we have our eyes to correct for the errors. Robots are generally controlled to a high accuracy, which makes their interactions with the real world difficult. The first place for robots to find real uses was in the production line and this allowed every part of every action to be very precisely defined. As Adam Spiers was looking to apply robots in less well controlled environments, he found these inaccuracies in measurement made many techniques he tried difficult to achieve successfully. He developed a video system to observe an arm moving, then an artificial neural network to interpret the motions and a hierarchical controller to move the arm (10.2.1). This is not that far from the way a human system works and it resulted in the elbow naturally pointing downwards.

This work fed into looking at robots working together (social robots), and then the way that surgeons move their hands. This showed the skilful efficiency of motion that the surgeons worked with their tools. It was not a big step to move from this to looking at the way that limb absent persons might use their tools. It was here that his earlier work gave him a different perspective from many in the field. Adam sees users' motions differently. The users of prostheses might wish to move their shortened arms the same way as their typical arm, but the motion of the prosthesis *"was not efficient because they were trying to solve a kinematics issue"* [115]. Simply, they had to change the way their arm moved because the arm itself could not do all that the natural one could.

By this time, the GoPro head mounted camera had become available. They were originally marketed to make recordings while the wearers undertake sports. In the standard form, the camera's capacity was limited to about thirty minutes. Adam extended it to five or six times as long with an external battery. Thus prepared, Spiers started. He took the equipment to volunteers' homes. He was able to set up the equipment and prepare the users. He began by using the personal touch because he was concerned that the position of the cameras on the subject's heads was critical to getting a good image of their hands and activities. It would be frustrating if the camera was not actually capturing the hands and their motions, if all the important information was just out of shot. In fact, the GoPro was made exactly to avoid such a problem. They are placed on the head of a wearer while they perform vigorous sports, so they have a lens with a wide enough angle that even if the camera was knocked while skiing or skateboarding they could still get a good shot of the action. The head mounted cameras missed little of the hand activities. Once he understood this Spiers was able to mail out the systems to participants, allowing him to get the systems to a much wider range of subjects, and acquire a larger pool of data. However, the early experience of going to the users' houses and talking to them he found *"...a really useful experience in getting to know the community better"* [115].

Subjects were able to record eight hours of activities at home over two days. If they used more than one prosthesis they were asked to split the time between them. From this, an observer has to view every frame, identifying and noting the activities. A trained observer must classify the activity by type and record how long they were used. To make this practical, a specific computer interface had to be written and a small army of students employed to work their way through hours of video footage to extract the data that could be used to understand the activities.

Any hope of simple analysis of the data proved to be in vain. Spiers learned what others had done before: Even a single real action (such as making a coffee, or feeding with a fork 3.4.3) had so many variations that it could not easily be compared. Conventional science experiments allow the scientist to control all the variables and change only one thing at a time. For example, two experiments to measure the rate of swing of a pendulum by two different scientists can use similar tests and compare the rate easily. In making coffee, the differences are so great it is hard to be sure they can be compared objectively. If the coffee makers are very different, the motions have to be very different, even if the basic process is the same, there are many ways to go about each action. Since the actions aren't the same, there is no simple way to compare the actions. A more fundamental framework had to be constructed into which the analysis could be done. Instead of how is a particular task performed, it became a study on how are all tasks performed. Adam devised a taxonomy of grasps.

He divided the range of grasps into:

Single handed The subject uses one hand (prosthetic or natural).
Bilateral The subject uses both hands passing the object between them.
Prehensile and nonprehensile actions The subject uses their prosthesis to hold an object or as a device to push, pull or clamp the object of interest.

This last one was the most crucial distinction. At a time when Chadwell was measuring activity without knowing if it was functional or otherwise, Spiers was showing that nonprehensile activities are important. He was also able to observe other interesting uses of the device, actions we might not see a person with two natural hands do, such as hanging things off the prostheses while working (hangers while putting out the washing).

Simultaneously, at the same lab, Gillian Cochran was looking at *"affordances in the environment used by subjects to aid manipulation'*. This is; using things in the environment to create ways to more easily handle objects. For example, picking up coins by sliding them to the edge of the table to allow a simple grasp across the coin, rather than directly trying to peel it from the horizontal surface [115,117].

When interviewed, Adam had studied four subjects. He has data for 27,000 individual manipulation actions and what he has found out is perhaps not a surprise; body powered devices are used much more actively than the myoelectric devices. The body powered users employed their devices for 20% of the time while the myoelectric users was only 17% of the time. Again, not a surprise, but it is useful to have one's suspicions confirmed.

The surprise was that for all of those with a transradial loss, the majority of interactions were *not* prehensile. Even with the body powered users 79% of all actions were nonprehensile. It was much higher than the 60% he had expected. While myoelectric wearers used their devices passively only 60% of the time, these large numbers can be compared with between 16% and 19% of nonprehensile activities in the intact population.

The work makes Spiers think about design and use of hands. He can see that body powered devices are faster in operation. *"It's really interesting. Watching the video [and seeing] the speed and agility that the BP users would do tasks it was really impressive"* [115]. These users aren't waiting for the hand to open, and they seem more confident in operation [115]. Most importantly they are happy to treat their devices robustly while the users of powered devices seem inhibited in their use as they are aware of the fragility and cost of the devices. This is something Mat Jury of Taksa Prosthetics would rather his users didn't need to feel [118]. Like many of us in the field, Adam hopes that his findings will get to the people who will benefit from them, whether it is in the design of new devices, or new occupational therapy interventions.

3.4.6 The future of assessment

The aim of assessment remains the same: To objectively measure the entirety of the design and use of a prosthetic limb. As with the rest of the world, the level of technology has changed the perspective by increasing what can be measured and how precisely it can be measured. The ability to then use these large amounts of data to change what we do, lags a little way behind. But given what we can record now, we are in a better position to try and find what are the right things to measure. We can avoid studying something simply because we can measure it. Instead we can measure aspects that tell us something important about the way a prosthesis is used, such as knowing the objective difference between wearing and using. We have moved from not being able to measure much at all, to being able to measure very many things, some of which may hinder the user. If we know that, then the first step is to remove the barriers.

Both Chadwell and Spiers see the same importance in being able to give the user a monitoring device and allow it to be used unobtrusively in the field: *"If we can develop something that they can stick on, send them away and have a wealth of information. I think there is a lot of value in that"* [112].

The control of all externally powered prostheses relies on the stability of the input signal used to control the prosthesis (how good the EMG signal is). This has an implication for what we might expect from a standard myoelectric hand. With the addition of more electrodes in pattern recognition this might well increase the unreliability by a greater amount than simply the number of electrodes used: Is four electrodes less usable than two? This is an argument for the external switches and accelerometers that some of the manufacturers are employing (6.2), as these instructions do not depend on a good electrode contact.

It is clear from the work of Alix Chadwell and Adam Spiers that not using the prosthesis is as much an important part of what users do with their devices as what they use them for. Nothing has changed. The clinical perspective is: As long as it matches the users' needs, it is acceptable for the team. However, the payer of the service might also want to be sure it provides what they perceive as 'value for money'. The danger is the assessment of value would be based on simple measures like wearing time. Radically, Chadwell asks if the *non* use of a complex device's capability should be assessed as part of the measurement of the prosthesis [112]. The work of the pioneers of assessment in the wild suggests that this sort of measurement should be extended for *all* devices, not just the complex ones. Chadwell sees the need to produce tools that don't need much time to analyse [112], which fits with Spier's experience of analysing his data. Computer based analysis of complex activities would require ways to automate the selection of the activity or action, so that a computer could rapidly classify hours of data and point to the interesting events. An important approach is to frame the problem in a way that makes the analysis easier.

3.5 Cybathlon

In 2012 a group in Switzerland led by Robert Reiner announced they would be holding a competition. It was billed as the competition for *Bionic Assistive Technology*. It aimed to raise the standards of the profession through a series of competitions for prosthetic arms, legs, wheelchairs and brain-computer interfaces, with the fastest in each category being crowned champion. Reiner is an engineer and son of a motor mechanic. As a young person, Reiner was torn between engineering and medicine. When the opportunity to work on Functional Electrical Stimulation (FES) for a PhD presented itself, he jumped at it. FES stimulates muscles to contract through external electrical signals so that some function can be restored. Its results are limited due to the complexity of the task, rarely does one muscle perform one task without the use of others to steady, adjust or oppose the motion. FES has enabled some persons with spinal lesions to be able to exercise, and restores some grasp function in people who have suffered strokes [119,120]. More recently Reiner had also progressed to work on artificial legs as well [121].

What struck Reiner, like others in the rehabilitation field, is that there was quite a disjoint between the solutions engineers and clinicians had proposed and the real problems on the ground. Reiner developed a long-term aim to try and create a way to get the two groups to talk together. The means to do so came to him in 2012 when he learned of a fund raising event created by the Rehabilitation Institute of Chicago (RIC). The RIC team were working on a design of powered knee and they got a user to climb the Willis tower in Chicago. This generated a lot of very good publicity. Robert felt: *"Why do the Americans always sell their research so good, we can also do it. We have good technology but we don't sell it"* [121]. Being Swiss he thought less of tower blocks and more of the natural resources Switzerland is famous for: He first considered getting a prosthetic leg user to walk up a mountain. However, concerns over the reliability of the weather jeopardising the event meant the idea moved to an indoor venue. Once indoors it rapidly became clear that any such event could host races for a much wider range of technical aids.

The championship sets tasks that encourage competition to improve the quality of the field, but importantly it aims to get engineers talking to users and clinicians. This is much like the early days of aviation with races and challenges between rival plane builders. Perhaps in this spirit, the users of the devices are referred to as 'pilots'. The intended competing organisations were to be universities and companies around the world. The idea was to allow them show off their latest mechanisms and controllers, and for them to be worn and used by the pilots. After getting the engineers to talk, the second aim was to influence the market in a similar way to Formula One car racing's impression on domestic motoring. F1's influence happens over the many years, slowly. The results start as gadgets for the most expensive cars. Then the technology trickles down to the family runabout in ten to twenty years. Reiner feels this approach is working: The first competition was in 2016, and two wheelchair designs used in the competition had already become products by 2019 [121].

The competition was designed by a panel of experts and users. The biggest challenge was to find tasks that were relevant, challenging enough for competitors, and interesting enough for spectators. The prosthetic leg competition was a form of assault course, stepping over obstacles and up and down ramps and stairs (all valid challenges for the prosthetic leg user). The wheelchairs were raced around a track. This is both easy to make into a competition and reflects the user's real life experience. To test brain interfaces they created a form of computer game played without the use of external paddles to control the progress of the competition. It was far from the real world, but such interfaces are even further from the usual experience of this group of users.

The arm competition is perhaps the toughest to make valid and interesting. *The Track* was a diverse set of manipulation tasks. Competitors needed to undertake various day-to-day activities such as putting a screw fitting light bulb into a lamp stand and opening a door with a round handle. More than half of the tasks were bimanual. All of the tasks were chosen so that they could be performed monomanually with the natural hand, but use of the prosthesis would give the pilot an advantage, it was worth trying to do them bimanually. There were a limited number of tasks where pilots were directed to use their prosthesis and not their dominant hand (something occupational therapists always dislike when it is suggested for assessment tools). Reiner admits this is a limitation, but he hopes that one day designs of prosthetic hands will match the capability of the natural hands and so this task in the competition will come into its own [121]. Teams entered a pilot who raced to complete the tasks in the fastest time. The first upper limb competition in 2016 produced a different result: A body powered prehensor won (see 5.5). This is less surprising when one analyses body powered devices (5.2).

Creating the Cybathlon required convincing sponsors that it was worth supporting the new concept. Potential sponsors were unsure if they were supporting a sporting event, a rehabilitation exercise, or a conference. The truth is, it is a little bit of each. A hurdle was that some sponsors only supported sporting events so had problems with the other aspects of the design. A Swiss bank would have liked to support the competition, except their website home page and their cash machine were not accessible by disabled clients, so to support the competition would have seemed hypocritical. The success of the first competition meant that the organisers were able to raise the money to support a bigger competition and more than six months ahead of the next event they already were fully financed.

Reiner's original aim was: *"To promote the long-term development of technology for people with disabilities"*. He was focused on the evolution of devices, not therapy. He later realised it could help with inclusion as he wanted to get both sides talking, engineers to users. Less than a year before the next competition, he believed it was working as the engineers were talking to the pilots earlier than they did previously.

It was only later he realised it could promote inclusion too, normalising the competitors. Following the success of the first competition they aimed to promote the work and aims of the Cybathlon with local regional events:

Cybathlon Experience These present 'casual' demonstration races at three or four rehabilitation fairs and shows. They do not have to follow the rules of the Cybathlon too closely.

Cybathlon Series These are much closer to the full Cybathlon. Competitions follow the rules of the main competition. The idea is that academic institutions will host it as this will be more credible than having companies acting as host. In the years between the first and second Cybathlons a powered leg competition was held in Tokyo and arm and leg races were held in Karlsruhe, Germany. Finally a BCI race was held in Gratz.

Cybathlon at School This aims to get the idea to the grass roots. The team developed teaching tools (hardware and software) that can be bought and then a person with a disability could present it to the pupils. The idea is that funding from donors will be raised to do this.

Reiner found the ambition for this part of the Cybathlon only partially fulfilled. The Cybathlon still sits outside the usual forums for prosthetics. Relying on external funding for each of the activities means the hosts need considerable investment in time and energy to pursue this. It is clear that having a university offering the Cybathlon Series competition gives it apparent objectivity. However, the level of funding required is significant, and the conventional source for universities is backing from governments or companies. The former does not usually support sporting activities through research and education institutions. It is more likely that if this part of the program is to succeed in the long term it will need a different approach to raise the money, such as a special interest group or charity. In any eventuality, the Cybathlon has shown that a different approach to innovation and fundraising can have an impact on the prosthetics field.

In the spring of 2020 the Cybathlon for that year was postponed due to the global pandemic.

CHAPTER

History of upper limb prostheses

4

Compared with many other areas of medicine and rehabilitation, prosthetic limbs have a long history. Surgery, the art of the physician, has a prehistory running until the development of anaesthesia and antisepsis [26]; modern practitioners would wish to divorce themselves from most, if not all, of the techniques and tactics of their forbears. For the engineer, this same era is a time of attempts at creating solutions. From then, we get hooks, peg legs and sleeve fillers which still have their own manifestations in modern prosthetics. The older devices remind us of our historic ambitions, successes and failures.

The Rig Veda, (a collection of ancient writings which is part of the foundation of the Hindu faith) tells of Vispala, who was given an iron leg when she lost her own leg in battle. Two Latin writers mention prosthesis users: Herodotus tells of a Persian soldier (Hegistratus of Elios) who used an artificial foot [122] and in Pliny the Younger's 'Natural Histories', he mentions Marcus Sergius, a soldier in the third century BCE. According to Pliny, Sergius was a formidable soldier. He was imprisoned by Hannibal for 20 months and managed to escape. He raised the siege of Cremona, and captured enemy camps in Gaul. Pliny reports he had a *"left hand made of iron"* with which he went into battle [123].

The earliest physical evidence of technology comes, as it often does, from ancient Egypt. There was an artificial foot made for Egyptian mummy [124]. It is uncertain if this foot was merely to fill the wrappings or was used for walking, although some wear suggests it might have been used functionally. A second mummy from sometime in the Ptolemaic Period (332-30BCE), now in the university museum in Durham, was X-rayed and it clearly showed that the person was a woman with a congenital absence of her right forearm. She had a hand and forearm prosthesis (Fig. 4.1). We know that, Egyptians believed in an afterlife. To prepare for it, they had symbolic food, servants, carriages made and placed in their burial chambers for use after death. Similarly, to be a whole person in the afterlife they would need replacement limbs to be made to make up for any absences. So the prosthetic arm's role for the mummy was to appear like a real arm, which would make it a 'cosmetic' prosthesis [124] in the modern era. According to *her* belief system, however, its function was to restore full manipulative capabilities in the afterlife, thus it would be a practical prosthesis. It was the most perfect and functional prosthesis ever made.

Most prostheses of the early modern era that we are aware of were made of metal and resemble armoured gloves. There may have been others made of wood and leather but they would have been less likely to be preserved. Even then very few would have been made as they could only be built for those who could afford them. Knights would have been one of the few professions likely to lose a limb and be able to afford a replacement. For this reason it seems the popular suggestion is that these hands were built by armourers for their warrior masters, and the designs were based on their expertise in making armoured gloves. I could find no direct evidence for this.

Early physical examples of prosthetic hands used in daily life are hard to find before the fifteenth century with the Alt-Rupin Hand (circa 1400 [125]). Documents describing earlier devices exist [126],

Making Hands. https://doi.org/10.1016/B978-0-12-820544-0.00013-5

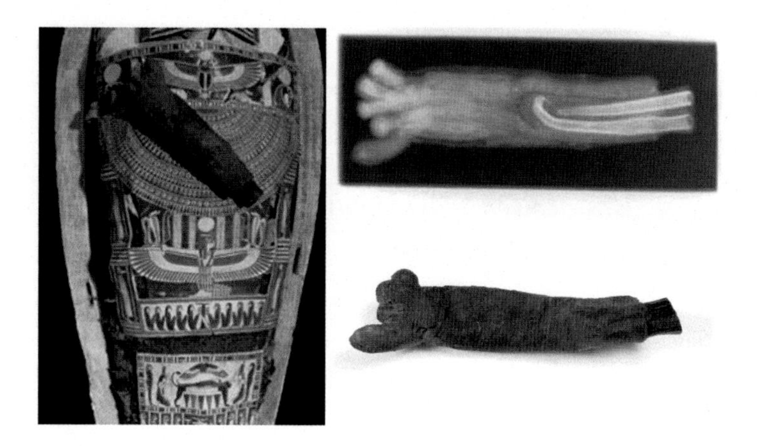

FIGURE 4.1

Mummy of a woman who would appear to have had a congenital absence of her left forearm and her mummy had a replacement made for use in the afterlife. (Copyright: Oriental Museum, Durham University.)

but they contain few details and no physical evidence survives. The most celebrated of the early modern users was the German knight, Götz von Berlichingen (1481-1562). In 1509 at the age of twenty-three lost his hand in the siege of Landshut [127,128] after being hit by cannon fire. Von Berlichingen's prosthesis seems to have set the pattern for a number of copies of the design (Fig. 4.2). For example, at one time the London Science Museum accredited the user of an artefact they possessed as Von Berlichingen until they realised it is clearly built for the wrong side [129].

The arms that survive generally have all fingers and thumb (and possibly depressions for the nails), but they also have frills and detailing, which is unnecessary for prehension, showing that appearance has always been a functional requirement. The digits may be multijointed, but the hand generally has a single axis to open and close, and a lock with a button release. So to employ the prosthesis requires the use of the other hand.

One of the most common popular images of an amputee is the sailor, this is not without justification. During battle canon balls striking the hulls of wooden sailing ships created a miasma of flying splinters below decks. Any splinter could penetrate the skin and trigger an infection leading to amputation. In reality, soldiers on land lost limbs due to blasts and infections too. Of course, the majority of persons with an amputation could not afford a prosthesis to be made for them, so there is a long tradition of the limb absent designing their own assistive devices. One recorded example is Shadrach Byfield, an English soldier in the War of 1812. He lost his right hand in the battle of Black Rock in New York State, August 1814. As a weaver he would have been unable to perform his trade when he returned to civilian life, leaving him with limited options to make a living. According to his autobiographical writings, he imagined a design of a prosthesis in a dream that would allow him to continue weaving and got the local blacksmith to make it up [130,131]. Sadly, no record of what it looked like survives.

There is always a cultural response to disability. Ancient societies such as the Spartans had little tolerance for the impaired [132], while the evidence of an adult Neanderthal skeleton with a congenital absence suggests that other societies would accept difference. Indeed, skulls have been unearthed with

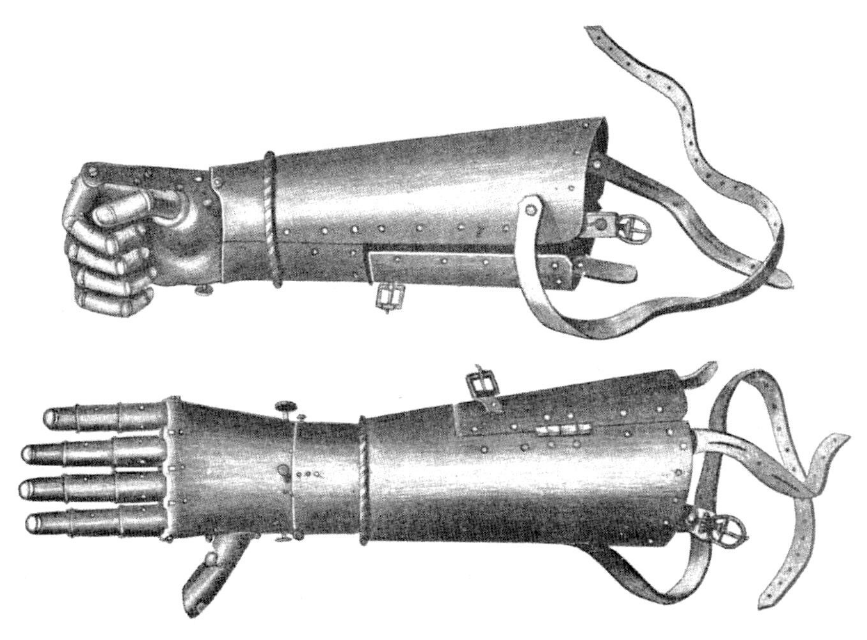

(a) The most well known historic hand design. The socket and the straps fit over the remnant limb and hold the prosthesis on the arm.

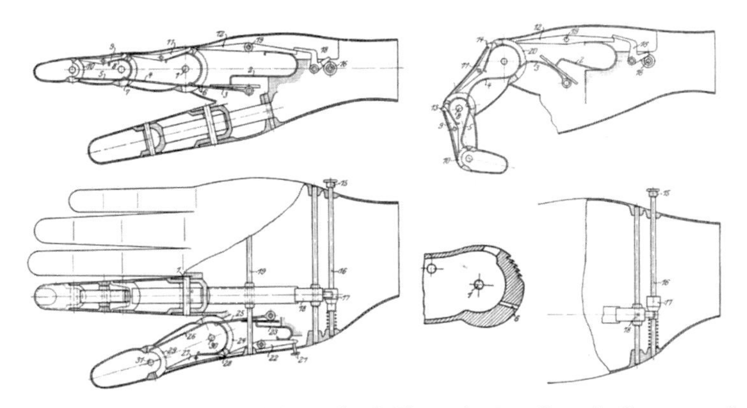

(b) Mechanism for the Von Berlichingen hand. The mechanism allows the fingers to curl around the target object. A release button (15) on medial side of hand is also visible on the drawing above.

FIGURE 4.2

One of the designs of two Von Berlichingen hands. (Both taken from [2].)

gaping holes in them. The holes are the result of trepanning, the boring of a hole to let 'evil spirits' out. What is interesting is that the edges of the holes have begun the process of healing, so it is clear that the person not only survived the operation, but lived with the treatment, something that would have needed help and acceptance [133,134]. Indirect evidence of long survival after amputation and prosthetic use in the sixth century comes from a burial in Logobord in Italy, the remaining bones and teeth, show changes that most likely occurred from the long use of a prosthetic limb [135].

The causes of limb absence or loss are either a congenital abnormality (an interruption of growth in the womb, not usually associated with any specific cause), or as a result of amputation. Until we learned to control infection, amputation was the option of last resort. In a time before knowledge of the bacterial theory of germs, any infection had the chance to grow and spread, rather than recede. The result was that the infection could take over the body and kill the person. Hence small cuts could be the cause for tremendous suffering and death. One of the few options open to the doctor was to amputate ahead of the infection line to prevent the spread. Unfortunately the surgeon did not have any form of control of anaesthesia or sterility. To perform the operation successfully the patient had to be physically restrained. It was not uncommon for patients to die either of the shock or the infection they received during the operation. Nor was it unknown for the patient to back out, unable to contemplate the pain of the procedure. Unlike today, the surgeon's most valued skill was speed [26], one such is the prolific French surgeon Barron Larrey, who performed two hundred amputations in the twenty four hours after the Battle of Borodino [136].

Only once better surgical techniques and the ideas of antisepsis and anaesthesia had been developed, and taken root in the medical profession, could amputation become more than the last resort. It took the first industrial scale war to create sufficient numbers of amputees to allow an industry to begin. This was the American Civil War. James Hanger lost his own leg at the Battle of Bull Run, and made a replacement out of barrel staves (Fig. 4.3). This proved popular with his fellow veterans and he founded a business. The Hanger company still trades as a prosthetics supplier today [137].

The progress of prosthetic limbs is influenced not only by technological advances but the people involved in their development; designers or users. In this prehistory I can only identify some of the significant influences on this progression before exploring the more recent pioneers.

A peculiar conceit of the field, is that external prostheses are known by the centre they come from (Edinburgh Arm, Montreal Hand). This is compared with internal prosthetic joints (hips and knees), which are more commonly known by the name of the surgeon who devised and put them in.

4.1 Early advances

While serving as an Army Surgeon, Ambroise Paré (1510-1590) developed better surgical techniques for many injuries and conditions he encountered [125]. One area he made significant strides in was in treating amputations. Paré applied ligatures to control bleeding. These are cords used to tie off the end of the stump to reduce or stop the bleeding. Prior to this, surgeons crushed the remnant or immersed it in boiling tar to staunch the flow. Paré also introduced follow-up care for his patients in which he aimed to improve their functional capabilities. He commissioned the construction of many prostheses for his patients. Production of the prostheses was under the guidance of a locksmith whom he called *Le Petit Lorrain*, the name the prostheses are known by today.

FIGURE 4.3

James Hanger veteran of the American Civil war. He lost his leg and founded a company to produce devices for fellow veterans. The company still operates today. (Creative commons licence.)

In 1812 a Berlin Dentist, Peter Ballif [125], made a great advance toward creating a functional upper limb prosthesis. He controlled the opening and closing of a hand by capturing the relative movement of some of the remaining joints in the same arm. This is the first recorded example of *body powered* actuation used in upper limb prosthetics. While a simple innovation, it allowed the wearer to use the hand independently of its contra-lateral, meaning that the user has some limited bimanual gripping function. Previous devices needed the other hand to actuate or lock them then users could not operate the prosthetic hand and employ their natural hand simultaneously, with this innovation they could. While the hand mechanism Ballif used was not particularly remarkable, the concept of the harnessing system was the breakthrough. The illustrations that have survived suggest the control of the harness comes not entirely from the shoulder, but also the chest (Fig. 4.4). The result of this complex harness is that it could not easily be worn under any clothing and would have been complicated to put on. In Ersatzglieder (4.1.2) there is a study on many designs of prosthetic hands to date, this includes a demonstration of the Ballif harness. The hand Ballif also show has multiple articulations and is pulled open and a spring for each finger closes the hand on an object, probably with a compliant grip (Fig. 5.1).

4.1.1 The Carnes Arm

The majority of recorded arm designs prior to the twentieth century are quite limited in scope. Many are passive and have limited motions. It is probable that they would only be of limited manipulative function. One indicator of this is the development of Carnes Artificial Arm which was patented in 1904 by William T. Carnes of Harmonsburg, Pennsylvania. It was seen at the time as a revolutionary

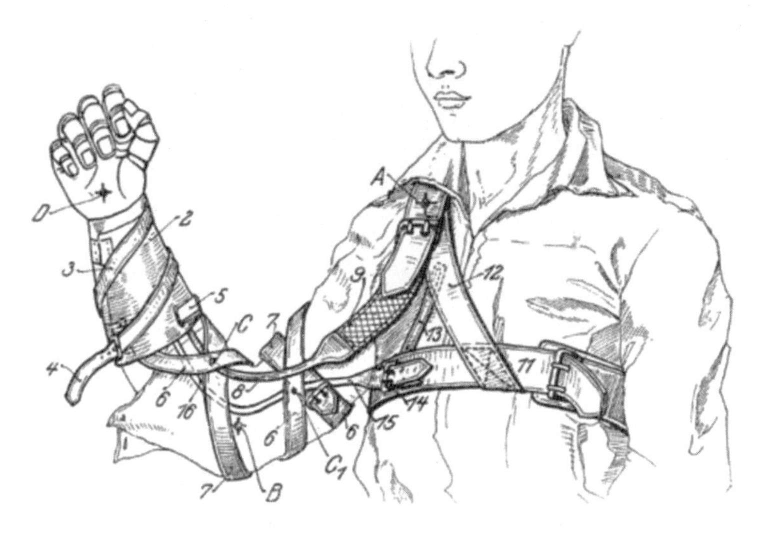

(a) Unlike a modern harness, the straps come from the chest and not just the contra-lateral shoulder.

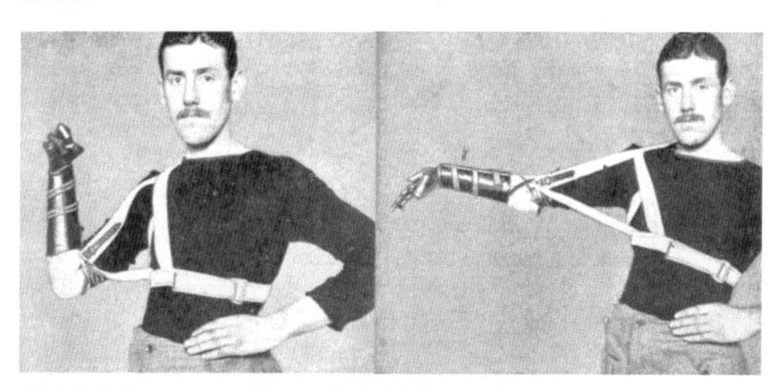

(b) A 1919 demonstration of the use of a Ballif-type harness showing it could not fit inside clothing.

FIGURE 4.4

The first recorded design for a body powered prosthetic hand, by Ballif in 1812. (Taken from [2].)

mechanical design. The original arm was cable operated from a shoulder harness, and a hand with articulated digits. Wrist rotation and flexion was passive and the joints were lockable. The patent [138] shows that the digit flexion mechanism was a cable operated rack and pinion system. The design was modified to a worm and wheel configuration in a 1912 patent [139]. A mechanism was added to rotate and lock the wrist.

In July of 1915, during the First World War, there was an international exhibition of artificial arms held near London, at Roehampton in England. Limitation on production (both men and facilities) in the UK during the war meant that the majority of the twenty-four exhibits were American products. The USA did not join the War until 1917. It still had a great deal of surplus labour and manufacturing facilities to respond to the need. At the exhibition, the Carnes arm was awarded the gold medal for the greatest advance in mechanism design, seen as able to deliver life-like dexterity. It was attached in the middle of the humerus in such a way, that the wearer had the finesse even to pluck out a hair [54] (Fig. 4.5). In the UK the arm gained the name the 'officer's arm', as it was not fitted to the lower ranks of military amputees after the war ended.

In 1915 America was a neutral country, so patents for the Carnes Arm were purchased by the German authorities in 1916 and the design was modified slightly to improve functionality of the mechanism. Meanwhile, the arm's popularity in the UK caused questions to be asked in the British parliament. In a debate in April 1918 Colonel Ashley directed his question to the pensions minister, whether

> "in view of the general opinion held by discharged soldiers who had lost an arm; that the Carnes arm was in most cases preferable to the Blatchford Arm, would steps be taken to issue such an arm in suitable cases?" [54].

Chas A Blatchford and Son is a major UK prosthetics manufacturer founded in the 1890s. It has a long history of innovation in prosthetics. In 1991 Blatchfords was the first company to introduce a microprocessor controller in a commercial prosthesis [140]. Since the 1940s it has only manufactured leg prostheses.

After the Great War a prominent British surgeon; E. Muirhead Little, described the Carnes Arm as:

> "a wonderfully ingenious machine; but the hands, which are made of steel, have the disadvantage of being heavy, weighing, according to type and size, from 16 to 24 ounces [0.45-0.68 kg]; On the other hand, they have the advantage of adaptability to various positions, which enormously increases their usefulness" [141].

Comments that could still be applied to modern prosthetic hands.

The ideas embodied in this arm have continued to reappear over the years. Carnes redesigned it, which allowed him to make new patents in 1942 [142]. It fell out of favour as it was expensive and heavy. Yet its appeal has continued. Recently an engineer with an amputation, Mark Lesek from Australia, built his own version, using newer materials [143].

4.1.2 Ersatzglieder - the first prosthetics compendium

Like Paré's experience in the 1540s, by the time the Great War had ended a great deal more had been learned about surgery, medicine, psychology and rehabilitation technology while treating the very many wounded soldiers. New classes of injury from high velocity shrapnel, gas attacks, fire and incessant bombardment had stretched the ingenuity and compassion of the people working with the wounded. In Germany a book emerged in 1919 which was a summary of much of the knowledge in the prosthetics arena. Its title is: *Ersatzglieder und Arbeitshilfen für Kriegsbeschädigte und Unfallverletzte* (Artificial Limbs and Aids for Disabled Veterans and Accident Victims) [2]. Alongside descriptions of socket design, amputation and forms of occupational therapy, the book contains a comprehensive

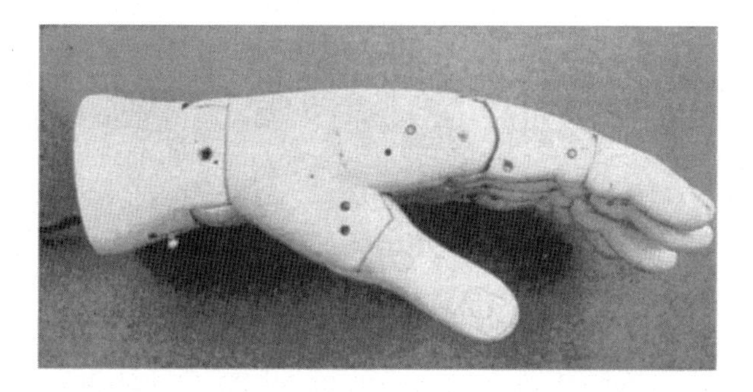

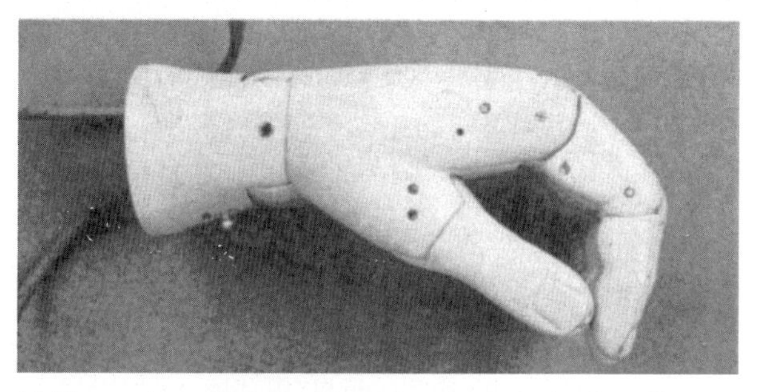

(a) The continuously curling finger tips oppose each other, allowing a range of grips to be achieved.

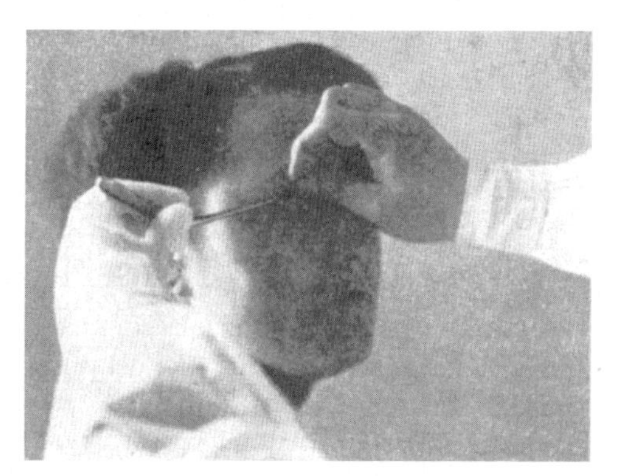

(b) Demonstration of the dexterity achieved with the Carnes Hand.

FIGURE 4.5

The Carnes Hand, was originally produced during the First World War and an American patent allowed it to be manufactured by both sides during the conflict. (Taken from [2].)

study of very many of the designs of prostheses up to that point. It remains a useful reference. Steve Jacobsen, the originator of the Utah Arm and the MIT/Utah Dextrous hand (7.2.0.1), firmly believed that Ersatzglieder had the majority of all arm designs collected together. So if a modern engineer approached Jacobsen and suggested they might have a new idea about a prosthetic mechanism, he advised then to turn to this book first, in order to find out if it has been tried out before [144].

One design not in the book is the worm and wheel mechanism used in the Touch Bionics' ProDigits design (8.2), but it is in the minority. Most of the other hand designs that one can observe being promoted today are variations on older themes. However, even if the design is in Ersatzglieder, it does not mean the idea should be abandoned. Simply one needs to consider why it failed to work last time, much as many of Leonardo da Vinci's designs had promise but perhaps lacked the materials or the power source to drive them. Some of the designs may be worth revisiting with a new perspective and new materials.

One thing that is revealed by studying Ersatzglieder, is two externally powered prosthetic hands in 1919, Fig. 4.6.

The first is electrically driven, but does not use rotary motors. Instead it employs electromagnets. It is probable that these were (at the time) more compact than rotary motors. The electromagnets pull on one lever, which appears to pull on a gear that drives the fingers to close in a group, whilst a second lever on which the thumb is mounted, is pulled into the palm to form the grip. The hand has rigid fingers that close together as a group. All of the mechanisms that work the hand are in the palm. The drawing includes a pair of terminals on the back of the hand, suggesting the power source and switches to control the hand are at the other end of the wires.

The second hand is pneumatically driven. It seems to have a separate cylinder in the palm for each digit. Every digit has three phalanxes which are linked together so the fingers will curl round the object as it closes. It is possible this would have given the hand a conforming or adaptive grip. Again the power source is external to the hand. This means it was probably aimed at being used at a desk or workstation.

While there is little evidence that these devices were built, they do represent the first steps towards the creation of externally powered hands and arms. For any such hand to be practical, it required great advances in technology to create compact motors, batteries and electronics.

4.1.3 The APRL Hand

Both world wars gave the governments of the participating countries problems of larger numbers of injured veterans to deal with, during and after the conflict. This was a problem mostly in Europe after the First World War, but the US government faced the same problem after World War II with approximately 17,130 men or 2.5 percent of all US wounded servicemen sustaining amputations [125]. In March 1943, the Surgeon General of the U.S. Army established six amputation centres in the United States (with another two added later). In 1945 a program of research and development was initiated, with the purpose of creating improved prostheses using the technological advances of the day. The program was created through the National Research Council and was headed by a committee of surgeons and engineers. It was originally known as the Committee on Prosthetic Devices. The program was supported by universities, private firms, laboratories run by the U.S. military, and the Veterans Administration. The program came to be known as the Artificial Limb Program [125]. One partner was the Army Prosthetic Research Laboratory (APRL). It was given the primary objective of developing artificial arms, with the emphasis on prosthetic hands. A result of this project was a body powered, voluntary closing hand. The APRL No. 4 Hand had a cast magnesium frame with a cam and a clutch

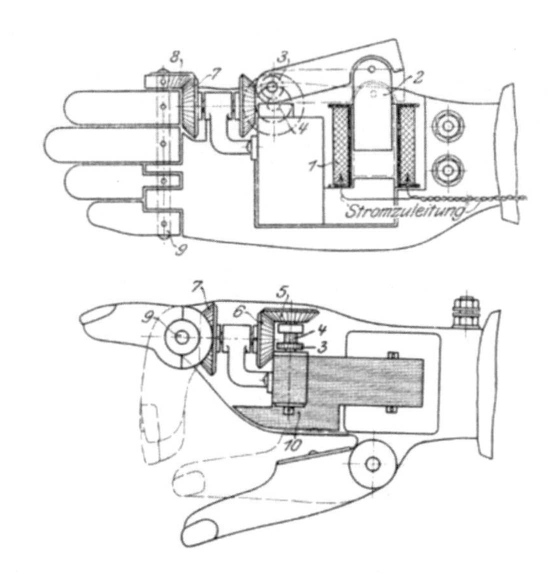

(a) The electrically driven hand uses electromagnets to close the fingers rather than rotary motors.

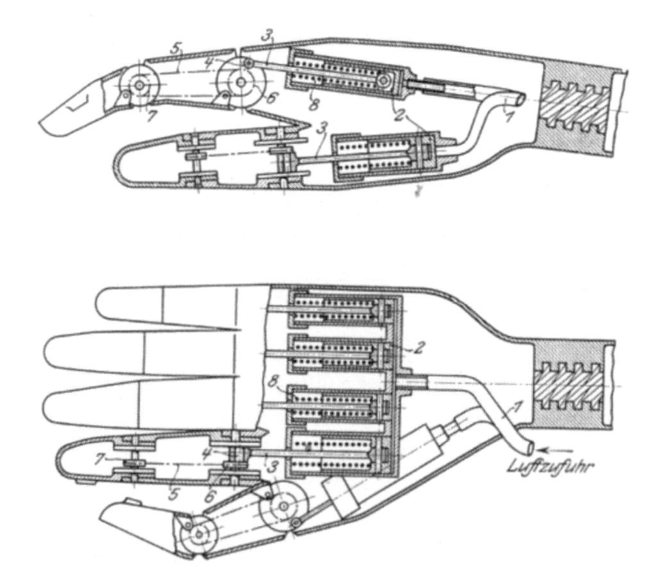

(b) All the fingers on the pneumatic hand close together, but have separate cylinders.

FIGURE 4.6

The first recorded externally powered hand designs. Both would have needed a large external power supply. (Taken from [2].)

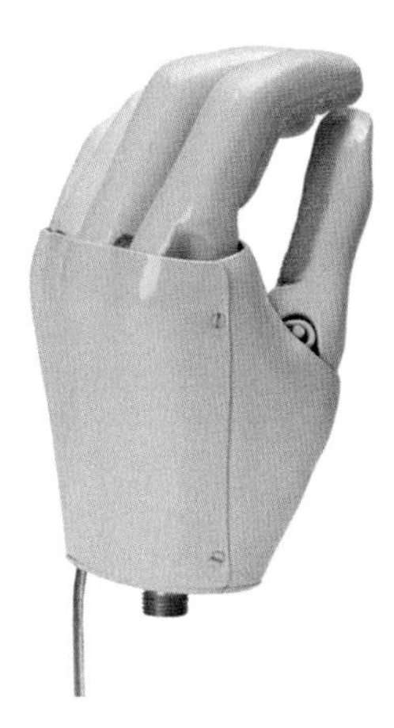

FIGURE 4.7

Although elderly in design, the APRL Hand is still on the market. One advantage the design possesses is a lower closing force than other anthropomorphic body powered hands [3], making it less tiring to operate. (Courtesy of Fillauer Companies.)

to lock the fingers in a firm grip. It was fitted with an external glove. The index and middle digit were flexed by the mechanism with the ring and little finger floating [56].

The hand was tested on 80 users over a 90-day period in 1951. The results of the test, shown in a report carried out by New York University, indicate that the majority of users found the hand to be preferable to both hook and existing hand devices [56]. The question of what factors they actually preferred is still something that plagues every engineer/clinician who is trying something new. Are the subjects really pleased with the device or simply happy to have the opportunity to be doing something to improve the technology? Will they become dissatisfied with it later and return to their old hand and ways? The APRL hand remains on the market (Fig. 4.7). When tested recently the APRL hand had the second lowest force requirement for closing the hand after the TRS prehensor [3], far outstripping the majority of other devices.

4.2 Development of powered prosthetic systems

An early powered device was the Vaduz hand [145], later known as 'The French Hand' after being marketed in Paris. It was controlled by detecting the bulging of the muscles as they contract by depressing a lever or by volume changes to pneumatic bags inside the socket. It had a gear shifting mechanism to give high gripping force. The digit flexion speed was considered good at the time. This is an effective compromise, and a form of gear shifting to move from grip speed and to grip force is used in the single axis Otto Bock hands (see 9.1.1). The design avoided detecting the contraction of muscles electronically which simplified the design at the time, this is this method that is associated with powered prostheses.

4.2.1 Myoelectric control

The most common form of powered arm control is via *myoelectric* signals. These are the electric signals generated by the action of contracting muscles [146] (6.1.3). These signals can be amplified and used to control prosthetic devices. The earliest work recorded in this field was by Reinhold Reiter at Munich University. He presented a prototype which used mains power and a valve (vacuum tube) amplifier, that was similar in size to an attaché case [147]. As a result it could only be used at a factory workbench as a form of workstation [146,148]. Reiter designed and built the system as part of his graduate studies in physics at the University of Munich from 1944 to 1948. The design captured the essential format of the powered prosthesis. It used the same form of controller with proportional speed control as is still used in commercial prostheses, but this last quality was only introduced in commercial prostheses very late in the 1990s. This hand could not proceed to a clinical tool in this form [147]. Reiter stated that in 1948, *"...the political and economic conditions in Germany were not conducive to further work on the project"* [149]. Germany was still damaged after the war and needed to re-establish its industry before it could tackle such work. It also required the development of solid-state electronics (transistors) before it could be made practical. Although published, the work was only rediscovered after the initial development of similar myoelectric systems in the 1960s.

Research in the late 1950s at the University of California at Los Angeles investigated whether electroencephalographic (EEG), electroneurographic (ENG), or myoelectric signals were the most promising for prosthetic device control. The authors favoured myoelectric control and outlined a number of concepts that would eventually be used in device designs [150].

Another early investigation into the possibility of using myoelectric signals from the residual limbs was made in the UK by Battye and others [151,152]. The work demonstrated the principle and concluded that control of a prosthetic device by this method was feasible. Other concurrent investigations into myoelectric control were taking place which resulted in several hand designs along similar principles. The so-called 'Russian Hand' [153] was the most influential of all the devices made at this time, while the Bottomley Hand appeared slightly later [154]. These early hands were simple designs with a single motion of all the fingers and thumb opening and closing together. This is a single degree of freedom (i.e. one motion) and these hands have nonadaptive grips, much as we shall see the most popular powered hands do (see Chapter 6).

4.2.1.1 The Russian Hand

As with rocketry, the most sensational developments in prosthetics in that period occurred in the Soviet Union. A joint group at the Moscow Machine Research Institute and the Central Prosthesis Research Institute first formulated the concept of using electrical signals from muscles to control a prosthesis in 1957 [155]. They described it as 'thought-controlled' [151]. It was a wearable, myoelectrically controlled arm system, this was the 'Russian Hand'. It was the first design to use transistors in a prosthetic hand [151,153].

The USSR's pavilion of new technological breakthroughs at the 1958 World's Fair in Brussels, showcased an arm powered by a miniature D.C. motor and battery pack worn on the user's belt. The hand used a 13.5 V motor and weighed 800 g. On top of this mass was a 120 g amplifier that was $100 \times 63 \times 37$ mm, and a 'feed unit' (batteries) of 320 g [155]. The grasp force between finger and thumb was up to 2.5 kg. This was achieved by pulsing the motors, much as the i-Limb Pulse does. Versions of this hand were demonstrated to people outside the Soviet Union and examples were sent to different centres to be tested and assessed. As a result it was arguably the hand that had the greatest single influence in the development of powered prosthetic arms in the following decades, until the i-Limb. The design was to have significant influence in the United Kingdom and Canada, where rights were licensed for manufacture. Worldwide, it raised expectations about what could be done for the limb absent and provided fuel for scholarly and popular science articles on the future of the man-machine interface. The direct impact of the Russian Hand can be seen in several of the stories in this book including Otto Bock, Boston Elbow and Institute of Biomedical Engineering, spurring development of myoelectric products in Europe and Canada.

In 1964, personnel from the Rehabilitation Institute of Montreal travelled to the Central Institute for Prosthetics and Prosthetic Development in Moscow and licensed the Canadian manufacturing rights to the Russian myoelectric arm. As is common with other examples of technology transfer, expectations were greater than realities. Although there were clinical products resulting from the Russian Hand licensing deal, commercial products had to wait a little longer. Initial investigation of the ten prototype arms the Montreal group took back with them from the Soviet Union [156] found that the arm was too heavy, the shape of the fingers limited grips and the electrodes had too much crosstalk [157] (interference of one signal on the other). The team recognised that the lack of harness was an advantage, as was the fact there was no need for effort to operate the device, they saw that the Soviet system was a good starting point. After extensive modifications and clinical evaluation, the Montreal group deemed them practical [157]. The changes included replacing the amplifier and battery components with North American hardware, redesigning the battery charger, integrating the components into the arm (except the battery), and adding an adjustable wrist unit.

4.2.1.2 Research in the UK

In the United Kingdom, the first research program into myoelectric control began in 1955 [152]. It was initiated at Guy's Hospital Medical School in London. The initiators were Drs C.K. Battye, Alfred Nightingale, and James Whillis. Battye and Nightingale were both members of the hospital's physics department. Nightingale would eventually become chief physicist at St. Thomas's Hospital in London, and the first editor in chief of the journal *Medical and Biological Engineering and Computing*. Whillis was a physician who never practised medicine, but instead explored a long-standing interest in human anatomy.

These three constructed an apparatus to perform a simple open and close action based on myoelectric signals from the arm of a test subject. The aim was to have the device open and close when the subject made a light grasp motion with the fingers. The focus was on proving the concept of myoelectric control, not on designing a workable device, and so the apparatus was bulky. The wooden box containing the electrical power supply, amplifier, and a 'discriminator' (what would subsequently be called a controller) was about ninety centimetres wide by sixty centimetres tall. Although the researchers experienced difficulties with the design of the apparatus because of electrical interference from electrical machines nearby, they deemed the preliminary work promising enough to suggest that a useful and dependable prosthesis could be developed [152].

It was about seven years later, in 1962, when the first British myoelectric prosthesis was designed and built by Alistair Bottomley. Bottomley was then a member of the scientific staff of the medical research council at West Hendon Hospital, London. The design of the device stemmed from the work Nightingale had done when he was at St. Thomas's, where Bottomley studied physiology. The progress made by this system was that it eliminated crosstalk so it could be used for proportional control.

Bottomley's prosthesis was subsequently handed to the British Atomic Energy Research Department for development [158], but it did not result in a commercial product.

The Bottomley Hand was the first portable hand to use proportional control of velocity [154]. Before this development, all that was required was that the EMG control signal exceeded some preset lower limit and then the motor would be run at a fixed speed, so the hand closed at one speed. This is difficult to use for anything except the most basic manipulations. Making the speed respond to the size of the control signal makes control easier and potentially more intuitive. In the Bottomley Hand, the stronger the user contracts their muscles, the faster the motor will spin and the faster the hand will open or close. This speed of opening and closing is directly related to the amount of perceived effort made by the user: Try harder and it opens faster. This is an easy to understand concept, yet despite its clear control advantages, it took decades to be adopted commercially. The Bottomley Hand mechanism had two moving fingers opposing a fixed thumb, similar to the Russian Hand and all commercial hands that followed it for four decades.

When the fingers closed on an object, the relationship between the muscle effort and motor drive was changed. A force transducer in the drive train was placed after a nonreversing lead screw and nut, and was it used to provide force feedback to the electronic controller. Instead of speed being related to muscular effort the grip force was made proportional to the effort. This too was easy to understand by a potential user.

4.2.1.3 Research in North America

In the US, the federal government responded to injured World War II veterans through the sponsorship of research and development, education and training programs and conferences. The US National Academy of Sciences decided that little modern scientific effort had gone into the development of artificial limbs, and in 1945 initiated a 'crash' research program funded by the Veterans Administration Office of Scientific Research and Development (OSRD). The state-of-the-art devices in 1945 were adult body powered artificial limbs with wooden hands. For children, there were only body powered hooks, as the hands had not been yet developed in small sizes [159].

In Canada too the research on artificial hands and prostheses began in a hospital facility. In 1949 the Sunnybrook Hospital in Toronto began work focused on making existing body powered and mechanical hand prostheses more useful through the use of new plastics and materials, novel suction sockets, and cosmetic gloves. Research on myoelectric upper limb prostheses at Sunnybrook would have to wait for over a decade. The major development in Canada during the period was the establishment of prosthetic research and technical units (PRTUs) in 1963 as a result of Thalidomide (4.4).

One of the major outcomes from the sponsored programs of the OSRD was the IBM arm. The difficulty in controlling the arm led the researchers to realise that users could not control the arm without conscious effort, and for most users that level of effort to control a prosthesis exceeded the benefits received. They suggested that future research should focus on electric arm control.

4.2.1.4 IBM electric Arm project

In the first half century of commercial computers the International Business Machines Corporation (IBM) was *the* computer company. This dominance ran from the 1950s through to the 1990s. They had created a position in the market where people knew that to buy their computers was a safe bet. Buying the next computer from IBM was an even more secure decision, with discounts for existing customers and support to create customer loyalty. A popular expression in the computer industry was that *no one ever lost their job buying IBM*. After World War Two, the OSRD recognised the disparity between the loss of function caused by an arm amputation and the limited function of the possible replacement prostheses of the time. To determine how to close this gap they commissioned IBM to investigate new power sources and methods of power transmission [160]. Two generations later, during the conflicts in the Arabian Gulf that were triggered by the Al Qaeda attacks on the USA, another committee came to similar conclusions.

The IBM project started in 1945. Pneumatic and hydraulic methods were investigated, but soon abandoned. Hydraulic drive was rejected due to the problems associated with piping power along the body by means of hydraulic lines. Pneumatic actuation lost out because it was found at the time to be *"grossly unsuited for artificial arms because of the weight required for such structures to be capable of withstanding the operating pressures"* [160]. The result was that electric power sources were pursued using electric motors to create a complete arm system.

The arm was based on the principal of one central motor for power with a series of clutches to take that power and deliver it to the specific articulations within the arm. Multiple-disk type friction clutches were used to transmit the driving force to the mechanisms, with the output force being proportional to the control force used to activate the clutch. The drives were then prevented from back-driving, employing the simple expediency of using worm gears with less than 50% efficiency [160]. The index and middle fingers were linked to the drive mechanism and could be operated to grasp objects (Fig. 4.8). The ring and little fingers were articulated, but unpowered. They could be positioned manually [161]. The thumb was locked in different positions to allow grasping of different size objects.

Different control strategies were used for the different levels of amputation. Control of high above elbow amputation (Model III-E) proved problematic and extensive research was conducted. The most successful control of the arm for this level of loss was with foot switches. Pneumatic bladders were placed in the sole of the user's shoe. The small toe controlled the drive motor, and the big toe and the heel were used to select the axis [160] (Fig. 4.9).

The stated aim of the IBM arm was to create control of the arm from 'movement primitives'. Much as music is made up of small blocks of tunes (melodies), the arm would be made up with simple motions, so that a 'symphony of manipulation' would result. While a worthy goal, perhaps one reason for its limited success was the truth that it took too much practice and skill to use. After all, not everyone is capable of playing a simple tune, let alone a symphony. In 1949, an evaluation was made by a team from the University of California, Los Angeles. They found shortcomings in the function. The elbow was the main source of erratic motions and the foot control switches were considered to be psychologically unsound because of their remoteness from the motions they controlled [160]. Further iterations of the arm were developed with the elbow design reverting to cable control. The resulting arm garnered some publicity and images of it suggest an interesting design [162]. Ultimately the program was discontinued. This was put down to the lack of sensory feedback and the inability to operate the arm without conscious thought [146].

A complete arm was tried by a user, John Sealey, which featured in Life magazine and filmed (Fig. 4.10). Sealey showes he can control the arm, but the limitations of the hand design of the hand and arm show through his compensations. The arm will raise only a little above the shoulder, but the film maker asks him to pick an object off a very high shelf so he has to bend sideways to raise his shoulder. He also has to pick the object off the shelf with the hand inverted as the fourth and fifth fingers are slightly flexed towards the palm. It was a long time before any other user can operate a whole arm over such a wide range.

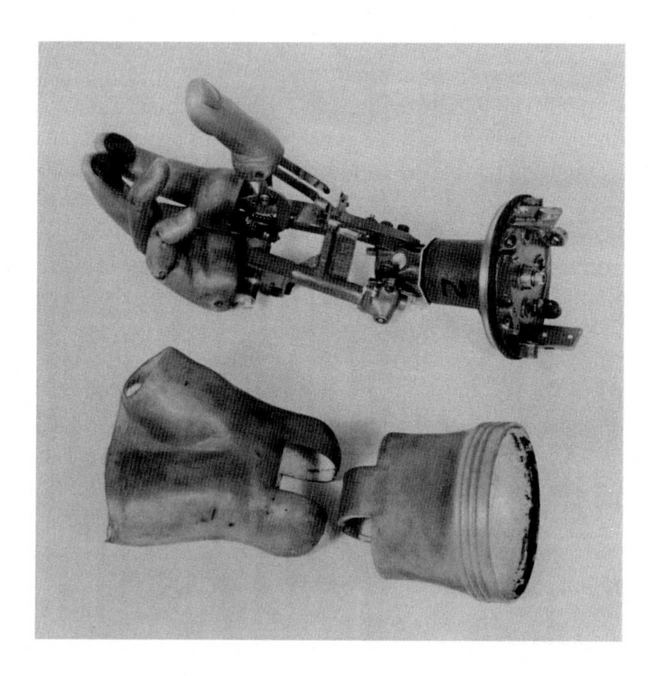

FIGURE 4.8

The hand mechanism for the IBM Arm, developed in the late 1940s. (Courtesy of International Business Machines Corporation, ©1948 International Business Machines Corporation.)

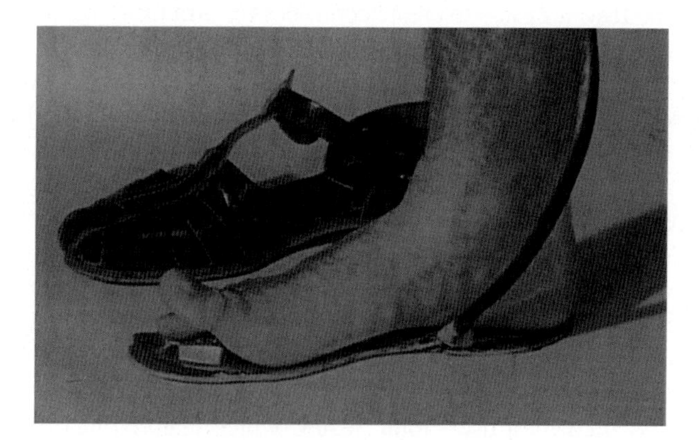

FIGURE 4.9

The control inputs for the IBM Arm. Bladders in the insole are used to switch the different functions of the complete arm. The air hose can be seen rising from the instep. (Courtesy of International Business Machines Corporation, ©1948 International Business Machines Corporation.)

(a) IBM Arm with test subject wearing the prototype arm on his right and a standard body powered arm on his left side.

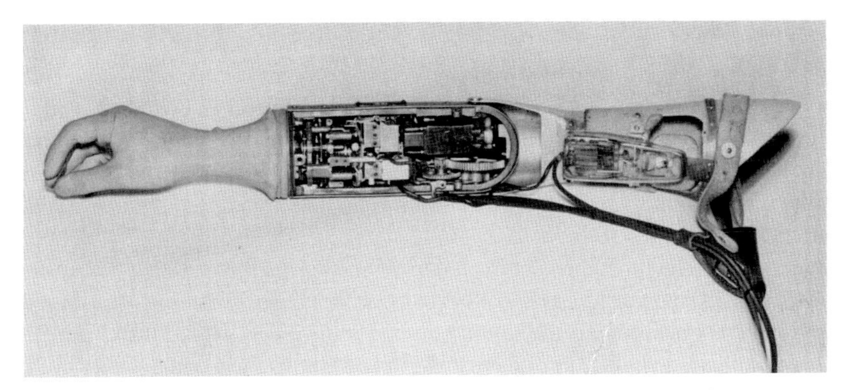

(b) The complete IBM Arm including powered shoulder and wrist flexion and rotation.

FIGURE 4.10

The IBM Arm. (Courtesy of International Business Machines Corporation, ©1948 International Business Machines Corporation.)

After the start of the wars in Afghanistan and Iraq in 2002 and 2003, there was a need not just to treat the wounded soldiers, but to be seen to be supporting the soldiers and allowing them to continue to serve despite their injuries. So the US government funded a program through the Defense Advanced Research Projects Agency (DARPA). The reasons for existence of the program were the same as the program sixty five years earlier. They focused on advancing arms rather than legs, projecting an image of little progress that ignored all of the work performed anywhere, but most especially that which was produced by the programs in the US over the preceding sixty years.

4.2.2 Innovations in the late twentieth century

Research and development was pursued in two distinct areas at this time. One was concerned with getting practical systems that responded to the needs of potential users and the other looked at more complex theoretical concerns.

4.2.2.1 Progression towards clinical applications

A centre that arose in this period was the Prosthetic Research Center at the Rehabilitation Institute of Chicago (RIC) (see 12.1). Incorporated in 1951, it developed one of the largest research programs on upper limb electric arms in the United States. The RIC was an independent body with its own board of directors and facilities. Subsequent centres were created at Guy's Hospital Medical School in London, the Princess Margaret Rose Hospital in Edinburgh, and Sunnybrook Hospital and Bloorview Children's Hospital in Toronto, and academic institutions such as the University of New Brunswick, Canada, and the University of Utah, USA.

From 1965 to 1976, research and development on myoelectric prostheses continued in the United States, Great Britain, Canada, and the Soviet Union, and new programs were initiated in Austria, Italy, Sweden, Japan, West Germany, and the Netherlands. These programs included research on electrodes, feedback, fittings, pattern recognition, and signal processing, as well as development of numerous myoelectric prostheses and control systems, spurred by the newly introduced transistor. It was near the end of this period that the first products began to emerge.

The teams looked at the design of hardware and control systems for one or more degrees of freedom (DoF) using signals from two muscles. These were design, build and test projects, and some resulted in products by the end the 1970s. The majority of persons with an absence below the elbow were potential users of these products. This is the biggest potential market. This bias continued to affect the field for another quarter century, with the first powered shoulder to be used clinically coming from the Edinburgh program in the late 1990s.

It is not uncommon to find that the later exploiters of an idea are the ones that become more successful. Although Kobrinsky had the advantage of developing the world's first transistor-based myoelectric hand and his group continued to improve upon it in the 1960s, it was the German firm, Otto Bock, that emerged as the dominant manufacturer of prosthetic hands, arms, and components in the late 1960s. Otto Bock benefited from collaboration with the Italian workers compensation organisation, INAIL (Instituto Nazionale per l'Assicurazione control gli infortuni sul Lavoro, the National Institute for Insurance Against Accidents at Work), near Bologna, Italy, and by the 1980s Otto Bock had emerged as the preeminent supplier of hand and wrist systems.

In North America at MIT, work began on the design of a system that become known as the *Boston Arm*, which would become the first myoelectrically controlled elbow, in the early 1970s (7.1). At the

University of Utah's Center for Engineering Design, work began on the Utah Arm (7.2). In Canada, UNB's IBME began development of its three-state controller for powered limbs (13.2), which was to become the first North American control system in 1965. In the late 1960s, at the Ontario Centre for Crippled Children, staff began work on the design of electronic elbows and hands for use by children.

There was little further innovation of the products until the mid 1990s. The companies showed considerable resistance to innovation, despite the growth of the application of microprocessors elsewhere. The first commercial prosthesis to use a microprocessor was a knee introduced by the UK firm, Chas A Blatchfords and Sons [163]. Based on studies made by Professor Saguchi of Nagaser University, a microprocessor was used to control the level of damping of the knee [140]. The concept of an advanced prosthetic knee was offered to Otto Bock and Blatchfords. Having initially shown some interest, Bock declined to take the idea further, and it was Blatchfords who lead the field from the mid 1990s [140]. No such innovation was reflected in the upper limb field until the turn of the century and then the progress was initially incremental. The only significant improvements to the commercial systems were to the electronics for the controllers. Initially the electronics were made smaller. The first use of microprocessors was to simply realise the same limited functionality in a single package. This is easier for manufacture, but it does not make any use of the potential of the technology, at a time when consumer electronics was changing rapidly. The first sophisticated use of micro-electronics was a digital signal processor, used in the Boston Elbow in 1999. The market changed as a new generation of prosthetists and users more used to the expanding high technology markets across the globe requested more functionality and were more ready to embrace mobile technology.

In Edinburgh a high-profile prosthetic hand with five powered fingers, designed by David Gow in the 1990s, was launched as the i-Limb by Touch Bionics, opening up the market for others to introduce similar products. Invented in the 1990s, its emergence at the turn of the twenty first century meant it could benefit from new innovation funding programs and maturity of the underlying technology.

One important nonproduct innovation in the field was the surgical Targeted Muscle Reinnervation (TMR) procedure (12.2). Todd Kuiken from the RIC first developed the procedure for persons with upper-limb amputations in 2001, and demonstrated it with patients by 2004. It has been implemented in many more than two hundred persons worldwide. TMR allows the user to move multiple joints of the prosthetic arm without having to use nonanthropomorphic actions to switch between axes, instead to move the electric arm more like a natural arm through use of the signals that would have controlled the arm prior to the amputation.

4.2.2.2 Basic prosthetics research

Basic research continued at universities. It was roughly in 1977 that powered upper-limb prostheses became clinically significant in North America [164]. At the same time the research funding streams dried up and the focus of research moved to other areas. A very few centres and researchers stuck to prosthetics research: RIC in the USA, UNB in Canada, Princess Margaret Rose Hospital (PMR) in Scotland, TU-Delft in the Netherlands, University of Southampton in England, INAIL in Italy. Often they had to make the prosthetics research part of a different or wider research program.

Topics included multifunction control systems for people with above elbow absences with only one functional muscle site. Three of the major research streams to the problem were to investigate and develop (i) endpoint control systems; (ii) sensory feedback systems; and (iii) myoelectric pattern recognition systems [164], but the research enjoyed little or no interest from the commercial companies.

Endpoint control systems promise users the benefit of less conscious effort to manipulate the limb and more natural movement. Research on these systems occurred at UCLA, Berkley, MIT, Case Western Reserve University, Rensselaer Polytechnic Institute, and the University of Southampton [165] (see 10.2.2). With funding from the Veteran's Administration, research at UCLA focused on developing clinical systems for users that provided preprogrammed movements such as opening a hand or hook when the arm reached out to grasp an object. An even more prescriptive device, the *Feeder Arm*, was developed at Case Western Reserve University. It allowed users to select complex programmed functions, such as the movement of a level hand for feeding oneself. This approach is appealing to engineers. It limits a large and difficult problem to a smaller set of solutions. These limitations are generally less well tolerated by the potential user who does not gain enough from the one action to find the device useful, except in very closely defined circumstances.

An alternative approach to feeding was developed by Mike Topping of Keele University in the UK, he developed a feeder robot that was only used during meals and is not worn by the user. Designed for a neighbour's child with cerebral palsy, the user selects the food they want from a dish and the robot picks it up and offers it to the user. The user then has to move towards the arm and take it off the spoon [166]. While not specifically designed for a prosthesis user, it has the advantage of delivering a service simply and predictably and could be used by people with many different impairments, including limb loss.

Sensory feedback research was undertaken at the University of Edinburgh, RIC and University of Ottawa. Research at the University of Edinburgh resulted in a pneumatically powered prosthesis that controlled the artificial arm by movement of the shoulder. It also permitted sensory feedback to the shoulder, providing users with position awareness of the arm (see 8.1). At the University of Southampton the feedback only went back to the electronic controller (10.2.1). Professor Jim Nightingale reasoned that it would only be possible to relay a limited amount of information back to the operator, so it was not practical for the user to drive the detailed motion of the arm themselves. The Southampton Prosthesis used the information from the hand and around the hand to automatically determine the grip shape, force, arm orientation and lead to the first microprocessor controlled prosthetic hand to be used in the field.

Pattern recognition of myoelectric signals was undertaken at Temple University and the Moss Rehabilitation Hospital in Philadelphia, Chalmers University Hospital in Göteborg, Sweden (see 11.2.1), and UCLA. Researchers at these institutions looked at the basic patterns with the contributions from different muscles to determine which grip had been used. It was difficult to reproduce this format repeatedly so later on at UNB's IBME, systems were developed for pattern recognition-based control using the information unique to the start of a signal as the person changes from one grip to another.

4.2.2.3 The DARPA program

From 2002, the US led wars in Iraq and Afghanistan led to the release of funding for prosthetics research on a scale not seen since the dissolution of the Committee on Prosthetics Research and Development (CPRD) by the National Academy of Sciences in 1977 [167,168]. Limb loss for the veterans of these wars was more prevalent than in previous wars because of the body armour worn. It protected the head and torso, but not the arms and legs [53]. This, plus improved medical treatments, meant more battlefield survivors with significant limb injuries. In response the DARPA *Revolutionizing Prosthetics Program*, was created following meetings where invited 'guests' paid to attend. Here the guests were able to input the sorts of ideas and conditions that would be explored in an upcoming program. The aim was to improve the state of the prosthetics art, shaping the details of the program. Following this, the call for project bids was issued in 2005 [169]. At the same time the routine provision for the US veterans was extensive. They were fitted with many advanced commercial systems.

Like the National Academy of Science's assessment of the field in the 1940s, DARPA in the 2000s found the state of upper limb prosthetic technology lagging. DARPA, however, responded differently. They targeted two projects focusing on the development of two prototype prosthetic arm systems, one by DEKA Integrated Solutions Corporation of Manchester, New Hampshire (which became a product in 2016) and the other by Johns Hopkins University Applied Physics Lab (JHU) in collaboration with the RIC. DARPA awarded $18.1 million to DEKA beginning in 2007 [170], and $30.4 million to Johns Hopkins University .

The overall goal of the program was the development of a neurally controlled arm and hand prosthesis that performed and looked like a natural limb, and would be ready for clinical trials within four years. Significantly they chose contractors with no previous experience in prosthetics technology. The result was that the Johns Hopkins' first whole arm system was not modular and could not be used except by someone with a loss at the shoulder [171]. They also chose a single main drive and clutches to power different joints, which had been tried by IBM in the 1960s. The second project, used foot switches for control (similar to the IBM Arm). Anecdotal evidence suggests that the teams were not even aware of the previous use of these methods.

The second generation JHU Arm was modular. The wished-for neural control did not occur in this project and DARPA returned to this with subsequent programs on implant technology. What will be the fate of these projects is unclear. It was stated at the beginning that it would create innovation that would benefit the field. Ed Biden, former director of UNB's IBME (13.2), believed that with the injection of cash there will be a trickle down of ideas from the very expensive arms of these two projects into devices in the field [172].

One of the veterans of the prosthetics field, who had been an adult during the Vietnam conflict, considered that one of the reasons for the DARPA program was the US government wished to avoid the problems of Vietnam, where they first lost the argument at home. Then they could not sustain the losses in the field, and so it became politically impossible to continue the war. For veterans of Iraq and Afghanistan, the US government ensured that the prosthetics provision and research and development budgets were kept high so that the injured veterans could not argue, as they had after Vietnam, that veterans were forgotten. The language used around these programs was carefully chosen. Injured soldiers were not 'veterans' but 'war fighters' a curious new construction [170].

Although his company was part of the DARPA projects, Bill Hanson, then director of LTI, was less convinced of the value of the program [20]. He felt that he might agree with the design if the first DEKA arms had been made more modular. Unlike the standard commercial arms, the DEKA arm did not break down cleanly into joints that take up as little space as possible above them on the arm. This difference would maximise the potential user population. The response of others in the prosthetics business was not positive. Sandford Meek (University of Utah) regards it as a *"Waste of 50 million dollars"* [173]. He suggested that it was an attitude of *"we can kill it with money, engineering and computers and not thinking what the fundamental problems are"*. Ed Iversen of Motion Control considered that the direction of the funds to aerospace companies rather than prosthetics companies meant that they would have to make a lot of mistakes in order to learn how to do it right, rather than relying on the experience of the prosthetic manufacturers [174].

The second DARPA arm was modular with the joints being stand alone. It was named the 'Luke Arm' after the character from Star Wars with an amputation. It has achieved a range of clinical fittings at levels from the shoulder down. While the foot switches remain in the armamentarium, since it has been placed in the hands of clinicians other control options have been added [175].

Additional work not part of the original specification included the assessment of the function of the person with the more complex devices [175]. Additionally, the advanced arms are encouraging researchers to investigate the impact of the cognitive abilities of the users, to understand more fully what is required to control a complex arm and ultimately match the arm to the user [176].

Bill Hansen sees a number of problems with the deployment of this technology: The first is price. *"Although no one has ever said what the cost would be, it is likely that it will be an order of magnitude above anything available at the time"* [20]. He later estimated it was more than $50,000 USD. So while they are having trouble with today's technology, who is going to buy it? Dean Kamen CEO of DEKA, said the arm would be purchased for the military. Hansen did not fully agree that this will provide a route to market. Even if Kamen was right, Hansen commented that it was a pretty small population compared with the general public: *"I think everyone is in favour of providing the best technology they can for those people"*, but such a small population will not easily bring the price down [20].

4.3 Powered prostheses today

The appearance of the Touch Bionics hand changed the market. Advanced hands had been regarded as 'impossible' by a very wide range of participants in the field. Suddenly this was demonstrated to be false. The reasons are likely to be the combination of advances in technology and changes in the attitudes of users and companies alike. A personal computer has moved from being a specialist technical tool to being part of a watch or a smart phone. So an interest in advanced hands that members of the user population have had for years [106] has become achievable in fact.

A problem with advanced prosthetic arms is that most prosthetics teams will not have the skills to fit and them successfully. As in the early days of myoelectric control, it will be centres of excellence which will have the skills and knowledge to fit such an arm. Since most smaller prosthetics shops will not fit many arms a year, they will not have practiced the skills enough to do an effective job. A poorly fitted arm is unlikely to be used to its fullest extent. Unlike the early days of myoelectrics (from 1980 to 2000) where innovation moved slowly, the expectations now are that technological change will mean continual innovations to the prostheses. The changes will be so frequent that next time a small clinic comes to fit an arm, it will be a new experience again. The staff will not have the chance to build up the knowledge and skills. It may therefore force a different model of fitting prostheses, with potential users travelling to a very limited number of centres with the experience to perform the fitting. This will clearly mitigate against those with insufficient funds or who can function well enough without the more sophisticated arms. This is similar to the Scottish National Health Service, which has a limited number of centres who specialise in the fitting of multifunction hands and microprocessor knees. This works well for a physically small country with most of the population served by a small number of centres.

An additional factor is the rise of the citizen engineers. They can use new tools of production to make custom devices previously difficult to obtain. An enabling technology in this change is the additive manufacture or 3D printer. This is widely forecast to be an agent for the democratisation of manufacture by allowing people to design and build objects at home. This is not a panacea. As in other areas of endeavour when the less skilled or trained have begun to adopt new methods, there is a sacrifice in quality either in the short or long term. Low-cost 3D printers produce low quality objects. More expensive printers produce a commensurate improvement in quality, but at a price. Since 2010 there have been many projects using the new technology, but one that arouses considerable enthusiasm is upper limb prosthetics. Many citizen engineers have sought to rectify the perceived need for cheaper prosthetic hands both in affluent as well as low-income countries. While the designs are varied, very few are an advance on the designs that Schlesinger collected in 1919. This means they can be made using existing machine tools and the 3D printing is only simplifying production while sacrificing quality. One of the potentials of 3D printing is its ability to make things that *cannot be made* using conventional machine tools. Once this is understood more widely it may unlock the true potential of the technology.

Two manufacturers who are beginning to use the technology to explore new areas are Open Bionics and Glaze [177]. The former is making personalised prostheses that match more closely the interests of the wearer. Glaze incorporates things that are not traditionally features of a prosthetic limb and questions what it means to have an artificial limb.

Externally powered arms are more prevalent today than in the past, but it is not clear if they are more readily available to the average user or are a small number of devices simply enjoying greater publicity. The statistics on how many arms are fitted or being used is hard to obtain, as it was twenty years ago. So we cannot be sure these solutions are more common. Advanced prosthetic arms certainly engage the public and it is the expense of them that many of the citizen engineers say they are responding to in their work to produce low-cost devices. An object lesson to us all should be how the pioneers of the past dealt with the same problems and overcame them, or failed to make progress. New technology creates opportunities to solve old problems in new ways. It is the aim of this book to provide a little insight to the problems encountered, so most of the rest of this book looks at the evolution of a number of the most important or influential designs produced since the Second World War.

4.4 The Thalidomide legacy

Most congenital absences are sporadic and have no known cause. There is, however, one notable example when this was not true. The development of a foetus is complex, with many stages having to occur in the right order and the right place for a healthy infant to emerge at the end of the gestation period. The process is easily interrupted by an external insult. For a long time it was thought that nothing could cross the placenta from the mother to the foetus, and affect the unborn baby. By the late 1950s this was known to be untrue.

Pregnancy can be a difficult time for the mother, and she is often held to be responsible for the baby's life, death or deformity. In truth she has little or no direct control over its development, a few simple precautions are all that she can take [178]. There is one event that has made sure the injustice of mother's guilt has been repeatedly reinforced in the last half century. It is also the one event that affects *the parents* of children born with some abnormality today, even when there is *no* link for today's parent, and there is no need for them to feel personally responsible. That event is the discovery of the teratogenic impacts of taking α-phthalimidoglutarimide (known as Thalidomide) while pregnant.

In the late 1950s, a new drug was marketed by the chemical company Grünenthal in Germany and then licensed in other countries (in the UK, for example, by the Distillers Corporation). The usual process in having created a new chemical is to proceed from development to a discovery program. This attempts to find uses for this new potential product. In the postwar period the manufacturers of sedatives found an increased market among the many people who lived through the conflict and were affected by the impact of the war. Many took sedatives and some overdosed accidentally, or deliberately. This is bad for the drugs business, so the companies focused on research programs that would identify sedatives that had less chance of being fatal even if ingested in larger doses. The initial identification of Thalidomide occurred when the workers at Chemie Grünenthal isolated α-phthalimidoglutarimide ($C_{13}H_{10}N_2O_4$) and found that, while it was unsuccessful as an anticonvulsant in epilepsy, it induced a deep all night 'natural' sleep. It was a hypnotic, inducing sedation and relaxation. These qualities are ideal for a sedative, as the person becomes relaxed and rested. Other attempts at sleeping tablets resulted in people gaining sleep but not being rested [179].

While Grünenthal noted some side effects, these occurred in only a few subjects, and the potential beneficial impact on the many was seen as significant. What was important to the scientists at Grünenthal was that in the tests to see how big a dose was needed to kill an animal, they could not find a dose large enough. It was apparently completely safe. This meant that if children found and took the drug, they would not die. This is a major benefit for any drug. Later on, the fact that the test animals (rabbits and rats) did not even get sleepy was a clue to the US Food and Drug Administration (FDA) that it only worked on primates. It was marketed in many countries and given many names (Distival, Talimol, Contergan) but it is now generally known by its original title: Thalidomide [179]. For people who witnessed the unfolding of the Thalidomide crisis, this one word has a jarring resonance that is hard to describe to those too young to know about it.

It is also hard to appreciate how different matters were in the drugs business in the 1950s. The change that has happened in the intervening seventy years occurred for many reasons, but many of them are specifically related to the impact of Thalidomide. The story is often shocking and surprising, coming as it does from a very different world. When asked in court, the workers at Grünenthal could not provide evidence that they performed even the most rudimentary checks on the safety of the drug. They reported they destroyed their records later. Instead of conducting extensive, well controlled trials, they distributed it for free to physicians and asked for feedback. Many tablets were dispensed and few came back with any problems. In fact, α-phthalimidoglutarimide was distributed both alone and with other medications, often over the counter, and under at least thirty seven names, for colds, flu, asthma, headaches and sleeplessness. As a result of its wide reach, it was later hard to reconstruct its path or its link to all the subsequent birth defects. What we do know is that it was sold in the UK, Canada, Australia, New Zealand, Europe, Scandinavia, Asia, Africa and the Americas and free samples were distributed in the USA [179].

The first recorded victim was born Christmas Day 1956 to an employee of Grünenthal. His wife had taken samples made available to them. Their daughter was born without external structures for her ears. This was ten months before the drug went on the market. Any side effects that were reported were not noted by the company. There were many problems including, giddiness, numbness, tingling and polyneuritis, but Grünenthal's response was to investigate the mothers and doctors to see if they were going to sue the company and to damage the company's reputation.

By December 1960 there was an article in the *British Medical Journal* postulating a link [180], but the struggle to get the drug withdrawn took until the following November in Germany. It was even later in the other countries who were distributing it.

The real danger period for taking the drug was later determined to be during the twentieth to thirty-sixth days postconception. Importantly, it is at a time when the mother was either unaware or uncertain about being pregnant. As a result the mother can never be apportioned blame for taking a drug she was told was totally safe, in a time when she was not even aware she was pregnant. What was worked out, forty or more years later, was that the drug interfered with the pathways that lead to the development of the limbs in the foetus. As a result, babies were born with malformed arms and/or legs, with digits on short arms and legs or at the shoulder and hips. There were many other body parts that were stopped developing or grew poorly, and as a result a large number of the babies did not survive very long.

In 1959 the drug had been on the market for roughly two years and the estimates of numbers of affected children vary between eight and twelve thousand, with five thousand making it past childhood [179]. The babies were born in 46 countries, with the countries focused on in this book: 105 in Sweden (5 post ban), 456 in the UK, 115 in Canada and 10 in the USA. In 1960 it was clear that there was suddenly and publicly a significant number of children with extensive congenital defects. It became politically expedient to attempt to do something about this issue. The details of each country's responses were different.

Sheila Hubbard from Bloorview children's hospital in Toronto suggested that the Canadian government felt some responsibility, having approved the drug, so that despite the fact that prosthetic provision is managed at the provincial level, the federal government founded a program with Prosthesis Research and Training Centres in Montreal, Toronto, Winnipeg and Fredericton (13). It was funded by a block grant from the ministry of Health and Welfare. Different centres worked on electronic controllers, lower limbs, and upper limbs [181]. The grant was continued until 31st March, 1975, exactly three years after the original intended termination date.

The drug's manufacturers dragged their feet for decades [179]. First, they did not admit any connection, then resisted an admission of liability when it was clear there was a connection. In each country the companies used the law to slow down the process [179]. Grünenthal dragged out a trial in Germany until they were able to claim that if the process persisted there would be no money left for the parents to settle on [179]. The tactics adopted leave the outside observer breathless with astonishment. The lawyers made different suggestions at different times that removed their client's responsibility, including that the foetus was already damaged, that the foetus had no stance under German law, that Thalidomide had preserved the life and that without it, the foetus would have aborted and lastly that the damage was caused by mothers attempting to abort their own children. These tactics delayed compensation for the parents and children. In the end it was settled, but without anyone being held responsible [179].

In each country the parents and children were left to fight the battle on local terms, and not to unite across borders. In general, all parents and children needed money to support the children. A tactic from both Grünenthal and Distillers in the UK, was to make a one-time offer of money, on the explicit understanding that they did not reveal the sum to outsiders (such as the press), or tell anyone that the deal even existed [179]. This isolated parents and ensured that those who tried to fight for a better deal were seen as preventing others from accessing the money. By dividing the groups, the companies could fight each battle separately. It is hard to guess the motives of the executives who made the offer and if the sum of money was really intended to cover the children for life, or just remove them from the battle, but inflation since the late 1960s would have ensured that if they had settled on the original offer, it would never have been enough.

In the UK the offer was to all parents, or none. This placed considerable pressure on all parents to settle, but at least six parents resisted this offer. The sum, while not small, was not large enough to last the person's life. For parents who had to look after a child with few intact limbs, it would have been the first money they had seen and would have been very tempting to take the offer. The tactic divided the parents and made the lives of the resisters even more wretched. When they did settle, years later, it was because the British company that supplied it in the UK, Distillers, was threatened with a boycott of their other products. Only with this threat were the shareholders mobilised to pressure the company into settling. The campaigners' advantage was that Distillers also made alcoholic drinks, so the pressure could be significant and precise: Stop drinking well known drinks. The campaign did not depend on the sick not taking their medicine to persuade the company to deal. No such leverage existed over Grünenthal.

While it was not marketed in the USA, many tablets were distributed ahead of the expected FDA approval, hence the ten children were born affected. The approval process was not as stringent as it is today. Dr Frances Kelsey, a Canadian-born pharmacist, new to the FDA, was given the task of what was supposed to be a simple process: Approve an already in-use drug with no reported side effects. She was unconvinced by the low quality of the evidence. The fact that it would not make mammals sleepy or kill them with an overdose did not mean it was a necessarily a good and safe sedative in humans. This worried her. While she could not actually refuse the drug, she could delay it while she considered the implications. What was important was that it was not being prescribed for anything terminal, so there was no urgency to save lives. It was a simple sedative and so it should need good evidence that it was safe before she would approve it. As she delayed its approval, evidence came from Europe about its neuropathic side effects. This was enough to delay approval further, and once it was withdrawn in Germany its chances of a launch in the USA had ended [179].

While it is an important story and the FDA learned a great deal about this, it is interesting to consider what would have happened under different circumstances. One scientist from the Johns Hopkins School of Medicine, Dr Helen Taussig, wrote in Scientific American *"If Thalidomide had been developed in this country, I am convinced that it would easily have found wide distribution before the terrible power to cause deformity became apparent"* [179].

The important question as to why it was allowed to happen at all, is hard to answer. People in the right places seemed not to know what to do or not to be diligent enough to follow their findings. Stephens and Bryner in their book on the tragedy [179], do not force a direct line between the callous behaviour of doctors in the service of the Third Reich during the Second World War and the entrepreneurs of Grünenthal. They do suggest that cultural influences existed less than ten years after the end of the war that led to the scientists in the company not following up all the procedures they should have done. What evidence survives suggests they did not carry out that many tests on animals, and when they did, they did not find the teratogenic effects, partly because it only affected primates, which were not the usual test animals. In fact the only test we have evidence that they conducted was the uncontrolled test on their customers [179].

The impact of Thalidomide on the growth of the children was very variable, depending on precisely when in their development it occurred. Some had intact legs and small arms that ended in one or more multijoint fingers. Others were missing legs, and some had poor development in their internal organs. So each person's need was different. The public desire to do something was considerable and a number of research/clinical centres were set up in the affected countries. I have worked in, or alongside, four such centres in my career.

Clearly, the mothers felt considerable guilt for taking the drug, but they trusted the system not to give them potentially dangerous drugs and there was no way that they could have known or prevented the damage. Unfortunately, the general reaction to the children was fear and even revulsion, including from their parents. This is a tale where the medical profession does not come out well. The children were often kept in hospitals for years. Surgeons attempted many different surgeries on the children, some life saving and some just, it seems, to 'tidy up', their limbs and make fitting standard prostheses easier.

Given the perspective of the twenty first century it is hard to comprehend the treatment of the children. They were isolated from their parents (willingly or 'for their own good') [179]. They were subjects of numerous studies and surgeries for cosmetic or 'functional' reasons no matter how misguided. The Thalidomiders were fitted with prostheses that did nothing, that they could not operate, or worse, that taught the children failure and frustration. From the perspective of the time, the work was generally well intentioned. This was the era when many children were taken away from their surroundings. All such children were in some way 'handicapped' and unable to defend their own position. At the same time in Canada and Australia, aboriginal children were placed in 'residential' schools to 'adapt' them to the Europeans' lifestyle, in a manner that was brutal and has been described as genocidal. The Thalidomide children were so strange and unusual they were bound to undergo treatment as 'different'. The shame is that the treatment was quite so poor. As with many other things, (drug validation, clinical testing, rehabilitation, prosthetics) we appear to have learned from the Thalidomide experience, but only to the benefit of those that will follow, not the affected people themselves.

The Thalidomiders have learned to find ways around obstacles. Often they see the obstacles as imposed by other people on them. They are as independent as they can be. However, as the Thalidomiders age, their joints and other bodily systems, worked hard all their lives and perhaps originally ill formed, are failing early. Now these people need all the technological help they can get, but I hope they receive a far more sympathetic response to the user needs than anything they experienced as children, partly (I hope) because society has learned from it. The assistance they need is not likely to be limb prostheses, but other devices to assist them in mobility and function. Much as many in the elderly population now have many tools to assist in their day-to-day lives. It is not at all surprising if many Thalidomiders distrust the medical and rehabilitation professions when the help is offered, and it is the profession that needs to work to earn their trust.

Thalidomide's influence on the profession was large and a number of the contributors to this book have made reference to their involvement in treating the children. The profession learned a great deal from the experience, but unfortunately the lessons the Thalidomiders learned were not so positive. As suggested, the professional response was to attempt to fit the children with prostheses. For example, David Simpson working in Edinburgh went to Heidelberg and saw first hand what they were doing there, before bringing ideas and techniques home to try. One, not unusual, reaction was to feel that the children needed so much help and intervention that in Toronto, Canada, they were admitted as inpatients for long periods, having schooling while there. The UK and Sweden had specialised centres for the children [179].

The need to provide the children any means to feed themselves meant that variants of the Michigan feeder arm was a device of choice for the innovators. It mostly took a fork or spoon from a plate area to near the mouth of the wearer. The usual fitting was to make one side the active arm while the other, a more passive device, hid the power source. In the UK, Henri Lymark did a trial of the devices for Thalidomiders at the hospital in Roehampton (London), where a prosthetist and a technician worked in a dedicated unit. The Thalidomiders were still children, so that even the smallest prosthesis was large, heavy and difficult to use, with many different joints to control. Robin Cooper, a prosthetist and leader of the research at the UK firm Hugh Steeper for years, noted the problem with the fittings [182]. *"All the kids gave up because the devices broke easily"*, and while it was possible to learn to use them it was: *"a clever feat, like getting a dog to balance on two legs, they could do it, but making it meaningful was difficult"*. It was possible, but not practical. Most of the children gave up with prostheses after a while.

The irony for the Thalidomiders is that despite all the efforts to assist them with high tech solutions, the majority of the people when they grew up still do not use any conventional prosthetic components, adapting or adopting other tools to assist them. This is due to many pragmatic reasons, such as the prostheses produced are always too complex and heavy to be of use. Many Thalidomiders have some form of hand at the end of their short arms or legs. These hands have great dexterity and sensitivity, close to or the same as conventionally developed hands. Thus they were able to manipulate sticks, buttons, etcetera with greater facility than any prosthesis ever produced. As a result, manipulating a simpler tool in their hands is usually far more effective. Occasionally, the surgical rearrangement of some of the digits to create an opposition digit (a form of 'thumb'), would add to their functional range, but as with the impairment, every solution was unique. Another difference is this was unlike a postamputation intervention. The surgeon was taking a person as they were and changing their form, something that could have significant negative impact on their self-belief or self-worth.

The response by the profession was well intentioned, but misguided. *"There was a feeling that technology could solve it"* Robin Cooper remarked. This was true in a number of countries including Canada, *"Everyone wanted to find a technical solution"*; it was like the USA's Defence project in 2005, *"there will be a solution, if we push enough money into this"* [182]. Sheila Hubbard regrets the experience of the Thalidomiders. Many left the rehabilitation system and have been working out their own solutions over the years. She feels that if a child effected similarly turned up in a clinic today, they would not be fitted with prostheses at all. The clinics would work on specialised devices and modifications to the home in the way that they do with other users today. *"They probably could have been helped more"* she commented [21].

The story of the Thalidomiders continues. It was only in 2012 that Grünenthal apologised to those that suffered. The drug Thalidomide has proven useful in many other disorders from AIDS to multiple myeloma, and its impact on Leprosy has meant that many specialist hospitals have been able to be closed, their patients cured. It is generally the drug of last resort, that is how the FDA have approved it in 1998 [183], and the work to find alternatives that do not have the same impacts as α-phthalimidoglutarimide continues.

The clinical use of the drug has been very heavily controlled as the thought of it being used at all disturbs those most directly affected by it. However, the truth is that many other drugs exist which would harm the developing foetus as much as Thalidomide. There are strict guidelines on when they can be used, partly as a result of *this* drug. Any further deployment of Thalidomide must be handled with sensitivity.

This is not to say that everyone impacted by the drug has 'suffered' all their lives, they are not victims. They are not defined by the mistakes of others. However, medicine and prosthetics have definitely benefited from the lessons of Thalidomide, while it is the Thalidomiders who continue to pay for it.

There should be no league table of recompense, but the Canadian Thalidomiders had one of the poorer settlements. In Europe, there have been funds which the companies have contributed to as the cost of living has risen faster than anticipated. This is not to say that the companies have ever (or can ever) make up for the damage, but it does mean that in some countries the adults can more readily get on with their lives. More recently, the Thalidomiders turned fifty and there was another settlement for the Canadian group, although they remain a long way behind their contemporaries in other countries [184].

Worse still, a combination of poorly developed joints and a lifetime's heavy use of their short arms and legs mean that the Thalidomiders are now suffering from considerable joint pain. Whatever settlements they get from government and company, they are going to need much help to make their last decades dignified and bearable.

CHAPTER

Body powered prostheses

5

5.1 Introduction

Body powered prosthetic arms use the relative motion of different parts of the body to pull on a harness or lever and move the hand or joint. The most common form of body power (BP) is a harness that runs around the opposite shoulder, pulling on a cable that links to the hand at the end of the remnant arm. Pull on the cable and the hand opens. This seems a pretty straightforward solution, but it took until the nineteenth century for the first design of a body powered prosthesis to be recorded. Ballif created the first such device in 1818 [185], shown in Fig. 5.1. There are many other examples of mechanisms triggered by cables going back many centuries, so assuming there are no undocumented precedents, why did it not get applied to prosthetics?

Bulliet in his study of the history of the wheel, considers why some societies had knowledge of the wheel but did not use it in surface transport for centuries [186]. He suggests it was not circumstances alone that make the wheel useful; the lack of draft animals or having a landscape suitable for wheeled transport did not cause the wheel to be adopted, but rather social and cultural forces. Historically, perhaps those rich enough to afford a complex prosthesis did not need to fend for themselves, so they did not need manipulative function, and could tolerate an ornate but passive hand. Those who could not afford any sort of hand were not in a position to ask for a functional device. Body powered harnessing could have been invented many times in the past without being recorded, it is Ballif who gets the credit.

Following this in 1818, innovation was slow and it was a little under a century before the most common design of body powered device, the split hook, appeared in the literature [187]. Prior to this a fixed hook would have been used. It can be very effective for performing many tasks of pushing, pulling and clamping objects. However, in the 1890s D.W. Dorrance developed an active device, the *split hook*, which he subsequently patented in 1912 [187]. In this case, a hook is split down the centre and one side is actuated using a body powered cable, so that it can be used to hold as well as retain. The hook is augmented by having the inner surfaces coated with a material which is less slippery than metal, improving grip. The shape of the hooks have varied to allow for different sorts of functions, but the usual formation is that at the pivot, the tines are separated making a space to grip large objects before coming together to form fine tips, allowing a precision type grip to be created. The Dorrance Hook is still in use today, though the design has been refined multiple times since the original version. Hooks are formed in symmetric and asymmetric formats, each has its own merits and users adopt the one that works for them (Fig. 5.2).

It was only nine years after Dorrance patented his design, in 1919, that the first designs for powered hands appeared [2]. This was part of a rush of innovations to prosthetic devices that were made following the First World War. There were some additional new designs produced after the next global conflict, but since then the pace of innovation of body power was not maintained [37].

Making Hands. https://doi.org/10.1016/B978-0-12-820544-0.00014-7

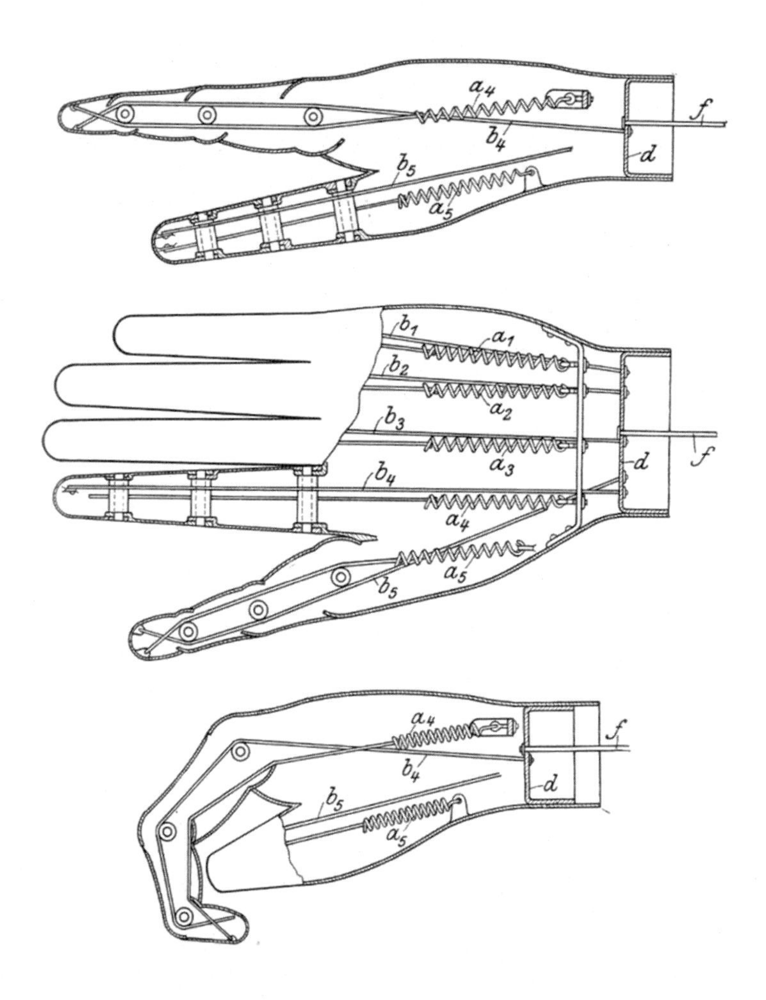

FIGURE 5.1

The design of the Ballif Hand. It is clearly voluntary opening and involuntary closing, with all the digits curling as they close to the tip of the thumb. (Taken from [2].)

Compared with almost any other area of technology still being used regularly in the twentieth century, very few innovations have been made with body powered prostheses in two generations. It allows those who have wanted to make a point to claim, as Dean Kamen did, that no changes have been made in upper limb prosthetics since the American Civil War. This is a gross misrepresentation of the truth, but it allows him to frame himself as a leading innovator in a neglected field, stepping in to save us from ignorance and inertia. However, compared with the money and intellect that has been devoted to externally powered prosthetics in the last fifty years, body power has been neglected by all but a small band of dedicated innovators [37].

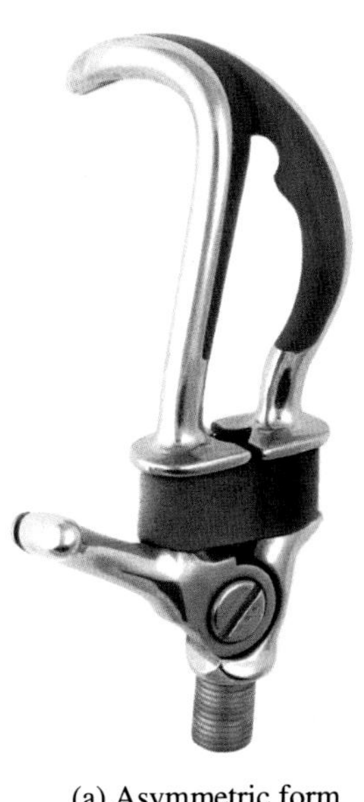

(a) Asymmetric form

(b) Symmetric split hook

FIGURE 5.2

The standard split hook formats. The rubber band closes the hook, a cable pulling the lever out to the side opens the hook. It is still one of the most popular designs of prosthetic terminal device. (Courtesy of Fillauer Companies.)

5.2 Advantages of body powered control

If we look at prosthetic control from the perspective of how someone controls a hand or arm, body power makes a great deal of practical sense. To control the position an object, such as a hand, we need to know about where it is. We use our senses (not just our eyes), but also the joint position sensors (proprioceptors). To control the movement of an object, we need to be able to change the speed or direction of motion fast enough to react to circumstance. If we wish to move a limb to a point in space, but we do not learn of its position until after it has got there we will be too slow to react, then a moving limb will overshoot the target. To get quickly and accurately to the target, we need to be able to sense the location and relay it back fast enough to allow the controller (be it electronic or human) to make a decision in time; to slow down and stop without overshooting. If the object moves more quickly than we can receive the information, we will always be too late to react and as a consequence the arm will always overshoot and bounce back and forward repeatedly overshooting, never stopping exactly at the target. In this circumstance the only way to get to the target reliably is to move more slowly. If we are controlling the position of our hand and the information arrives slowly, then the hand must move slower than the information. If the transfer of information gets quicker, then the hand's speed can increase.

Body powered prostheses use a fixed link, such as a cable, to pull on the prosthesis to move it. This means the user can 'feel' the movement as they pull on the cable, and so react quickly to changing circumstances (in roughly 40 ms). Electric devices use the signals from muscles, which are slower, so

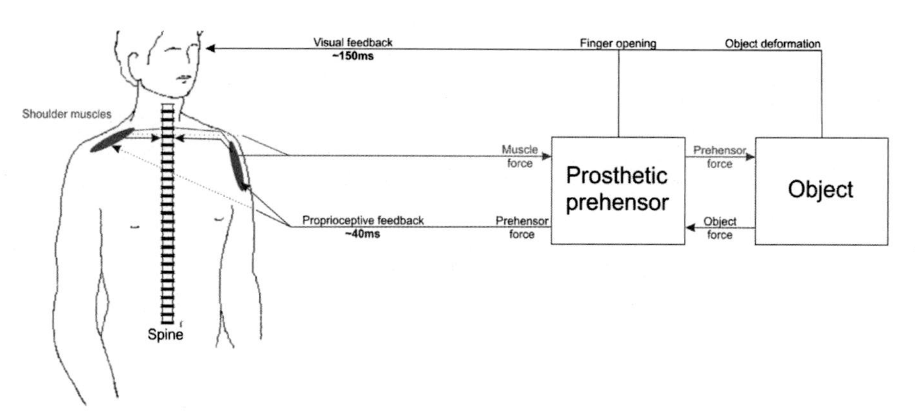

(a) Body Power - The delays in initiating movement or feeding back information come from sensing the movement of the muscles that pull on the harness (proprioception) and is roughly 40ms. Visual feedback takes much longer (150ms)

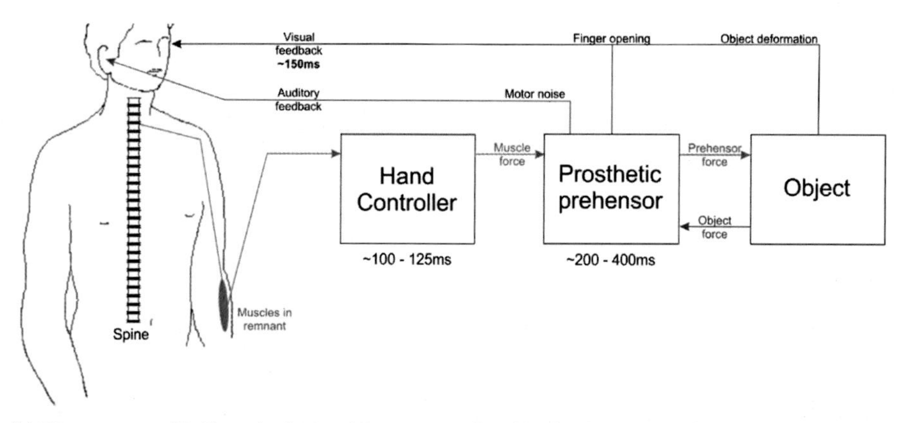

(b) Electromyographic Control - Greater delays are introduced by the slow nature of EMG and the prosthesis mechanism (up to 165ms).

FIGURE 5.3

Delays introduced in prosthetic control. (Adapted from [4]).

the prosthesis must move more slowly. The only way the user can detect motion is when they can see it has happened (in about 150 ms). Add the slow nature of the muscle signal to the slow motion of the prosthesis itself, and the person will see the movement more than 305 ms after they initiated the action (Fig. 5.3). From this it is clear that a powered prosthesis cannot move as quickly without the operator losing control of it, at least temporarily [4]. The best they can do is based on previous experience. They 'launch' the hand in the direction of a target, in the same way one might throw a ball at a target, with enough effort to get to the target. This skill can be learned, but once we let go of the ball we have no more control over the ball itself. If the target moves we cannot change where it ends up. Similarly one can learn to control a hand to open or flex an elbow, but changing circumstances cannot be as easily accommodated. The body power user of a hook can feel how much it is open, or if an elbow is flexed, by how the cable and harness moves, and so be more confident in their control.

Both forms of control rely on seeing the movement (visual feedback) to confirm where the prosthesis is. Tests have shown that persons with intact hands tend not to look at their hands often. They look where the hand is going to be in a few milliseconds time. By the time the hand has arrived there they are looking to the next place the hand will be [104]. Users of prostheses tend to look at their hands and follow them as they perform their tasks, even those with body powered devices. Other experiments show that if we combine two sources of information about something (position of an object or force on it), we place greater confidence on the one we know with the greater precision [73]. When some form of artificial feedback has been applied to prostheses, it has never been as detailed or as precise as vision, so we tend to believe our eyes over other senses. While other forms of feedback for hands have been tried, they have tended to fall short of the success that people have hoped for them. Experiments that use abstract tasks [188] show improvements. When the tasks are practical or in the real world [189,190], there is no recorded functional advantage.

In addition to the better quality feedback, body powered prostheses can have many other advantages over more sophisticated devices. Body powered devices can be made simpler, lighter, cheaper and more robust than an electromechanical hand with a computer controller. These are highly desirable traits for a prosthesis.

The most usual form of body powered control/actuation is a cable running from a harness across the shoulders down the arm to the terminal device (Fig. 2.4). This cable is a stranded metal cable in a coiled wire outer sleeve, often coated in plastic. It is a 'Bowden cable' and is exactly the same technology as a bicycle brake cable. It is cheap and easy to manufacture, very robust and is easy to repair. However, if the cable has to go round many corners then friction in the cable builds up and it can take a lot of force to operate a hand at the end of the cable. When tests were performed on cables running to terminal devices in the form of hands they found the friction in the cable and the prosthesis mechanism required up to 131 N to be imparted to get a 15 N pinch force [3]. As the cable is pulled by a harness on the shoulders of the operator it is uncomfortable for the operator to generate this force, especially to do so repeatedly [191].

The standard harness generally uses a loop that runs round the opposite arm (an 'Axilla loop'). One of the consequences of the large forces required to pull on the cable is that the pressure on the skin and nerves at the base of the arm can be considerable. Long term use can damage nerves in the shoulder and arm causing loss of sensation in the arm and hand, and pain along the arm. One novel solution comes from a user, it is perhaps a solution that only a user would think of. Debra Latour devised an anchor that sticks to the skin on the shoulder blade of the opposite shoulder [192]. This means the same motions can create the force to move the prosthesis, but the pressure can be spread across the skin of the shoulder. Modern adhesives mean that they can be applied and left for many days.

The cable only works in one direction, if it is to work in two directions it is usual to add a spring, so it is pulled back in the opposite direction. For a terminal device, such as a hook or hand, this means that the hand is either sprung closed and pulled open (so call *Voluntary Opening, VO)* or the reverse *(Voluntary Closing, VC)*. Each have their advantages, but the market has focused more strongly on the former designs. When tests are conducted by users employing a prehensor that can be either voluntary opening or voluntary closing, the subjects' choice depended on the task, and neither form dominated [193]. Not unreasonably, it seems that the choice depends on the task needing to be performed.

One flaw with voluntary opening devices is the trade off that has to be set between closing force and controllability. If you make the hook powerful (large closing force so a large grip force), then it is hard to open and it is hard to control its pinch force [194] and operation becomes tiring. Alternatively, you can make it easy to use with a light closing spring and you lose all the closing force.

The drawback of voluntary closing devices is that they require the user to continue applying the force for the entire time they want to hold an object. Alternatively the user needs to apply a brake, which has to be engaged and released easily somehow. Clearly if they need to use their other hand to do this then it is not a particularly useful device. Bob Radocy of TRS's solution is, as usual for him, simple, practical and elegant; a small wedge is added to the outside of the arm and when the cable runs in it, the wedge grips the cable. It is not however perfect. It can slip and has a limit to the force it can retain in the jaws of the device. For a prosthesis used in support of one typical hand, the forces needed are usually sufficient.

A body powered prosthesis works by using the motion of one part of the body relative to another to control and power the device. For example, the forward or backward motion of the tip of the shoulder relative to the chest, or the level of elbow raise (shoulder abduction) relative to the body. The problem for someone missing more than one joint is trying to capture multiple relative motions to enable more motions to be used to control more prosthetic joints.

Hence this leads to the prosthetic contradiction: *The greater number of joints lost, the more need to be created, but there are fewer opportunities to control them.* This is true for the externally powered devices as well.

Someone who is missing a hand can have a socket over their forearm and a cable that runs from their hand to their shoulders and back with two loops (a *figure of nine harness*) between the shoulders. The harness both holds the prosthesis on the short arm and allows the opposite shoulder to pull on the cable and open the hand. If the user is missing their elbow, then a second control cable has to be added to the front of their upper arm so that it can engage a brake on the elbow through suitable movement of the shoulder.

When the brake is on, the elbow is locked and the force on the cable will open the terminal device similar to the way a below-elbow device is operated. The other cable is activated when the person shrugs forwards. This unlocks the elbow which can now swing freely. The first cable is now controlling the amount the elbow is flexed, raising the hand. To perform this easily and unobtrusively requires coordination. The adjustment and alignment of cable lengths, and where they attach to the harness relative to the natural joints, is crucial to ensure that it is possible for the user to operate the device. If the brake requires too much motion to be engaged, it may tire the operator, so it is not used. If more motion is needed, the operator may not be able to create enough movement to trigger the brake at all. If the link is short, then the brake may come off too easily, so items are dropped or operations are filled with mistakes and repeated attempts. To create the correct form requires the prosthetist to carefully set up the harness correctly, and user practice to perfect its use.

If the loss is further up still, then a shoulder joint is required. This can only be unpowered as the forces required to lift the arm through bodily motions will be prohibitive, even for a very light arm. The joint is held still by a brake but with no other independent motions to actuate the brake, it will need to be operated by an external switch. One option is to place the switch where it can be reached by the operator's chin, so it is placed around their neck. To operate the arm, the brake is taken off and the arm moved up by placing the hand on a surface, such as a table. Then the person bends forward so the arm rises relative to their body and when it reaches the correct angle, the brake is reapplied and the person stands up with the arm now in a salute. The alternative way to raise the arm is to bend forward so the arm swings in front of them and then apply the brake. Of course if the other arm is in some way intact the person can use that to move the arm. Chances are that if they do have one functional arm, they will not use the prosthesis, nor ask for an active joint. They will instead leave the prosthesis hanging and employ the sound arm for all tasks.

People with very high-level losses can control many axes in their prostheses, mostly in a serial fashion, i.e. each axis at a time. This is time consuming, but many can be very effective with such high-level losses. Some individuals with two missing arms can lead mostly independent lives, but every operation takes longer and requires thought and practice.

If the person has lost both arms at a high level, it is difficult to find and isolate enough body power axes. An effective solution is to fit the dominant arm with the body powered joints and put electric joints on the nondominant arm, which are then switched using muscle based signals [195]. Jesse Sullivan the golden user from Chicago has this arrangement, (Fig. 12.2), with body powered devices on his right, two chin switches are visible on his right shoulder. This can create an extremely effective user. Jack Uellendahl describes a farmer with a high level bilateral loss, who mostly operates his own farm without help [195]. Videos of him working are awe inspiring as he carefully arranges himself and items around him so he can manipulate them. This includes tractor, trailer and farm equipment.

The current range of body powered components is quite broad with many different companies making different devices. It is not clear why much research has not been done in this field in the last fifty years; Bob Radocy of TRS speculated that perhaps it is the perception that it is all solved, or that it is somehow an old-fashioned technology not worthy of study [28]. This is, of course, false. Many other older systems and inventions have benefited from new ideas, techniques and materials. Body powered prostheses are no exception. For example, one can employ newer materials, that are lighter or easier to use. Harold Sears of Motion Control suggests there is no way to recover the investment in the development of body powered devices (5.2.3).

5.2.1 Terminal devices

Existing designs of terminal devices fall into the four categories: Hands, hooks, prehensors and specialist devices [19]. The first are nonactuated anthropomorphic (or nearly anthropomorphic) hands. These overlap with the so called cosmetic hands, being lifelike in colour and shape, but there are some that are only somewhat like natural hands and can be used to cup or hold items, or used by infants for crawling. These are soft rubber and bend when the wearer rests on them.

5.2.1.1 Hands

There are many designs of anthropomorphic hands. The majority have some limits set on the number of moving fingers. For instance, some have the same format as the simple powered hands. They are set in a precision grip where the index and middle finger are part of the structure and their flexion is linked to the thumb so that they open and close together and oppose each other. These may even share their external gloves with the powered hands, which is a convenience for the manufacturer (Fig. 2.5).

The Becker hand is different as it has all fingers actuated by a flexible mechanism inside the shell. The fingers have coil springs as their main external structure. The mechanism pulls all the fingers closed in a precision grip. When it is opposed the mechanism locks and can be unlocked with a pull on the control cable. The spring-like fingers allow the hand to conform round many different shaped objects.

The majority of these hands are voluntary opening. There are some similar hands that are voluntary closing. All of these hands are stiff and hard to actuate and so require considerable effort to operate.

5.2.1.2 Hooks

A major group are the 'hooks'. These are generally of the same form, with fine tips for precision grips, opening out at the back for larger power grips. The gripping surfaces are coated with a durable plastic, but generally have a small surface area and which is also flat. A stable grasp is the result of spreading the grip over as wide a surface as practical. This is not possible with a hook. The gripping surface is small so the force that is be applied to the object to hold it has to be very large. The hook-like shape also allows for it to be used passively to pull objects towards the user, or pin them down to be manipulated by the contra-lateral hand. There are two sorts of tines on the hooks, either symmetrically bent outwards to wrap around an object or with one straighter and one bowed outwards (Fig. 5.2). Each have their advantages. Harold Sears estimated that of the 20 or 30 thousand hook users in the US, most hooks are either Hosmer or the APRL design [196].

The majority of standard hooks are VO with a spring (rubber bands or metal springs) which determines the firmness of the grip. As the gripping surfaces are generally small, the gripping forces have to be large to create a stable grip. This means using a strong spring to close and making the device harder to control, as the user has to pull against it every time they open the device. This can be taken to extreme, one of the users I have worked with employed up to twenty elastic bands to hold their hook closed. They had to push very hard against the harness to pull on the cable to open it. The cable was pulled by a harness across the top of other arm and this pressure tends to cut off the blood supply or compromise the nerves in that arm. Thus the cable was arranged to be slack and only pinch the nerves when the device was activated. This meant to operate it required a lot of physical action hence it was not used as much as he might because driving it was an effort. For comparison; Bob Radocy from TRS can control his prehensor with little or no effort and is very effective.

5.2.1.3 Prehensors

Two nonanthropomorphic devices that are not hooks are the CAPP (Child Amputee Prosthetic Project) and the TRS GRIP Prehensor (more of this design 5.3.1). Parents and providers might be concerned about providing a hook with sharp points to a small child. It could be used as a weapon. One solution is to make the hook soft or use broad points, but this would tend to make it less useful as a manipulator. Prehensor designs get around this while creating a practical device. The CAPP is alligator-like in appearance. This appearance can be emphasised to the young children to encourage its use. They decorate it with eyes added to the top half. The device grasps objects rather than hooks them.

5.2.1.4 Specialist devices

The final category of devices are those designed for specific tasks such as sports or occupations. The history of commercial sports prostheses is short. In general, it was only in the last few decades that manufacturers made devices specifically designed for different sports [28]. A large number of the most interesting and creative come from TRS (see 5.3.1), but other companies have also produced devices. As a category, it is an extension of the devices designed for specific activities. These designs go back much further into history, with prosthetic users making their own specific terminal devices. There have been knives and forks interchangeable with the standard hook or hand, for many years [19]. These have been added to; tools for the kitchen and the workshop, such as knives, hammers, and rings [19]. The majority of devices use the same interface at the wrist to swap in and out devices [19]. These are simple and have not changed much in years. To release the hook, it requires unfastening the driving cable and fixing that either on the new terminal device, or parking it on some extra anchor so it is not lost up the sleeve. Then the terminal device is released and the new one added and used. The truth is that the level of innovation that now is used for tool changers for power tools is so much greater than this, but this sort of thinking has yet to find its way into prosthetics.

5.2.2 Wrists

Wrists form part of the terminal device's fitting. They can incorporate this quick release/tool change facility. They may also rotate freely or be braked, or have a series of positions and resistance so the user 'clicks' to the next position and the TD is held there. Some wrists allow flexion and extension. This motion can be passive, set by pressure on the hand resting on an object or by manipulation with the other hand. The TD can be moved towards the body to allow the user to get objects to the mouth for feeding or backside for toileting. It is these off-axis motions that are most essential for personal activities, especially if the absence is on both sides [19]. Operating the TD with a flexed wrist takes more effort to actuate, so these wrists are more often used by persons with a two-sided loss, where they cannot compensate with their other arm.

5.2.3 Elbows/shoulders

The elbow joints that are available may be passive (simply set in a position and locked), or pulled up via a cable and lowered with gravity, with the brake activated by manipulation of the harness (shoulder shrug). The designers of elbows try to take up as little of the upper arm (humeral space) as possible, so that they can be fitted to the widest range of users. The smallest space needed use hinges that fit either side of the extant elbow structure. These hinges can be used with a damaged natural elbow that cannot support the prosthesis, or for a person who has lost the limb at the elbow joint itself so there is no humeral space to accommodate the full prosthesis.

There are three elbows that compensate for the force of gravity. The first in the market was the Ergo arm from Otto Bock. It uses an elegant counterbalance and lever mechanism that creates an almost massless forearm. The lightest flick of the shoulder causes the arm to rise. It then has a brake actuated by a cable, so that a simple shrug can catch the elbow at the top of its swing and lock it. A skilled user of an Ergo arm can use it dexterously, with the operator using it swiftly and precisely. The design incorporates an ability to easily alter the amount of gravity compensation by rotating a wheel on the underside of the forearm. This can compensate for the mass of the hand or the stiffness of a coat. Subsequently Otto Bock added a motor to this design to make the Dynamic Arm, an obvious but elegant refinement (see 9.2.1). Motion Control also produced their own counterbalanced arm that was a physical match for their powered elbow, adding a gravity compensation mechanism [197]. This innovation was in the opposite direction to Bock, taking a design with active motion and converting it to passive. Finally the Espire Arm, developed for College Park and now marketed by Steeper, was designed from the start to have a passive, counterbalanced variant (7.1.3.1).

Commercial shoulder joints are passive with a 'manual' brake (as outlined above). If unlocked this allows the arm to swing forward and back while walking. The majority also have a second passive axis to allow the arm to swing away from the body sideways (humeral abduction). If the loss is at this level, but from one side only, it is not uncommon for the wearer to be unable to tolerate the mass of such a long arm. The result is that they will opt for an entirely passive arm, made from tubes with passive joints in the appropriate places and then a foam cover to pad-out the arm and a shoulder cap to ensure that their shirts and coats are filled out and fit.

It has been observed that there have been limited changes in this area for many years, Debra Latour has contributed to progress in the field through her cable anchoring system. She has a congenital absence of her right hand and is an occupational therapist by profession. She was a practising OT for more than twenty years before she worked with fellow users. At this point she realised that the problems she encountered day-to-day were quite usual, and based on their experience she refined her belief that harnessing and socket fit were the two most important aspects of a prosthetic fitting [198]. Her idea was that an anchor stuck on the same side as the limb difference could reduce the chances of damage to the arm. She had been working on the idea for many years. In her early attempts as a new therapist working on herself, she used velcro to attach the cable to her brassier in a similar way to the usual chest strap arrangement. While initially effective and allowing her to wear formal dresses without the unsightly harnessing, at one occasion she had a *"fashion faux pas"* [198] which set her back. Once she was working with users she revisited her ideas and the solution came to her in a dream; she saw her own back and the harness attached somehow to her scapula. Following her epiphany, she employed modern glues. These are easily applied and the result is clean lines and good function [192].

While the lack of research in this area is hard to fathom, it is clearer why the companies have not produced any new designs, especially in North America. This market is characterised by low-cost products, strongly influenced by insurance companies' reimbursement limits. Even when not bound by the system (private payers), other providers in the USA tend to use this as a framework. Currently, the system is based on a description of the device and no attempt at functional assessment is made (although this may change in the future). Harold Sears of Motion Control, an advocate of body power, commented that the financial margins for simple body powered devices are so narrow that it is hard for a company to recover the costs of developing a new device in any reasonable time. Thus they are more likely to devote their energies on more expensive devices where the margins are at least a little higher [14]. This only explains the lack of commercial research, not the paucity of academic study. One reason for the latter might be that academic study requires financial support and governments may be looking for more than incremental progress; a slightly better hook design may not make for what is seen as 'high impact' research.

5.3 Innovation in body powered prostheses

Despite these barriers there are a small number of enthusiasts that have continued to look at body power. Two of the most significant contributors form the first two case studies: One is a successful prosthetics company and the other is a research group that aims to make a commercial device.

5.3.1 Therapeutic Recreation Systems

Therapeutic Recreation Systems of Boulder Colorado (now Fillauer TRS) is a very different prosthetics company. If you look at the TRS website you will see their wares laid out as a series of photographs of people using their products. There are children hanging from overhead ropes, and teenagers playing ball. Amongst them is an athletic looking man with a full head of wavy hair, with images of him curling weights and drawing a bow. Most impressively, he is crossing a rock face using a TRS product. This person is Bob Radocy himself, the CEO and founder of TRS. He knows his products and trusts them.

Bob always liked sports. As a child he played many, from baseball to hockey, from basketball to football (soccer). In his seventies he still has the build of a sports person, and any time you see him in a prosthetics conference you feel you just called him away from taking part in sport or that he is on his way to a game of something. His career path was pointing towards marine biology or engineering, when his calendar and that of one of his closest friends clashed. The friend was getting married in Chicago and Bob was in Miami needing to sit some exams. Bob chose to revise hard, then take the exam, and at noon climb into his car and drive to the Windy City. Unfortunately the long hours of concentration caught up with him, and in Kentucky he drifted to sleep and across the centre of the road. The truck driver did everything he could to avoid the car, moving as far right as possible, but he ran out of road. At least Bob did not run out of luck. While the side of the truck tore the side of Radocy's car off, he only lost his hand. The driver was quick to his aid and put a tourniquet on and made sure that he got medical help. Radocy credits him with saving his life. He was only twenty-two.

Radocy was at least partly ambidextrous (batting with his right hand, kicking with his left foot), and he had lost his left hand. The degree he was studying was predominantly biology, but included some engineering. His father had been an engineer, working on high quality recording tapes (the company made the first successful 'Compact Cassette' tapes). Providing solutions for problems was something Radocy considered in his everyday life. It is the nature of an engineer that they will see something wrong and worry about a way to fix or improve it. During recovery in a hospital, a long way from anywhere he might call home, he started sketching ideas for devices he might use. The surgeon who had performed the operations to stabilise Radocy, visited his bedside and asked what he was drawing. When the medic was told it was prosthetic devices he countered with *"You don't need to do that they have everything you need"*. Radocy reckons this put him back about five years [28].

Certainly for a number of years Bob persisted with the prosthesis he was given. His loss was pretty much ideal; reasonably short but strong trans radial amputation, with little additional damage. This meant a good, well fitting socket could be made, and his control was precise. However, Radocy felt that while he *"could use a split hook as well as anybody, [I] was being limited by the technology of the hook in things I wanted to do"*. Radocy's goals were distinct from other users. He lists climbing and archery as two of his interests. With conventional devices he could not hold or pull, so he did not feel secure climbing.

> "To me, I think, gripping strength is the most essential thing to replicate in a prosthetic (hand), split hooks just don't have the prehension capability to mimic the hand, they are far inferior".

He thinks they don't do well on cylindrical or palmar grips either,

> "They don't come anywhere near replicating human capacity when it comes to palmar prehension, so handling tools... doing things in general I was always frustrated with the split hook" [28].

For Radocy the hooks were good at fine pinch, precision style grips, but as soon as a larger object was placed in the hook the two jaws pivoting from a single point at the distal end of the hook formed a wedge that pushed the object out of the grasp. Radocy likens it to trying to pick something up with a pair of scissors. This is a significant flaw in the design of the standard hook. The Dorrance Hook

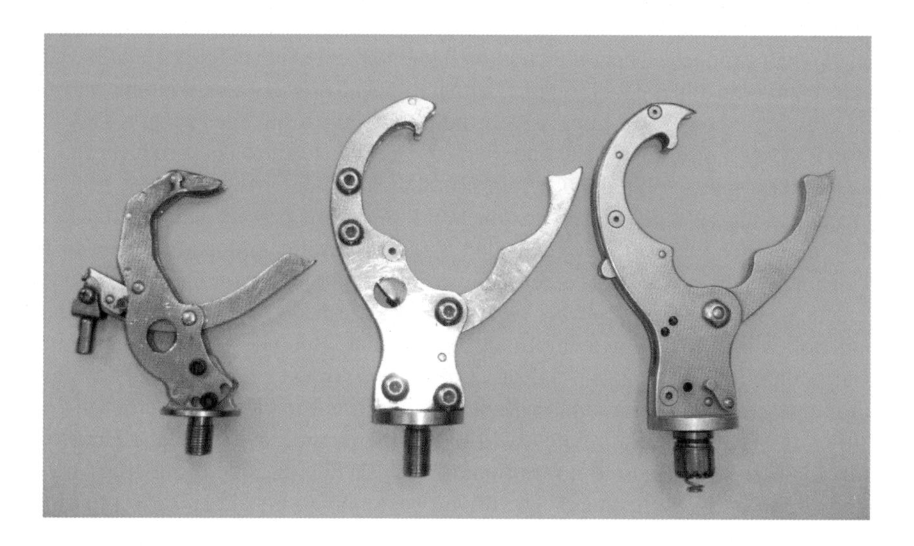

FIGURE 5.4

Evolution of the GRIP Prehensor. Left to right - the modified Mollenhauer device, first version of the 'Prehensile Hand', an early version of the GRIP1. (Copyright: Fillauer TRS, Inc.)

(Fig. 5.2) is typical of many other designs of hooks. Its jaws start at a fine tip, open up and then come back together, so the jaws are in the most part in parallel, but only ten millimetres apart, once it is opened it creates a wedge, which combined with the closing of the jaws tends to eject the object [28].

The second flaw Radocy saw is that the majority of body powered prosthetic devices offered were voluntary opening. There is a clear explanation as to why the market had chosen to focus on this as the single solution. It is probable that the choice is made based on the assumption that the prosthetic side is most often used in a support role, holding objects lightly and precisely. For example, a nail is held in a precision grip in the prosthesis, while the hammer is held in a power grip in the natural hand. The prosthesis then holds the object with a light force and the user can hold it with a VO hook, without thinking about it or exerting any significant effort. If a user wants more from their prosthesis, then the voluntary opening hook will not do. Radocy certainly did not want to use his prosthesis for this application. All over the world, broken, worn and damaged hooks come back to clinics, suggesting it is not the only use that others put their devices to. Bob found voluntary opening devices frustrating and it led him to look for alternatives [28].

Radocy found one German designed hook (Mollenhauer) that was voluntary closing [28]. Its design clearly had little to do with appearance and everything to do with manual function (Fig. 5.4). It had a ring sticking out to one side for fastening to hooks etcetera for carrying objects. It was a small device, and Radocy really appreciated its functionality. He used it until it wore out.

After his accident Bob moved away from marine biology. After working for several years in the air pollution industry, he returned to college to study to be a physiotherapist or occupational therapist. Ironically, his loss was seen as a disability and the transfer to a liberal arts college affected his grade point average (cumulative academic score), driving it down. He could not get into the schools he

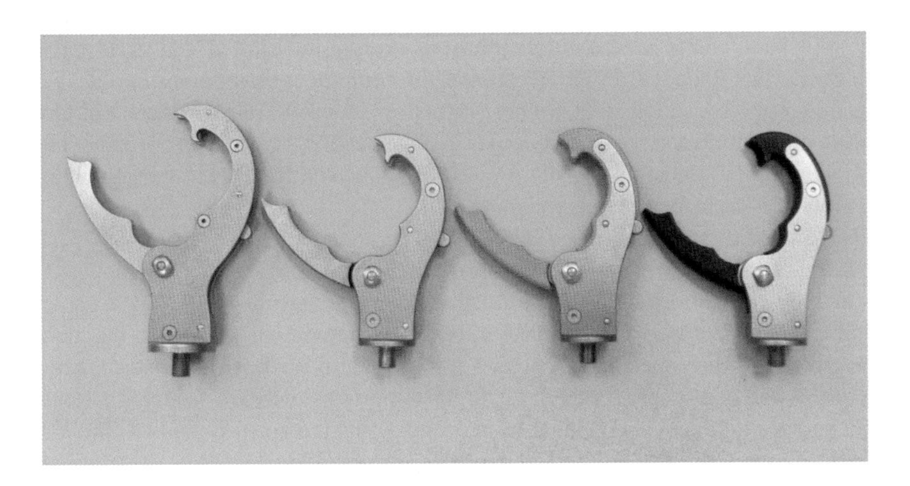

FIGURE 5.5

GRIP Prehensor - An elegant solution to providing a series of different sized cylindrical gripping surfaces for a body powered terminal device. Later versions (right side) have rubber based gripping surfaces. (Copyright: Fillauer TRS, Inc.)

wanted. Instead, his master's became one in physical education and recreation with an emphasis on therapeutic recreation. It is this emphasis that shapes TRS's approach.

Bob's epiphany came when he was complaining to a friend about the devices he had used and the friend asked him why he did not try making a better one himself. This seemed to give Radocy back the permission to think about designs for himself that the surgeon had taken away from him in the hospital. Bob went to the library of the university and trawled through all the patents he could find. He found very many different devices. Some very elegant, but he felt that all of them would be totally destroyed the first time someone wanted to use it forcefully. *"Being a prosthetic user, I look at those and I go I pick up a hammer with one of these things, I swing it once and it goes in the trash"* [28]. It was clear that no matter how neat devices were from an engineering perspective, none would be useful for the sort of work he wanted. This is still true and the majority of 3D printed hands produced in the last ten years would fail this test.

Instead Bob designed one himself. His first design choice was for a voluntary closing device. Having worked on the design he scraped together enough money to pay for the parts to be machined. When it was finished he started to use it, it did break, and so he revised the design. Over the years Bob's device has been strengthened, and its grasping function improved. However, the prehensor that TRS still sells essentially reflects his original design.

Radocy is very particular when he calls his device a *"prehensor"* and not a hook, because he finds hooks an ineffective way to firmly grasp an object. His solution is a pair of curving jaws with a series of concentric enclosing cylinders on their inner surface (Fig. 5.5). What he is mimicking is the action of the fingers of the hand wrapping around an object. It is, after all, the function he is trying to restore. The obvious extension of this is a fully anthropomorphic hand with multi jointed fingers curling as they flex. Indeed, this has been the design sought repeatedly throughout this history. However, Radocy's philosophy (like any engineer's should be) is to make it simpler, lighter and cheaper. Any anthropomorphic hand would need multiple joints, as every joint tends to add mass and expense and reduce reliability. Thus Radocy's design overcomes this with the inner gripping surface being a subset of cylinders that provide solutions that are near enough to be useful in a diverse set of circumstances, holding a wide range of objects, roughly circular in cross section, more securely than a hook. He does admit that holding a flat object, especially a piece of paper is difficult, but that is low on his list of priorities, particularly as he has one natural hand [28].

Once his first design was being used he began to promote the prehensor. At his first prosthetics conference he met Jerry Leavy, who was a bilateral user who worked for the US supplier/fitter Hosmer. Leavy was a *"bilateral split hook user for thirty years so he wasn't going to convert over"*, but he saw a use for it and encouraged Bob to make it, telling him there was a place in the market for it. In 1978 he did try selling the concept to Hosmer but they were not interested. Instead, he started a one product company, and was incorporated in 1979. Sales started slowly (he nearly went out of business in 1983), but he got a lot of encouragement from occupational therapists to build a child's version. It is clear one of the reasons for this; children tend to be more physical than adults. Adults reserve specific times to undertake many activities, such as going to the gym or the playing field. Adults can expect to rely on specialist devices, changing when they get to the workshop or the gym. Children are more spontaneous and use their hands for many different purposes, so their device must be more general purpose and rugged and this made a great deal of sense to him. Convinced of a market, Radocy pulled together $130,000 of venture capital, sold off part of the company, and developed the *ADEPT* line, which turned TRS around. These were the first voluntary closing devices marketed for children.

The approach that TRS takes is a therapeutic and sporting focus, producing specialist devices to perform different sports and activities. Radocy aimed to establish TRS as a company to provide good quality, reliable, lower cost devices, while the rest of the industry has been on a more than fifty-year quest towards expensive and sophisticated prostheses. He could see there were still big gaps in the market to fill. *"We want to provide a lower cost alternative to bionic type limbs"* [28].

His first sports device was the 'Super Sports'. For his own recreation, he used to tape a tennis ball over his prosthesis when he played football (soccer), to protect himself (and others) from the sharp point, but with it he could not fully take part in the game and do throw ins. He developed a cup shaped hand with flexible elements to hold the ball and allow it to be thrown. The only other sports device manufacturers at the time were Hosmer, with a bowling ball adapter, a baseball glove adapter and RobinAid in California who were making one-off devices to hold golf clubs.

The design of the prehensor used moulded rubber, which opened up Bob's knowledge of polymers. He learned how to use polyurethanes in different ways to create complex structures with different properties in different parts of the structure. This allowed him to investigate the idea of replicating the overall biomechanics of the task in passive structures. The structure bends or flexes in directions that are useful to performing the task while it limits motions in unwanted directions. By using polymers the device can be cast in a single piece and does not need hinges or joints which are likely to be more expensive, heavier and reduce reliability. An example is a device to hold a golf club. It is flexible in the direction a wrist bends to swing the club, while it is stiffer in the direction away from the body. This means the terminal device will tend to go in the direction of the club's swing. Polyurethanes are being used in many places in prosthetics and he thinks there is a lot of scope for combining purely mechanical design with polymer science, to replicate the elasticity of human joints more accurately. Working in polyurethanes is one of the major skills of TRS. Bob believes he is in a good geographic position to do more polyurethane production in the future. One reason is that Colorado is dry and polyurethanes are

FIGURE 5.6

Freestyle Swimming TD - A simple solution to providing a reliable means to push against the water in the power stroke, while folding out of the way during the return stroke. (Copyright: Fillauer TRS, Inc.)

hydroscopic (absorb water) and so the quality of his finished product is better than his competitor's in more humid locations [28].

One of the things Radocy is particularly proud of among TRS's achievements is that they can build something quite sophisticated at a low cost to allow people to do sports and recreation.

> "Probably the thing we developed that was most rewarding for me personally was the Cobra, even though we haven't sold that many" [28].

The Cobra is a throwing arm, that holds the ball in a cup and can be controlled to direct the throw precisely.

> "I used to throw left handed and so I hadn't thrown left handed for forty years... I am able to throw a fifty plus mile-an-hour fast ball, which is not big league wise, [it] isn't that fast, but certainly fast for an average individual".

He reckons he can throw precisely to someone thirty-five metres away. What impressed him was that the first time he tried the new device, the hand eye coordination came right back. He picked up a ball and threw it exactly to where he wanted it to go [28].

5.3.1.1 Creating the Cobra

The development of the Cobra is a good illustration of the collaborative way that Radocy and TRS work. Ideas come from everywhere. He sees a need and adapts an existing device. About a quarter of the products come from users themselves. Radocy then attempts to make them commercially. Ideas also come from prosthetists. Robert Gabourie had an idea for a swimming hand, the patent he owned was very complicated with valves and four fingers and Bob felt it was too complex to be made easily and cheaply. He offered to pay the licence even if he could make it another way. Instead of a complex shape he thought of butterfly wings. The *Freestyle* Swimming TD was the result (Fig. 5.6) It is two hinged flaps that close on the back stroke allowing the water to rush past. They are sprung to open up and are blocked in the open position during the power stroke so the user gets to push against the water. The product has lasted thirty years mostly unaltered.

The story of the Cobra began with a previous sporting device: the 'Hoopster'. It was developed from the invention of Hector Picard who had lost both hands; one at the shoulder and one at the wrist (and also a foot). He was an electrical linesman and had touched the power lines. The electricity shorted through his body and burned away parts of both of his arms. Wanting to play basketball Picard took the ring off the bottom of a galvanised pail, welded a thread on to it and covered it in a rubber bicycle tube and used it to play basketball one handed. The hoop held the ball in one side and he was able to move about and play quite bimanually. Bob asked Picard if he could evolve the design. It was not patented, but nonetheless Bob paid him a royalty for the idea. To create a safe product Radocy needed to remove the metal ring to reduce the chances of injury to the user and fellow players. Thus it became a simple rubber hoop. Once he used it Radocy realised that the hoop stores energy in the rubber which helps the throwing. He next wondered about other size balls to throw. His aim was to see if he could use it to throw a baseball.

The first consideration was to make sure it would fit within the rules of the game. To do so it had to be compact. An arm longer than the natural one would give the user an advantage. When he was brainstorming ideas the vice president of TRS asked Bob if he had seen the ball throwers made for dog owners. The device is a cup on the end of an arm that allows the owner to repeatedly throw the ball often and far, without tiring. Bob had not, so he ran down to the local pet store, bought a couple, and began experimenting with them. The original device is a half metre rubber beam with a cup on the end. To use it the owner places a ball in the cup and whips it forward. When they stop the cup moving abruptly, the ball carries on and flies out of the cup. The beam gives the user a longer arm and stores a little throwing energy, so the motion will throw the ball much further than a simple flick of the wrist alone. If the owner chooses to employ their entire arm, the result is the ball carries a long way. Bob took the device and reduced its length, padded out the cup to take a baseball, and then used a belt sander to adjust its performance. He removed parts of the rubber beam to give it a particular whip, plus he changed the cup's overall shape. The first attempt at a throw was a disaster; *"it went straight to the floor"*. He realised that the 'wrist' had to be adjustable to allow it to be angled correctly. This worked well enough to encourage him to build one from scratch. Once he tried it, he rapidly realised it needed to be adjusted left and right as well as up and down. The next iteration had a flexible 'forearm' and an adjustable 'wrist' to tune the trajectory of the ball. All this took about a year of experimentation to get it right. Once it is adjusted Bob can throw consistently along the line of sight as naturally as with his original arm (Fig. 5.7).

From this we can see that Radocy has a very practical approach. He will make items himself and get an outside machinist to build the complex parts. He is based in Boulder Colorado and he thinks it is in a good position for this sort of work;

> "If I lived on the east coast I'd probably be going from anywhere from New Hampshire to Georgia to get the same kind of [manufacturing] services I get here in Denver" [28].

Bob realises he can not solve the problems of everybody with a limb absence. The needs of someone with a unilateral loss are different to someone without a hand at all. For example, the first evaluation of his prehensor made by the USA's department of Veterans' Affairs (VA), which looks after the needs of all servicemen and exservicemen in the USA, was undertaken by a person with a bilateral loss. It did not go that well. This was an important lesson, in that it showed that it was designed for someone

FIGURE 5.7

Cobra throwing device - A rubber cup on a flexible beam holds the ball and releases it at the end of the swing of the arm. Its orientation is adjusted to the individual thrower's biomechanics. (Copyright: Fillauer TRS, Inc.)

with one good hand so it was good at supporting the actions of the user, but not necessarily initiating them. He believes that before him, no one had actually targeted unilateral loss. The manufacturers had focused on those who needed the most function. However, he is in the majority, with only one missing hand, so has a bigger market. He sees the company existing to create solutions for persons with a unilateral loss and he is not forcing the products to fit the second market. He regrets not being able to do more for persons with a bilateral loss, but he realises that at the same time he also has a business to run. He knows that rehab of a bilateral is *"far more daunting than someone like myself"* [28].

Radocy's philosophy is that people can work single handed, but TRS provides the tools to allow people to excel if they want to operate bilaterally. His philosophy for TRS is: *"For those who want to use a prosthesis it allows them to be truly competitive and functionally bimanual in the activities they want to take on"*.

5.3.2 The ENER JOINT elbow

A second example of TRS's ability to respond to market interest is a passive elbow mechanism named the ENER JOINT (EJ). It was designed by Bob Radocy, in response to a request from prosthetists at the company Advanced Arm Dynamics. They were looking for an elbow for people performing vigorous sports activities such as mountain biking. When riding a bike over rough terrain we use our elbow to damp out the up and down motions of the handle bars. Prior to the EJ's introduction prosthetists typically tried to apply conventional shock absorbers across the elbow joint to absorb the motion. This creates a bulky elbow which does not fit easily under the rider's clothes and is awkward to adjust for the riding position. Bob chose to build the resistance to motion inside the joint, using the company's skills in the fabrication of rubber. The result is a spring like response to small motions that resists more as the pressure to flex the elbow increases. *"It feels more like a real arm controlling the bike"* says Radocy. The resistance can be matched to the user, with four different levels. It is simple to operate; the elbow position is adjusted and it is clamped with one hand using a quick release lever made for bicycles. Terminal devices have been made for specialist applications for many years and are now commonplace, but the EJ was the first activity specific elbow on the market.

5.3.3 JAWS

A more recent development is JAWS. Its Bob's first foray into voluntary opening devices. He thinks they will produce more of them than any other product, because there are so many suppliers of VO devices that it is an easy change for a user. They do not have to learn new dynamics or a new system, just put it on and use it, Radocy suggests it: *"Does everything your split hook won't do or can't do"* [44].

The JAWS was designed for the many users who use an electric hand but find it cannot not grasp firmly enough cylindrical objects like handlebars of bikes, snowmobiles, all terrain vehicles and mowers. JAWS can be used with or without a cable to open it. In the latter case it is pushed on to handlebars. It is lightly sprung closed and uses a lever to increase the grip force. A lever on the other side of the prehensor, changes the direction of force of the spring. At the lowest force the line of action of the spring is in line with the pivot. The lever moves the line of action to oppose the opening of the jaws. This means the effort to open the device is not against the full closing force and the force required to change the setting is also low. The lever can be set at four force levels and uses a 265 N spring [44]. He is marketing to both the prosthetics profession and the individual users [44]. He is not using it himself, his GRIP 5 is all he needs but he thinks it is *"Kind of a fun product"* (Fig. 5.8).

Radocy is a strong advocate for body powered solutions, his life and activities are a good advert for the effective use of the technology and how elegant solutions can be achieved with simplicity. However, the best technical understanding of the technology comes from The Netherlands.

5.4 The Dutch perspective

The region of Europe now referred to as The Netherlands is an area of land that has been part of a number of world empires: Spanish, French and ultimately their own. Their history includes wresting land from the North Sea though drainage. Their relative influence on world art and commerce is well above the size of the home country. The Dutch celebrate their reputation for being more relaxed about rules and being 'careful' with their money [199]. Outside engineering, their ability and the quality of their engineering is less well known, but being able to drain the sea to create up to 17% of the current surface area of their country shows the quality of their ingenuity. That work has to function reliably year after year so it is perhaps unsurprising that a practical Dutch engineer would see the value in investigating body powered prostheses.

Delft is a port city facing the North Sea, and has a long history. It is famed for its fine pottery particularly with delicate pictures in a strong blue on a white background. As one of the home ports of the Dutch East India company, Delft has attracted wealth and innovation for centuries. The Technical

FIGURE 5.8

JAWS terminal device - designed to firmly grip objects such as handlebars and paddles. It uses a lever to increase the grip force once the device has grasped the object. (Copyright: Fillauer TRS, Inc.)

University of Delft has buildings across the city, the main campus being a few minutes cycle from the main train station.

In the preceding sections reference has been made to the limitations of body powered prostheses, and some of the advantages they have over externally powered limbs. Despite these disadvantages, much of the effort and money devoted to prosthetics research over more than the past half century has been towards developing the more sophisticated devices. A look at the literature over the last fifty years suggests that for most of that time there were fewer than three new papers on body powered research every year, while there were far more than that number per month on powered prostheses [37]. In the last few years this has begun to change, but many of these papers are from the Delft group. There are a very few people over this time who have spent any effort on body powered research. Some, like Dudley Childress of Chicago, have directed part of a much wider program of effort into prosthetics in general on body power. Others, such as Maurice Leblanc, have spent time looking at the problem. However, no one in the last couple of decades has shown as much dedication towards solving the problems as Dick Plettenburg and his team at the Technical University in Delft, the Netherlands (TU-Delft). In particular in quantifying the differences between body power and external power and finding the limitations of the current technology.

For a Dutchman, Plettenburg is only mid height, rather than the towering postwar men and women who characterise their country in the modern era. He is a gentleman who quietly presents his ideas at conferences with a dry and understated wit. In 2002 Dick was asked to present a keynote address at the location of the heart of myoelectric research; the MEC, Myoelectric Controls conference in Fredericton, Canada. These conferences were set up to promote the use of myoelectrics as the control format of choice for prosthetic limbs, therefore externally powered limbs. Dick unashamedly called his talk *'Batteries Not Included'*.

The origins of the interest in prosthetics at Delft started with Jan Cool (pronounced *Cole*). Cool studied Mechanical Engineering in Delft in the late 1950s. Like others featured in this book, he became involved in prosthetics as a result of Thalidomide (in 1962). At the time Professor Boiten was head the laboratory for Measurement and Control, and he had links with a physician in Nijmegen. They had about twenty five Thalidomiders in the area and they wanted to help. Similar to a number of other groups, Nijmegen started by importing the Heidelberg arm, and in common with the other groups, they quickly found out the arms were very difficult to control and unreliable. The physician asked Boiten to help to improve the control. They investigated building miniature pneumatic components, first trying to improve the Heidelberg equipment and then making their own. They rapidly came to the common conclusion that to undertake significant clinical work, one needs a strong connection with the clinicians and especially with the users. In 1967 a multidisciplinary team was founded at De Hoogstraat in Leersum. This was the first such team in the Netherlands, and it was shortly followed by one in Nijmegen at the Sint Maartenskliniek [200].

There were four groups in the Measurement and Control department. The one devoted to instruments (including pneumatics) was logically the right group to become involved in the prosthetics work. The group title: *Werkgroep Instrumenten Laboratorium Meet En Regeltechniek (Working group of Instruments Measurements and Control)* in Dutch has the acronym WILMER. When prostheses were built and handed over to the clinic, there was a tendency to refer to them by the acronym i.e. *"This is a WILMER device"*. Over the years this has become increasingly confusing, because it is the name of a group, and became the name of the company that produces the prostheses, *and* the name of a foundation to help technological transitions out of the university group. More recently the research group was renamed the *Delft Institute of Prosthetics and Orthotics* (DIPO), and is called the *Home of WILMER*. The spin off company called *Delft Prosthetics* marketed the WILMER products until the end of 2017 [200].

One facet that can often be observed in the products from Delft is a tendency to find the simplest solution and realise it in an elegant manner. By the late 1960s or early 1970s attempts to keep prostheses simple drove the team to what they saw as an obvious conclusion: Body powered devices were simpler and easier to control. Looking at control theory and the methods of feedback, it was clear that there is potentially a better and faster link through body power (5.2). At the time, the attitude prevalent in the Netherlands was that children do not need prostheses, that they should be allowed to grow up before they need to be fitted. This meant simple and cheap prostheses were the best option, it also meant that there was a gap in the provision and there were few designs for children, which the group at TU Delft moved to fill. The designs proved popular, the consequence was that the team had to follow the children as they grew, and the designs got bigger. However, a full adult version did not work satisfactorily and the designs did not scale correctly. *"So still something to do there"* Plettenburg remarked [200].

The WILMER products were always elegant. They provided simple solutions to existing problems. For persons with a below elbow loss the WILMER products included five innovations that stand apart from the conventional body powered solutions.

The first is an open frame suspension that goes over the olecranon process (elbow) of the wearer. As the frame locks in place around the elbow once the remnant arm is in the socket, it is cooler than the standard socket. It is also easier to make a firm fitting while allowing donning and doffing to be easier. Conventional prostheses have to fit round the elbow to prevent them slipping off, while being loose enough to allow the person to take it on or off. The lock means greater space can be made round the end of the arm to place the residuum into the socket before pushing the support in and locking it in place.

The second innovation gets around one of the flaws of the cable operated hook designs. In a conventional prosthesis, the cable has to run from the shoulder down the outside of the arm and out to the lever on the outer side of the hook. This means it tends to damage cuffs of shirts worn over the arm. Also when the hook is inverted, as it might be to pick up some objects, the cable has to run around the outside of the arm, increasing the resistance to actuation. A WILMER prosthesis is operated using a push rod that runs along the centre of the wrist axis. There is no interference with the shirt and it does not matter which way up the hook is oriented.

The way to capture the actuation force is the third innovation. It is a lever built into the elbow of the socket, the elbow extension is used to drive the hook. This removes the cabling and creates the same clean lines of a myoelectric hand with a self suspension socket.

The fourth innovation is to create the hook that can take a rubber boot to protect the mechanism (and the user's clothes). This boot is easily interchangeable and can be made in many bright colours. In the past two decades customisation of the entire prosthesis to match the person's tastes has been a growth area. However, in most other versions of this the technology it is a one-time choice of colour or pattern, much like a tattoo. The WILMER device can change the boot colour any time the user chooses, and it was available long before customisation was more widely adopted.

A final product of note is a small toddler's crawling hand. It is made of rubber and is introduced early enough that the small child gets the idea that the prosthesis is part of their body image. Its softness means it is gentle on anyone around the child who might get swiped by them [201].

Plettenburg obtained his BSc and MSc in Mechanical engineering at the University of Twente in the Netherlands. During his master's program he spent the first half of 1979 in the US working on the Boston Elbow. His job at Liberty Mutual was to repair the elbow when it was returned. He also upgraded its two-way clutch at the same time. From this he was able to observe what happened to a product when it was exposed to the real world [200].

Back in the Netherlands, his graduate project was to design a device to be used in open heart surgery. After obtaining his master's he was looking for a job and saw TU Delft was advertising for an eighteen month post on upper limb prosthetics research. It was a turning point, once he joined he never left, retiring in September 2020. Half way through the time Jan Cool, the head of the prosthetics research group, asked him if he was interested in a PhD project on a pneumatically driven prosthesis. He readily took up the post, and was working on the design when he was asked to join the faculty to replace someone who had left. Dick said he was honoured and would do it, but asked *"what about his PhD?"* Cool said it would be ok, he would just combine teaching and research and it would take a little longer. This was 1986. Plettenburg finally finished his PhD in 2002, by which time he was well established in the profession, having already been invited to present his keynote at MEC in Canada.

It is usually a big mistake to take on additional responsibilities during a PhD. It is often a researcher's only time in their life when there is a chance to concentrate on something truly theirs. Distraction by teaching or other projects can interrupt, slow down or even stop a student's study, as it did for Dick. He could have easily lost momentum and drifted, never quite to complete the study. It is a mark of his determination and resolution that he did not. It is this doggedness that had made him pursue an unfashionable aspect of prosthetics in the face of disinterest and neglect, because he believes in it, and can justify his stance, carving out his unique niche in the profession.

Plettenburg synthesised the many years of experience of prosthetic design and use by the TU Delft team into the Delft philosophy of prosthetic fitting: Prosthetists need to recognise *"the 3 Cs: Cosmesis, Comfort and Control"*.

If the *cosmesis* is right, then people might consider wearing the prosthesis. For some users the cosmesis is to look as natural as possible, while for others it means looking cool. If they do want to wear it they will want to wear it all day, so it cannot be heavy and it cannot rub the skin. It has to be *comfortable*. Finally, if it is worn all day it would be nice to be able to use it in a subconscious way. Meaning that it needs proper feed forward and feed back signals, which you do not get with other forms of *control* [200].

One dominant theme that stems from the work at Delft is the improvement of the efficiency of body powered systems. It is reasonable to expect that if one is capturing useful force from the movement of the body, then the least amount of energy that is devoted to moving the prosthesis must be better for the comfort of the operator. This is especially important when the damage caused by the amputation may leave the user with less force or range in their remnant than an uneffected person. The trigger for his research was when the team wanted to use elbow movement to control the prosthesis, and not shoulder motion. As suggested above this can lead to a much cleaner and simpler looking prosthesis, but for some, especially children, the elbow muscles were far weaker. Some users had to summon all their strength and they still could not open the hand more than one centimetre.

This happens because in a voluntary opening device, the operator has to overcome the resistance of the return spring (and the glove in an anthropomorphic hand). This spring needs to provide enough closing force to hold the object. Radocy's response was to make a hand that is sprung open and pulled closed, but the Dutch investigated a different direction. The result was an entire philosophy of making prostheses where there is little or no energy expended to operate it. Ideally, when someone is opening and closing the hand there should be no real force required, even against the resistance of the glove. Once the hand has closed on an object, when the object is gripped (as long as it is stiff), there is no movement, so no work. In theory there is never any input work done by the operator. Of course there always is some work to overcome resistance in the mechanism, but it can be seen that well designed prostheses could lower the energy requirements for an operator considerably.

The first phase of driving a body powered limb is the capturing of the effort, which is then transmitted via the cable to the terminal device. When Plettenburg's graduate student, Gerwin Smit studied the force required to operate hooks and hands [87], he found that the cable contributed enormously to the resistance of the effort. Combined with the design of the hook or hand mechanism it could lead to the results already mentioned, 131 N of effort at the shoulder is needed to generate a 15 N force at the finger tips. Smit tried many devices and each device tested had a different characteristic. In general, the mechanical 'hands' required between 1.5 to 8 times more mechanical work than the 'hooks'. The resistance of the glove contributes a great deal to this inefficiency.

This is easy to appreciate. If a thick work glove is used on a natural hand, it is hard to grasp objects firmly and it is very tiring. The material of the glove absorbs the energy when opening and closing the hand. The energy lost when opening and closing is called hysteresis. It means that there is more effort needed to move the hand in both directions (rather than getting most of the energy put in to opening the hand back on a sprung return). For the gloved prostheses this was between two and twenty-seven times higher than in hooks [87]. Perhaps unsurprisingly, the most efficient and easiest to use was the TRS prehensor, as it needed the smallest activation force (33 ± 0.2 N) and had a very light spring to open it. The Hosmer Soft hand needed the largest (131 ± 0.7 N). The only similar study was performed some twenty-four years earlier by another pioneer of body power, Maurice LeBlanc [202] (who worked with the Delft team in 2012). What was observed in the new study was there was no improvement, and indeed some of the newer hands were less efficient than before.

The closing spring is an important part of the device. It controls how firmly the hook grips the object. If a person uses a stiff spring for a strong grip they then cannot easily control the prosthesis [194]. They find it hard to resist the spring and will tend to over grip the object.

Clearly this demonstrates a need to improve matters. One of the next steps taken in Delft was to consider a new hand mechanism that was more efficient. Different finger designs were evaluated and the more efficient design was chosen [3]. When they studied the glove it was found to contribute significantly to the resistance to opening and also to the absorption of the effort [203]. For anyone who has tried to flex the inner glove of a prosthetic hand, it will come as no surprise that the silicone gloves tested showed less resistance and lower hysteresis than the PVC gloves. One elegant aspect of the work at Delft is the attempt to overcome such problems with simple mechanical solutions. Two have been attempted:

The first uses the same principle as the Anglepoise lamp, which uses springs to create a device that is easy to move throughout the range of its motion, even with gravity pulling down the lamp. Each joint has a spring that counterbalances an external force, in the lamp this is gravity. Similarly, the resistance of the prosthetic glove acts as a spring, and the designer can compensate for the resistance of the glove with a second spring on the other side of the joint to oppose the force of the glove material. It is simple and elegant. The practical difficulty is that the glove's properties are not the same or as simple as a conventional spring. This is difficult because the structure of the glove may mean that it does not flex the same way throughout the range of motion of the finger inside (while the metal spring will easily be very much the same throughout the range). The second potential problem is the glove's behaviour (characteristic) changes over time. When the glove is new the compensation makes the hand move easily throughout the range. If the glove became stiffer with age the hand becomes more difficult to move. This is a problem, but it should not insurmountable. If the compensation makes the hand easiest to use when new and only slightly less easy to use as the glove ages, then it is still better than the hand without the compensation throughout the life of the glove. This change will still make any such design easier to use than existing hands.

The same improvement could be created for *any* of the existing gloves with limited modification of existing designs. This includes electric hands, so that they could use less power or have greater pinch force generation for the same motors or batteries. So far manufacturers have shown little interest in using this technique, except in a limited extent, with a bow spring in the Otto Bock Z6 derived hands and a single carefully positioned spring in the Motion Hand. The biggest barrier to the adoption of this solution would seem to be that gloves are moulded not at one end of their range, but in the middle, half open, so the spring needs to act in both directions depending on how open the hand is. The alternative of new moulds for the gloves has a cost implication for manufacturers. The cost of making moulds is considerable and at one time the manufacturer Steeper chose to change the shape of their prosthesis mechanism to match the mould, and not the other way around [204].

A second approach is a 'biphasic' device. Here the mechanism changes with contact of a held object to make opening low effort, while gripping can generate larger forces. Like the gearbox in the Otto Bock hand it switches between two states; fast and easy to open and then on to a higher gearing on contact with the object. A standard device is pulled closed with a spring. This closing force needs to be opposed to open the digits. The Delft group moved the direction the spring's force so it was acting through the pivot of the digit instead of simply stretching the spring. If it is arranged properly it requires very little force to open the thumb as the spring is not being stretched. Once the thumb closes on an object, the structure changes so that the spring is moved to a different position and it is used to retain the object alone. Gripping requires no force and opening the hand very little effort. So far, the design has not succeeded as change in the line of action for the spring needs to be different for different size objects. Again, a structure that works to remove some of the force over all of the range still makes the operation easier. The JAWS prehensor uses a similar change in the line of force of the closing lever to increase its gripping force, without needing a large effort from the operator to apply the lever.

5.4.1 Recent developments

The knowledge gained from the glove experiments has been incorporated into a new hand design. The first prototype was constructed as part of Gerwin Smit's PhD [89]. It is a return to the origins of the Delft prosthetics group. It uses micro hydraulics rather than pneumatics to drive the fingers of a multifunction hand. The design has gone through design iterations and has seen limited use in the field *"which resulted of course in modifications again"* [205].

Plettenburg saw a new way to look at prosthetic hand use. From the level of damage of unactuated hands observed in the clinics, it is clear that so called 'passive' hands are used for prehension. As observed by Adam Spiers, the hands are used to hold down items, and they are used to push or pull objects (3.4.5.4). The fingers break often enough to indicate that they are bent by users to hold objects. The team have looked at a passive prehension system. When an object is forced into the palm of the hand, the fingers flex around the object and the mechanism locks. A second inward pressure causes the fingers to unlock and release the object. Objects can be placed in the prosthesis by a contralateral hand, or by pressure from the arm onto the object. It is possible to make a five fingered hand that wraps around an object and grasps it [206]. At the time of writing, the device was beginning to be assessed for its use in the field [205].

The next frontier for Plettenburg is powered prostheses. It is clear that some users cannot generate enough force or excursion to operate a body powered arm, so it needs a power source. Plettenburg is not talking about using electromyogaphy to control the hand. He wishes to combine the virtues of body power with the energy generation of external power. Instead of using a myosignal he wishes to use a harness across the shoulders which measures the movement between two body parts, like a body powered device. The body movement is then measured with a potentiometer, so the users will get the feeling of movement and the proprioception but without needing to generate enough force to open the hand. This movement is relayed to the prosthesis, which feeds back its movement to the harness, like a body powered system. The user can feel the actuation. It could be regarded as 'power assistance' for body powered devices.

Looking at the profession, Plettenburg is always surprised how people seem to work on things that he can see will not work at all, *"trying to do the wrong thing better and better"* [200]. He is of course referring to myoelectric control. He admits that there is a lot of good and intelligent work, but feels it is still going in the wrong direction, particularly pattern recognition, which he suggests needs 101% accuracy, which of course is an unattainable goal. While he is not alone in this opinion, he is rare in not liking myoelectric control. Simpson from Edinburgh also disliked it (see 8.1), but the difference is that Simpson worked in the early days of myocontrol whereas Plettenburg still holds this view at a time that might be considered myoelectric control's golden age.

FIGURE 5.9

Bob Radocy as the champion of the 'Powered Arm Prosthesis' competition at the first Cybathlon in 2016. (Copyright: Fillauer TRS, Inc.)

For his own work, Dick hopes that in ten years they will really have improved body powered devices so that the forces needed will be far lower and the users will be able to benefit from proprioceptive feedback. Echoing the titles and claims of the USA's major prosthetics initiative of recent years; he hopes that they will *"revolutionise prosthetics, but now for real"*. In this way he believes there will be more users of body powered devices in ten years than there are today.

5.5 Body power at the first Cybathlon

The first Cybathlon was held in 2016 (3.5). For this competition, Dick Plettenburg invited Bob Radocy to be their pilot and they went to Switzerland to take on the world. Reiner and colleagues had imagined competitors using multifunction hands perhaps using neural control and a variety of grip forms. DIPO entered a standard body powered product. However, Reiner did realise this team stood a really good chance of winning [121].

Unsurprisingly, Radocy made the finals in the competition. He is a sportsman with considerable kinesthetic talent and he should do well at most competitions. For the Cybathlon he had more than thirty years of practice. While the other competitors were going as hard as they could, fiercely working their advanced hands to manipulate the objects, running between the different stations, Radocy was strolling through the competition. He dominated the field and won comfortably (Fig. 5.9). One of the most challenging tasks was the 'wire-way'. No other competitors completed it twice in the time allowed. The test was much like an old fair ground competition: Competitors must run a loop along a wire on a circuitous path without touching the wire. The loop must be rotated left and right to negotiate the course. Touching loop to wire was a fail, and competitors had limited time and limited number of repetitions allowed to complete the task. Bob held the loop in his standard *Grip Prehensor* and used his body and arm to move the loop around. He said he could have mounted the loop straight into his socket and moved it that way without the prehensor [44]. The other competitors tried to use their wrist rotators and when they failed to stop the prosthesis over rotating, the loop touched the wire and they were out. Bob felt that if the other pilots had approached the competition the same way, they could have achieved similar performances [44].

Following the win there were suggestions from some quarters that it was hardly a significant victory, as Bob only won by seconds. The world champion of Formula One only needs to win by split seconds. What an F1 driver is not able to do is stroll through the entire race in second gear and still win. There was clearly enormous capacity left in Radocy's performance. Any advanced device had to be much better to win. What is significant is that for the second competition intended to be held in 2020 the organisers added to and changed the tasks. While the wire-way is still there, changing a light bulb has moved to a ceiling fixture from the lamp stand. The bulbs screw into the fixture. Previously, the light was the most challenging task for Radocy, he needed to get the mounting thread on the bulb started by feel alone. Plettenburg thinks that the new fixed configuration will not be impossible to solve with a body powered hand. If the first thread in the bulb was started on the screw fixing, then body compensations would allow the body power competitor to complete the task [205]. Radocy is not sure he can get the prehensor with a harness in the new position [44].

A new task is a 'haptic box'; the competitors will need to feel with the prosthesis in a box and based on the feedback they get from the prostheses they must identify objects without seeing them. The aim is to use high levels of good quality sensory signals from an electronic prosthesis to allow users to identify the object. In contrast, Dick thinks that someone with a voluntary closing body powered device will still be able to sense a great deal from the cable connection [205].

Bob Radocy has studied the tasks and, given that no one had time to finish all the tasks last time, he is sceptical that anyone will finish the tasks the next time [44]. The next competition will not involve Radocy. He considers it is time to retire from competition. Even so I hope that his influence on prosthetics is not yet ended.

CHAPTER

Externally powered prostheses

6

6.1 Introduction

The application of an external power source opens up a number of possibilities for the prosthetic engineer. First, the control signal and the power source are separated (unlike in a body powered device), which means that control sources too weak to be used as motive power can still be used as guidance. These signals might be small excursions of the remnant limb, or movement of the nubbins in congenitally shortened limbs, or simply the electrical noise generated when muscles contract. Secondly, it means that more axes can be actuated than would easily be achievable with body power alone. The drawback is that the power source needs to be carried as does the mechanics to convert the power into movement. Additionally, the feedback path to the operator is lost and the device will be slower (see 5.2). In electrical devices the mechanics are the batteries, motors and gearboxes. These features will add bulk and mass to the eventual arm. One of the roles of the designer is to find ways to reduce these as far as possible.

6.1.1 Power sources

The distinctions between the pneumatic, hydraulic and electric have been outlined (3.3.2). The former two have the right velocity range, being better starting from stopped, and slower movements than an electric motor. Stefan Schultz and Christian Pylatiuk of the Institute in Karlesruhe, Germany, investigated one solution to the problem of motor efficiency and recharging [207]. Electric motors are best when spinning at a fixed speed, and not doing cycles of stopping and starting, while hydraulics are good at stop/start and poor at continuous motion. Schultz and Pylatuik worked on a hydraulic device in which had a compact electric compressor mounted in the palm of the hand. The compressor used an electric motor, not constantly slowing down, stopping and reversing, simply turning at the most efficient speed, or off. The pressure built up in a reservoir drove the fingers open and closed. This is not unlike an electric hybrid car, where the petrol engine only runs to charge up the battery, which in turn drives the road wheels. As a result the petrol engine runs at the most efficient speed for fuel use, without constant changes of speed that burn fuel less efficiently.

Schultz and Pylatuik produced a promising device that was built into a novel hand design with multiarticulated fingers and it was used a little in the field [207]. Schultz later stopped development to move on to the Vincent electric prosthesis (see 14.1.1).

As we have seen, the idea for one of the first powered hands recorded used solenoids (electromagnets) to move the fingers; this was in Germany in 1919 [2] (4.1.2). This odd solution may well have been because there were no rotary motors small enough to fit within the hand at the time. The majority of electric prostheses have employed conventional rotary motors to drive the hand. The modern

Making Hands. https://doi.org/10.1016/B978-0-12-820544-0.00015-9

alternative is 'brushless' motors. Here a microcontroller switches on and off different electromagnets arranged in a circle to pull the rotor onwards. The brushless motor is more efficient as the switching is exactly the right amount at the right time and there are no brushes to wear out, but the entire device increases complexity. Brushless motors are becoming commonly used including in prosthetics; the earliest adopters of brushless motors in prosthetics was LTI (7.1.2).

To be useful for the slower movements of a prosthesis, the drive must be reduced in speed through a gearbox. Gears can be used to increase or reduce the rotation speed through the size of the gears. Generally, reduction is used in hands. This means a small gear meshes with a large gear, the relative sizes determining the reduction. The result is all the load of the drive must pass through the small gear and a small contact area between gear teeth. If this load is greater than the material strength of the gear, the load will destroy the teeth. For this reason, there is a limit to the amount a single gear stage (large and small gear combination) can transmit. To reduce the speed further requires a second stage. To obtain sufficient reduction may require several stages in a gearbox. Every stage add mass to the hand and more inertia to the drive. During normal operation, the hand changes direction between opening and closing. To get all the gears moving and then to slow down and perhaps stop or change direction requires more effort. Each stage in the gearbox makes the eventual finger, or hand, slower to react to commands to change speed or direction. Each stage is also likely to add to the noise the gearbox makes. So the smaller the reduction, the better for overall performance of the hand.

The latest multifunction hands have five or more motors and so are not lightweight. While each motor needs more capacity from the battery, advances in battery technology are keeping ahead of this limitation. Electric motors remain the most popular solution as they are easy to control and power.

One way to improve the hand is to make the precise design of the motor match the needs of prosthetics, rather than using a general-purpose motor. This is much like specialist rally or Formula One cars; they are exceptional at their sports, but would make a poor car for the family shopping. Otto Bock of Germany has taken this approach with the motor for their DynamicArm; it has to be made just right for the application. Even their head of innovation, Hans Dietl, called it: *"One of the most crazy designs you can imagine"* [208], but the result is perfectly tuned to its job.

There are other, interesting ways to convert electricity to motion, and occasionally they are suggested as ways to drive a prosthesis (see 3.3.2). Regrettably, none so far have proven to be even as efficient as the standard electric motor. Indeed many are very much more wasteful of power and lower in force generated.

6.1.2 Control

Once the power of the battery is converted into motion, it is necessary to control this action. An electronic controller can take many different inputs from switches to muscle signals to instruct the controller. The best way to make a flexible prosthetic controller is to standardise the input form (say a voltage between zero and five volts) with the understanding that this signal is then converted into the controlled output. This output might be the voltage to the motor so that the motor gets faster with the increasing input signal. If the input is standardised it means that many different input devices can be used to generate a signal. As a result, the prescription of the prosthesis will be a matter of determining what the person can voluntarily change to control the limb and adjusting the socket to take the relevant input device.

The possible input signals are various: The relative motion of one part of the body to another can be used to pull on cables to generate electrical signals, or use a switch to select a different action. Any part of the body could be used, but trailing wires from opposite arms or legs increases complexity and the chances of something breaking. The prosthetist will choose a balance between complexity and reliability. This means using inputs that can be sensed close to or inside the socket of the prosthesis. The new multifunction hands have more grip options, the user must switch between the grips before opening and closing the hand.

If the person can move parts of the body separate to each other, similarly to body powered control, then the motion can be detected by a potentiometer. This is an electronic device that can measure, rotation or linear motion. These devices can be returned to their starting point with a spring. If a potentiometer is placed within a harness so that the pull or rotation can be measured, relaxing it returns to the starting point. This can be used as the instruction for the amount of flexion of a powered elbow: No effort into the sensor is interpreted by the controller that the prosthetic arm is straight, while increasing the input is converted into an increasing angle at the elbow, raising the hand.

A joystick is a pair of potentiometers linked together at right angles to each other. Usually, motion up/down is one signal and left/right is another: The two potentiometers give two distinct, but linked signals. If a part of the body can voluntarily move in two directions, then it could control a joystick. The shoulder can move in two directions relative to the neck (forward/back and up/down). The first user of a shoulder controlled joystick was an early user of the Utah arm in 1981 [86,173] (see 7.2.1).

The prosthetist and therapist will attempt to identify all the possible motions that can be capitalised upon and then employ those which are easiest to use or do not interfere with (or are affected by) any other actions of the person. For instance, a common problem with myoelectrically controlled arms occurs when raising an arm causes the prosthesis to shift, it will inadvertently trigger the hand to open (see 6.1.3.1). If a potential user can move the end of their residuum (or their shoulder), within a specially designed socket and touch the inside of the socket then a pressure sensor called 'Force Sensitive Resistor' (FSR) can be mounted there. Pressing against it can be read as intent. Sometimes the control of the residuum is not very fine, so the operator cannot generate progressive signals, it is simply all or nothing. If this is the case then a switch can be used to activate a state change (i.e. hand or wrist) or turn the motor on and off. When controlled like this the motor turns at a constant speed when or not at all, confusingly referred to as 'digital control' [209]. Alternatively the motor's speed can be in proportion to the amount of control signal (*proportional control*). So the greater the input effort, the faster the prosthesis moves. This input may include pushing harder raising their shoulder higher or bending their elbow. This is called *proportional velocity control.* With a more sophisticated controller, the angle of flexion of the elbow can be made proportional to the signal instead. At a low signal the elbow is extended and as the signal increases the elbow flexes more. This is *proportional position control.*

Once you have the signals you need to do something with them. The conventional control format is to take two signals and make one drive the motor (and so the joint) in one direction and the other signal maps to the other direction. If they are derived from two muscles in opposition to each other, then, flexion and extension in the operator, can generate flexion and extension in the prosthesis. For the majority of potential users, there are unlikely to be enough pairs of controllable signals to be sufficient for every absent motion. The simple solution is to switch between the different axes. For example, if the person wants to use a powered wrist, elbow and hand, and they have one pair of usable muscle signals, they can use those to flex and extend one of the joints. Then a switch (say in the harness) is triggered by shrugging the shoulder. This switches between the joints, first from the hand to the wrist. Another shrug transfers control to the elbow, before a third returns it to the hand.

All of these controls can be achieved using any number of different possible control inputs, but the form most associated with external power is electromyography.

6.1.3 Electromyography

Electromyography is the study of muscle function by investigation of electric signals. The word 'myoelectric' combines the Greek word 'myo', meaning muscle, with 'electric'. The end of the word, 'graphy', is commonly used to be the study of something, such as photography ('drawing with light'). It is one of the quirks of English that we get the two forms; myo-electric and electro-myography, presumably because the word 'myoelectric-ography' is inelegant.

The electromyographic (EMG) signal is small (a few millivolts) and it is associated with the contraction of a muscle. Any muscle will generate a signal, and these are very useful signals for diagnosis of certain disorders and study of different processes from muscular degeneration to understanding walking (gait analysis). In the past, the small size of the signal created many challenges for engineers and biologists. The rise of myoelectric control mirrors the rise of electronic technology and the rise in EMG as a diagnostic tool [210,211]. The challenge has always been to separate a tiny electric signal from the sea of external electrical noise that we are bathed in. In the early days it required an entire shielded room to isolate the subject from the rest of the world to see and study an EMG [212]. These days electric circuits exploit the fact that the noise signal is the same everywhere on the body: Two electrodes on different places on the same muscle will see different signals from the muscle, but the *same* noise signal The latter can be subtracted from the former. To work, this requires electric circuits of considerable precision. Once upon a time this meant very expensive circuits costing thousands of pounds. Today one can buy an EMG electrode amplifier for a few pounds, giving rise to a whole new area of hobbyist activity. These circuits are not as good as the commercial prosthetic ones and will pick up more interference, but they are good enough for noncritical applications.

Many of these new investigations like to use EMG as a control input for very many different activities, but with physically typical subjects. It seems almost magic; with a wave of the arm it causes a change in a robot or computer. This creates something with an appeal and looks cool. The truth is, EMGs are difficult to use consistently, being prone to interference and fatigue of the control muscle. Often, activities that are proposed for EMGs could be achieved some other way, more cheaply or more easily, if not as coolly.

Electromyograms (EMGs) as an input to an artificial arm have been used because the signals are associated with the control of the absent limb. There is no hand to hold the joystick, but there remain the muscles that would have controlled the hand. The EMG signals are available on the surface of the remnant limb. The amplifiers can be built into the socket so they do not require cables to run from one to another part of the body (like a body powered arm), and the prosthesis can be more compact. However, EMGs are a difficult source of signal because even with high quality amplifiers they are prone to interference from outside electrical sources and movement of the socket.

The precise signals are only associated with muscular effort, but there is no simple, direct relationship between the level of the signal and the muscular force generated by the operator. Any electrical noise is not simply added to the signal as road vibrations might be added to the motion of the steering wheel, but it *multiplies* with the signal. Anything but a really small noise signal will have a large impact on the total signal. Attempts to control motion smoothly with myoelectric inputs require significant levels of special filtering to reduce this impact.

As the signals are the by-product of contraction, David Simpson of Edinburgh likened EMG control to managing the engine of a car by detecting the emissions from the exhaust pipe. Needless to say, he was not keen on EMG control of prosthetic limbs and found other ways to control limbs (see 8.1). It is dealing with these conflicting qualities that has kept research in electromyography alive for over half a century.

6.1.3.1 The generation of and use of EMG signals

Muscles can be regarded at many scales. At the largest, the muscles are attached to the bones via tendons. The tendons run across one or more joints and provide drive to the bones. Few muscles provide a single action. The movement of one part of the body is generally the result of a mix of several muscles to create a motion. The control of the hand is not simple; there is not one muscle per finger. Different muscles pull the fingers and their joints differently. For example, several of the muscles in the forearm cause all the fingers to flex or extend. If you wished to pull only the middle finger down (flex it), its neighbours will flex towards the palm too. To isolate the middle finger requires the other fingers to be pulled way from the palm (extended) as well, which requires additional effort.

Each muscle is divided up into bundles (known as fascicles). Each fascicle is a bundle of fibres, and each fibre is a bundle of myofibrils [213,214]. It is the myofibrils that are the engines of the muscle. A myofibril is composed of sets of filaments. Each filament can bond chemically to its neighbour. The fibres bond then shorten before letting go, lengthening and bonding to a different place further up their neighbour. In this way it pulls itself along the fibre, like a person climbing a rope, grasping, pulling and then releasing and extending the arms to grasp again. The more bonds that attach fibres to their neighbours the more forceful the muscle contraction is. The faster they bond, shorten and release the faster the contraction. A normal contraction is a combination of the number and frequency of these interactions. The top contraction speed depends on how often the fibres can bond, shorten, and release. It means that when the contraction is slow, many bonds can form at once and the force is at its greatest. As the speed goes up the molecules need to grab their neighbours and shorten their lengths for shorter and shorter times and so the output force drops.

This is why the force that a muscle develops is greatest at slow speeds, and gets less and less at increasing contraction speeds. Ultimately it can contract quickly, but generate little or no force. This is useful for animals as most muscular action is at slower speeds. Electric motors are roughly opposite to this. They get more powerful as the speed increases (below some limit). At no speed they generate very small forces indeed. Consequently electric motors are not generally that effective at directly replacing the action of muscles (Fig. 6.1).

The other consequence of the force generation mechanism is that muscular motion works in one direction. A muscle can only actively contract. To lengthen again, a second muscle on the other side of the joint must contract and stretch the first muscle to its precontraction length. This is why muscles are arranged in 'opposing' pairs (see 2.6.1). When muscles are used to command a prosthesis to open and close the tendency is to use the signals from a pair of opposing muscles as they correspond roughly to this sort of action.

The muscular force is the result of molecules changing shape as the chemistry of the cells changes. This change needs energy to occur, which comes from the food we eat. One result of the chemical reaction is a small change in the electrical field in the muscle fibres. This is the electromyogram and is detected by electronics. The smallest level is a small twitch of a single muscle fibre [214]. To increase the force more fibres must twitch more often. To sustain the contraction this must happen frequently,

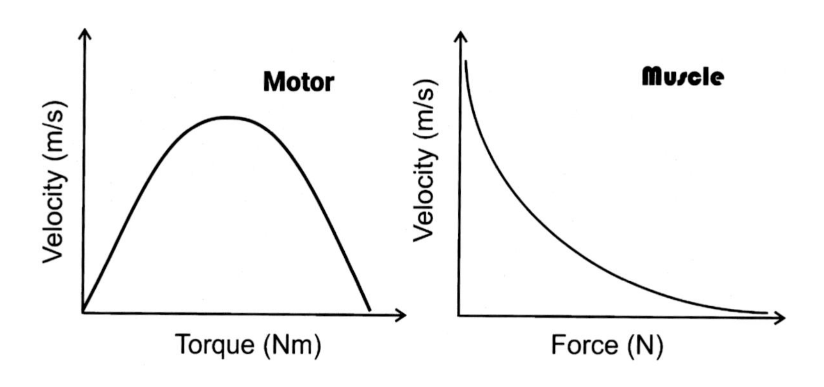

FIGURE 6.1

Characteristics of motors and muscles. The faster the muscle contracts the less force can it generate. A conventional electric motor has low torque at low speeds and then peaks at higher speeds. For this reason their use in the driving of limbs is less efficient than muscles that evolved to perform this task.

over a longer period. Individual fibres need a little rest between twitches to recover their chemical balance, if they cannot, fatigue sets in. To be a smooth contraction, the individual twitches must happen many times and in a random manner. The arrangement of these twitches is complex and still the subject of investigation (and debate), but as they occur randomly in time and across the structure throughout the muscle, a smooth force can be produced [213]. The overall force of the muscle depends on the number of fibres in a muscle. This number of fibres can increase as a result of the body reacting to regular exercise, but they will be removed again, if they are not used for a time (such as if one stops exercising).

The control of the force depends on the number of different bundles that are in a muscle. If there was only one bundle, the muscle will be all turned on or off at once. Adding more bundles gives more steps to increase the force to the maximum. Many bundles means that an individual fibre can turn on and off frequently and can rest between twitches. The increase in tension of the muscle can be subtle and progressive. Eventually, the reserves of energy of the muscle become depleted. One result of this is the tendency for all the fibres to begin to fire at once. This means they will tend to all go on then off together, which is why when a person is fatigued their muscles start to tremble.

The individual twitches come from individual places spread throughout the muscle. When an electrode is placed on the surface of the arm, it covers many of the fibres and it will detect a lot of these little twitches at the same time. As they need to turn on and off randomly to give the smooth signal, the electrode will see many small spikes at different times. The electrode adds all the little twitches together and the result is a spikey low frequency random signal (all below 1 KHz). This is the raw electromyogram that the prosthetic controller sees.

The more the muscle twitches, the more effort the muscle can generate and the more energy (amplitude) of the signal is observed by the electronics. This means that the level of the electrical signal roughly corresponds to the amount of effort put in by the person contracting the muscle. This is why it can be used so easily as a control input. The relationship is not precise, so it cannot be used to say *how much* force this particular muscle is generating. For prosthetic control the electronics generates a signal

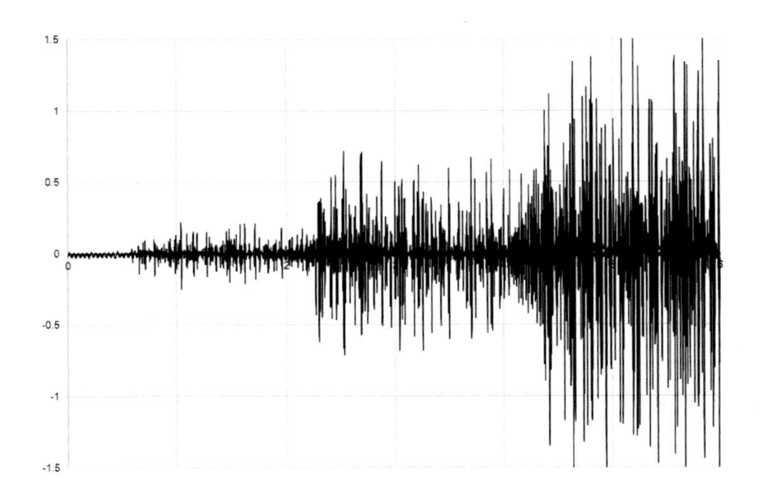

FIGURE 6.2

Three contraction levels detected using surface electrodes on the forearm. The signal is noisy and the amplitude roughly corresponds to the intensity of the contraction.

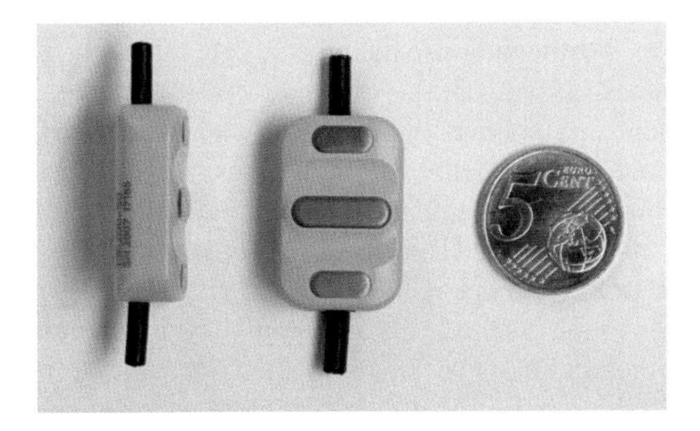

FIGURE 6.3

Active EMG electrodes. The packaged electronics amplify and filter the signals so that it can be simply used to drive a prosthetic limb. Like a number of subsystems in prosthetics the package size and shape of the electrodes has become standardised, these are electrodes from Otto Bock. (Copyright: Ottobock SE & Co. KGaA, Duderstadt.)

that reflects the average of the raw signal. This signal looks even more like a graph of the muscular effort. It is *not* exactly like it, just enough like it to allow a person to feel the control is intuitive. Three contraction levels of EMG are shown in Fig. 6.2.

The majority of commercial prosthetic products called electrodes are actually combined electrodes (the metal bit that contacts the skin) and amplifiers (Fig. 6.3). They include electronics to perform some basic signal processing, so that their output to the controller is the rough average. These are sometimes referred to as 'active electrodes'. The signal displayed on a myoelectric training tool is this rough average. When a trainee user observes the graphical display they see this level responding to the effort they are putting in contracting their muscles. It feels right to users; the harder they try the higher the signal. The precise signal depends on many factors. Place an electrode on the skin over the muscle and you are sampling a part of the total signal. A large electrode on the skin will sample many of the fibres in the muscle and give a general measure of muscle effort, but the signal is impacted by the local fibres a little more than the distant ones. So if the electrode is moved slightly across the skin, the signal will be a bit different, as it will be closer to different muscle fibres.

The precise signal is also dependent on the environment of the electrode. Most of the energy of a muscle is along the long axis of the muscle, if the electrode is rotated away from the axis, the signal will get smaller. If there is a lot of dead skin under the electrode the signal will also be smaller. If there is moisture (sweat) the signal may get bigger as the water transmits the current more easily. If the sweat forms a bridge (short circuit) between electrodes, the received signal will then disappear. As the person fatigues the signal changes (11.2).

The result of the sensitivity to changes in circumstances is that every time the electrode is removed and placed back on muscle, or after a little time when the person has sweated a little, or they are tired, the signal will be at least a little bit different. Reproducing *exactly* the same result after an electrode has been taken off and replaced is unlikely. To compensate, the controller must be designed to not let the small changes impact on the resulting control. The changes in the muscle signal are detected, and the prosthetic movement still follows the effort sufficiently for a user to feel they are in direct control of the hand. The muscular effort is reflected *roughly* in the speed of the motor or the position of the arm. This uncertainty makes it difficult to use the precise patterns of contractions to create a complex command signal for a prosthesis (see 6.1.3.3).

Good modern semiconductor amplifiers built to detect and clean the signal can almost completely remove an interference signal that is roughly a million times bigger than the muscle signal. However, this only works when the big signal is identically common to both electrodes. The identical nature of the common signal is not true when one electrode loses contact with the skin ('electrode lift'). Then the difference between the two electrodes is huge and the amplifier just gives the maximum voltage possible (the supply voltage to the electrode) as an output. The problem is that without knowing this movement has happened, the output signal seems to show that muscle is contracting fully, when it might actually be relaxed. If this happens a false open or close command is issued. Similarly, if there is a source of interference which is closer to one electrode than the other, then the amplifier will boost the stray signal. The most likely source would be a mobile phone, or perhaps from the spark plugs driving an internal combustion engine (if the hand is within the engine bay).

Each electrode has its own local environment of sweat, dead skin and underlying tissue, which change the local signals. When the electrode is first put on the skin it needs to settle down, to wait for the electrodes' environment to become the same and the interferences to become common and so subtracted out. This takes from a few seconds to a few minutes. Once it does this, the electrode should work predictably. If the electrode then moves this local environment changes and the output from the amplifier will change for a while. This is 'motion artefact', which might be interpreted by a hand controller as a signal to open or close the hand. Knocking the electrode, or moving the socket holding the electrode, might cause this form of transitory interference. This interference can happen because the socket shifts on the skin when a heavy object is picked up in the hand, or the hand is raised above the head. Depending on which electrode has moved, a person might drop the object or be unable to release the object while the hand is above their head. This is frustrating and can lead to the user rejecting the arm as being unreliable. If one can detect what is the source of the wrong signal, it can be ignored and the control of the arm made more predictable. It is possible to detect increased force against the electrode by adding force sensors to the back of the electrode. With this information it is then possible to cause the controller to ignore the signal, until the contact force returns to normal [215,216], this has not yet been adopted by any prosthetics company.

Once the signal is large enough to be useful, what can be done with the signal is limitless. It can be combined with other signals to extract meaning or remove interference. A computer can convert the signal to numbers (digitise it). Once in the computer, the numbers can be converted into different forms, such as sounds or colour to help humans analyse them. With these tools we can use them to see meaning, such as in the order the muscles contract. Or we can see changes in the way the muscles work over time and potentially track the progression of a disease. In modern terminology this is 'signal processing'.

6.1.3.2 Myoelectric signals used in a prosthesis

The principle of control using a myoelectric signal can be quite simple. As outlined before, one muscle signal is used to switch the motor to spin in one direction. The other muscle signal can be used to spin the motor in the other. The speed of the motion is made roughly in proportion to the tension of the muscles. So if the prosthetist uses a pair of muscles that would have flexed and extended the natural wrist or opened and closed the original hand, then the user may find the arrangement very intuitive. In their mind, they can make movements of their missing hand which are similar to the motions of the prosthesis. People are often sufficiently adaptable that other combinations of motion are possible, but generally the first thing to try with a potential user is the solution closest to the natural.

There are some things that make this solution more complex than it might first appear. One is that the myoelectric signal is quite noisy. The designer generally creates a minimum signal below which it does not trigger any motion. This is called the *dead band*. Greater effort needs to be expended to generate a signal larger than the top of the dead band. This is why the size of the dead band should be as small as possible so the user does not have to use much muscular tension to open or close the hand, as this is tiring. The muscles are rarely entirely silent so without any dead band these little signals will cause the hand to jitter; it will open and close a little all the time.

This is the conventional form of the myoelectric control. The effort of the two muscles are mapped on to a straight line, flex to one side, extend to the other, with the middle for both muscles relaxed. As a muscle progressively contracts then the point shifts along the line to the left or right, depending on the muscle. This can be seen in Fig. 6.4. The mapping of the pair of muscles is then in three sections: the dead band in the middle and the flex and extend two regions (often from an opposing pair) either side.

A second aspect of control to consider is what should happen if both muscles contract together (cocontraction). The hand cannot both open and close at once. What the controller actually does will be the result of complex interactions between the muscles and the design of the controller. This cocontraction might happen when the user starts to pick up an object and 'stiffens' their arm. If it does, the controller must be able to ignore this, not follow one muscle rather than the other. If it did, it would either to open the hand and drop the target, or close more tightly and crush the object.

If the person never inadvertently generates a cocontraction, and they can perform it reliably, at will. Cocontraction can be used to switch control to a different powered joint, instead of using an external switch. This is a much cleaner solution. The arm will need no harnessing or extra movements

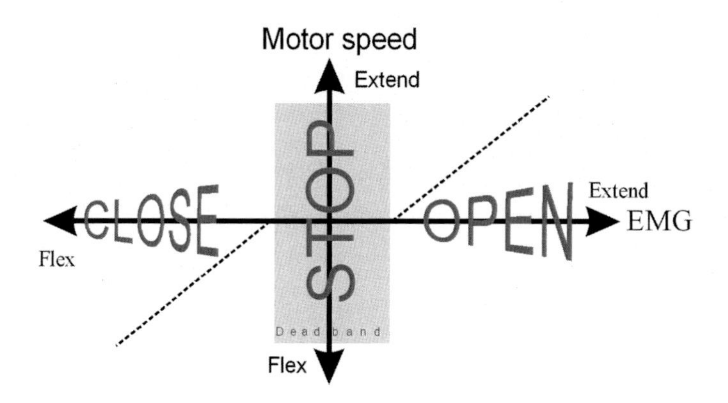

FIGURE 6.4

Conventional two muscle EMG control. The effort from fully flexed (to the left) to fully extended (to the right) is a continuous line with both muscles relaxed in the centre. A deadband in the middle where the controller does not react means the hand will not open or close with small signals from the user. Only larger, intentional, muscular effort creates a command to open or close the hand, motor speed being proportional to muscular effort.

to operate, but can only be used as long as the effort required to contract the two muscles at once is not too tiring, otherwise the user will not employ it often. For example, when the second axis to be controlled is the wrist, it can take time and effort to cocontract and switch the EMG from hand to wrist. It would be simpler to use their shoulder to move the elbow outwards and rotate the wrist. The user may adopt this strategy, even if it imperils their shoulder in later life due to overuse (see 2.3.1).

Every potential user is unique and one variable is the structure of the remnant limb. One person may have damage to their arm following the original trauma and may only have one viable muscle. Another may tend to contract the two opposing muscles almost together, so the prosthetist can then only use one signal. There are three formats they can use with a single muscle:

Using the single muscle as a switch. When it contracts the motor is turned full on to open the hand. A spring or the electronics are used to pull/drive the hand closed when the muscle is relaxed. This is the so-called 'cookie crusher' that is used in children's hands. It is simpler to learn and often is used with the first prosthesis while the children develop their cognitive and kinesthetic skills. Once they are older and can control two muscles separately, they go on to more sophisticated controllers.

The second controller is an extension of the cookie crusher. It is analogous to the body powered devices. The hand is opened in proportion to the effort. When the user relaxes, the hand closes again in proportion to how much the operator has relaxed.

The final single muscle arrangement divides the range of the muscle up into three sections. Very low effort equals stop (the dead band), some effort closes the prosthesis and a higher effort means the hand opens. This *three-state controller* was developed at the Institute of Biomedical Engineering in Canada [217–219] (see 13.2.2). It was the first alternative controller developed after the standard format was pioneered by Reiter and the Russian Hand. As the EMG signal has noise on it, which is to some extent dependent on the user, studies at UNB worked out at what levels of effort to set the tension switch between the states to give the best performance (Fig. 13.1). The lab at UNB made the controller as an add-on to existing commercial units. It was used until microprocessor controllers replaced the earlier devices.

The ultimate goal for prosthetics has always been to detect the intention of the operator and convert that into prosthetic movement. One line of research has been to attempt to intercept the signals in the nerves that would have driven the arm and use those to control the prosthesis. This is not easy as the body tends to reject any implanted device. It is also true that the nerve signal is even more complex than the EMG, so interpreting the signal has been the subject of much research. At the time of writing, there are a small number of places that have some encouraging results [220–222]. The application of this technology is not simple or straightforward. Aside from the technical barriers, there may be many people who do not want something as invasive as electrodes permanently in their arms. So this may not be a solution for every person who could use a prosthesis.

6.1.3.3 Pattern recognition

If you grasp your forearm a third of the way down from the elbow, and wiggle your fingers you will feel a pattern of activity that is mostly your muscles working together to create the pattern of movement in your fingers. This pattern of activity in the muscles does reflect the movement. If a person with an absent hand still has many working muscles, it is possible to study these patterns of contraction on their arm and relate the patterns to the intent, and so teach a computer to recognise the patterns and control a more complex prosthetic limb. This is referred to as ‘Pattern Recognition’ (PR) and over the last 50 years there has been considerable effort to turn this idea into a practical system. One of the important labs that has conducted PR research is the Institute of Biomedical Engineering (IBME) in Canada. This will be explored further in the appropriate chapter (13.2). Some of the earliest practical work was conducted in Sweden (11.2).

The idea of a PR system is that there are multiple electrodes placed on the forearm to record the activity of more than two muscles. Some applications have the electrodes over specific muscles. More recent efforts have concentrated on electrodes evenly spaced around the arm. This distribution can only be achieved in an ideal situation because a real prosthetic deployment will also need to avoid areas which cannot tolerate hard contact points such as electrodes. These areas include any scarring on the remnant limb or places where there may be insufficient cover of flesh on a bone. The placement of the hard surface of an electrode on these parts of the limb will be painful. In these cases an electrode will feel like leaning all one’s weight on a hard table top on the very tips of ones elbows. The prosthetist has to take into account the user’s anatomy when they are designing the socket.

A pattern recognition system needs to be customised to the user. Once the electrodes are in the correct place, the next stage is to teach the computer what the specific patterns look like. For example: Open/close hand, rotate wrist, flex/extend wrist, form a fist, and do nothing. The person is instructed to perform these motions, perhaps going through a few repetitions of each motion, with the arm in different positions. The computer then extracts the underlying patterns in the signals and applies a way to match any incoming signal to the patterns. If it determines they are close enough, it will cause the hand/arm to perform whatever grasp it thinks it is being instructed to perform.

Here again some compromises must be made. How close is *close enough*? Applying strict matching criteria when even the slightest deviation from the general pattern will mean the controller will not accept the majority of instructions given by the user. The controller will not move the arm, this will lead to the user becoming frustrated at the arm’s inactivity. If the PR controller is more generous in its interpretation then it may instruct the hand to form the wrong grip or move the limb when it was not commanded to move. This will also be frustrating to the user. It may even start one grip and then briefly switch to another. If this occurs as the person is reaching for an object it would clearly be unacceptable.

The PR system is taught what the patterns look like and then interprets the signals accordingly. The original idea was the same as a conventional fitting: Training in the clinic and then adjustments on return trips. There are some potential challenges that all designers have to overcome if theirs is to be a practical system.

- **How do the patterns change if the socket is removed or shifts on the remnant limb?** Since the signal coming from a muscle depends on the location and orientation of the electrodes, the pattern is a subtle collection of muscle signals. These include the signal sizes and timings. Moving the socket will change these subtle relationships. Taking the socket off and replacing it will change the signals.
- **How do the patterns change with time, or when fatigue or the muscles get stronger with use?** These are longer term shifts over hours and days. As muscles fatigue, the signals get smaller. As they are used muscles may get stronger. Both these changes will affect each muscle differently. The patterns learned by the computers will then change too, making the original training out of date.
- **How do the patterns differ between people and practice?** As users practice, they get better at generating the signals. They may pick an action when first using the system that becomes hard to do, or they may get better at reproducing them. Again, the patterns change over time.
- **What do you ask a person with a congenital absence to do?** For those who once possessed a hand, asking them to close their fist to generate a pattern makes perfect sense. They knew what it was like to move their hand. For those that never had a hand, there is some learning to be done. This is even true for the simpler two muscle controllers, but is more easily overcome. If there was nothing ever attached to the muscle then the simple act of contracting the muscle can be hard as it makes no intuitive sense. The person never had a hand so asking them to move their hand is meaningless. As one potential user most appropriately put it: *"It is like being asked to wag your tail"*. What do you think about or try to do when asked that question?

The response at UNB, was to focus on the problems of high-level amputations while looking at the patterns of myoelectric activity based on ideas of the anatomy and physiology [223]. Only later did a different approach occur to Blair Lock that promised a solution (see 13.2.5).

This approach to reading the patterns begins with the idea that if you know the anatomy, how the bones and muscles are put together, and the physiology of the muscles, you might be able to create a mathematical model to predict what the patterns will look like. A mathematical model is simply an equation or set of equations that produce a result that is similar to the thing we want to model. For example, if we want to model the population of an animal, and every pair of animals (say meerkats) has four babies during their lives, then every generation doubles the population of meerkats. This would make a very simple model and its results would quickly be different to any real population of meerkats. If we wanted to make the model more sophisticated, we could then build in the effect of predators on the population, so if eagles eat one meerkat per generation, then the new model would be to double the number of meerkats then subtract one from the number of breeding pairs. Then we can add many other factors from the amount of food changing, to how many babies survive to breed, to the occasional baby being trodden on by a cameraman making wildlife films. Each change gets closer to what happens in the real world, but to make this model we have to notice that these factors exist in the real world, and work out how big a contribution they have on the population. The same is true for the model based approach in PR. Not only do the modellers have to work out the contribution to any activity of each muscle, but also the variations between people and time, and then the changes due to the movement of the electrodes. This makes for a complicated process.

An alternative approach is to find only the approximate result. The hunt for a pattern is like looking for the location of an object in a landscape. One can find out precisely where it is or one can simply find out it's 'over there' somewhere. This can work for prosthesis control as one can assume that there are only some motions that are possible once the arm is moving. Then the user can correct for the differences as they occur. This is also true for other computer interfaces. The Kinect system and other games interfaces detect the patterns in movement of the arms and body and control a computer interface. They are of the simpler 'over there' form.

The signals a Kinect interprets are based on the general idea of a person who has two arms and two legs. To play the game, the player waves their arms, and if the avatar moves a little to the left from the player's real position, or perhaps does so with a little delay, the player can rapidly adjust to the difference without it affecting their ability to play the game. In fact, if the player is someone with part of the arm missing, this does not necessarily affect their ability to play the X-Box. There are prosthesis users who play Kinect based games without problems. If you look at the details of the results you will see that the alignment of the person and their avatar is not as close to the actual motion as it is for a person with four intact limbs, but it is not bad enough to make playing impossible (or even to give the player any disadvantage). Most important is the relationship between the motion of the player and the avatar. If it does not change much, then the difference is predictable. The differences are acceptable for a game. If an EMG Pattern Recognition system had a performance similar to the Kinect, would it be reliable enough to drive a prosthesis for everyday use? A mistake, where the user wants to command a hand into a lateral grip and the controller moves it into a power grip, is frustrating. The designer of a PR system has to decide how often is *too often*? One instruction in ten? One in a hundred? Never?

6.1.3.4 Evolution of pattern recognition

The first manifestation of this control philosophy was developed in Sweden, in the Sven Hand project (see 11.2). This was a team from Sweden who created a multiaxis hand system with a controller compact enough for the user to carry it around. It ultimately led to the first commercial multiaxis hand system, the ES Hand in 1981 [224] (see 11.2.1). This hand system was only used by a limited number of wearers. The development of pattern recognition systems was a significant outcome from this line of biomedical research. It is instructive to realise that the next commercial system after the ES hand broadly based on these principles did not appear until 2014.

A second approach to pattern recognition, adopted in North America in the 1960s, was to look at the actions in other parts of the body which are associated with performing a task, and detect patterns in that. For instance, when a person wishes to raise their hand from beside the leg and up to their chest (as when lifting a dumbbell), the back and legs also need to stiffen to steady the load. There are also contractions of the muscles attached in the shoulder blade (scapula) and the back. The hope was that the detection of these signals would give the controller a clue to what was intended by the user. Unfortunately this method proved to be very complex. Having worked on this problem and demonstrated it in the laboratory, Steve Jacobsen of Utah believed it was impossible to render this in a practical system, requiring too much computer processing power (see 7.2.0.1). While that might not be true any longer, it is complexity may continue to prevent it being a practical system. Electrodes would need to placed precisely on the back of a user. This is hard to do alone. It will require some ingenuity to create a viable clinical system based on these principles, unless the electrodes are implanted, which creates a different set of challenges.

One of the barriers to PR being a useful clinical system was the idea that the system needed to be the same as a standard fitting: Set up in the clinic and fixed until the next visit. This would include all the training of the computer. Unfortunately, as indicated, the signals change subtly over the course of a single day, or even an hour, it was considered to be very unlikely to work in the field. The breakthrough came when it was accepted that the signals would change and the hardware was capable of relearning the patterns in situ, (13.2.5). A second problem is that the majority of pattern recognition systems can pick up one pattern (and so command) at a time. The result is that operator must rotate their wrist before they open their hand (or vice versa). This slows the operation, potentially making the prosthesis less useful.

6.2 Control of multifunction hands

Once there are multiple fingers to control, providing the means for the operator to select the appropriate grip form becomes urgent. Touch Bionics struggled with this problem for years and so have the other manufacturers of multifunction hands. With five digits moving separately or together, the number of possible grips rises rapidly to over 100 poses and grips. While a great many of these are not useful, there are still too many to be easily selected at any one time. The task of the clinic is to help find the subset of grips that will be of most use to the individual.

Conventional two-site EMG can be used for a hand with five fingers if the digits are all driven together, solutions need to be found to change the grip. The simplest method is to engineer some sort of extra actions with the EMG, but this is generally not easy to perform repeatedly. For example, contracting both muscles together can be used as a switch. However, as noted, there are circumstances when this cannot be used, such as when picking up an object.

One solution requires resorting to other coding schemes, making different patterns of muscle contractions. A commonly suggested solution is tensioning and then relaxing muscles in a sequence like a form of Morse code. Since it is possible to contract muscles quickly or slowly this can be used as a switch. Otto Bock, for instance, uses fast or slow extension for different states, especially for switching control between the wrist and the hand. Touch Bionics adopted a form of this; when the hand is fully opened, the person relaxes and then applies two or three quick opening pulses to switch into a different grip form. In all of these, the challenge is learning how fast is fast and how slow is slow. For the i-Limb it also needs the operator to be sure to relax the opening command muscles without accidentally triggering the closing muscles, otherwise the user will have to start again. It is possible to learn these, but they are slow to execute and hard to repeat, taking time and concentration, which might not be tolerated by the user.

Another way to change the grip posture is to use external physical switches. The bebionic hand has a switch in the middle of the back of the hand. This requires a second hand (or an object) to push against, but once triggered the action is unambiguous, there can be no mistake. There is no frustrating false triggering that is possible with EMG switches. A second internal switch in the bebionic hand records when the operator rotates the thumb round into the lateral grip. This causes the controller to close the fingers to create a platform that can oppose the thumb. It then maps the EMG command exclusively to open and close the thumb.

Manufacturers have tried other forms of external switch. The need to find different solutions, matches the differences in approach individual users take to their lives and any list is not comprehensive.

One solution is to create an app for use with the wearer's smart phone. This is straightforward. However, it takes time to initiate, it is hard to do spontaneously, and is only practical for switching into a mode for a particular circumstance, like when using the hand in the kitchen to perform a range of tasks with the same grip. It also requires the other hand to be involved in holding the phone and switching the mode. The user must also have a phone with them, and be able to operate it with their prosthesis, although this is unlikely to be a great barrier for many.

A more hands-free operation uses RFiD tags placed in the environment. Radio Frequency Identification tags (RFiD) were developed to allow remote identification of objects. The first and primary application was in retail. A worker can identify a product, the amount in stock, the location and the price, simply by waving a detector near the product. Once you have this contactless means to identify objects, it can be incorporated into other things; it has moved into many forms of credit, debit and access cards. It can also be used with a hand. The controller can be programmed to identify a particular object and the grip most appropriate to pick it up. The tag will be associated with the object (perhaps stuck to the bottom), and putting the hand near the object will trigger the right grip. The user's world can be seeded with these tags. When they are away from their preset environment, they can wear some of these tags in their clothing, and by waving of the hand near the tag the hand will respond appropriately.

While it requires thinking ahead and an inflexible response to circumstance, this can be an advantage. We all have fairly stereotyped responses to things and will do them the same way without thinking about them. Someone with any motion impairment, including needing a standard prosthetic hand, has an even more restricted set of actions and will tend to think further ahead than most of us. Users tend to have learned stereotypical approaches to situations. The subtle change of shape of the finger tips of a prosthetic hand by the manufacturer can be an unwelcome change if the user has already got used to the hand they have. From this, it can be seen that the tags may no drawback at all.

A third option taken by Touch Bionics is to incorporate accelerometers into the hand. If the hand is held still in front of the body the controller is programmed to become receptive to an accelerometer in the hand. Accelerometers record the movement of the sensor and have become very cheap now that they are incorporated into mobile phones. If the hand is then flicked left, right, forward and back, one of four different switches are engaged. The choice of grip can depend on the needs of a user, the choices are aimed to make the motion match the grips to the user's requirements. For example: An inward flick (medial direction) will close some fingers into the palm creating a two jaw chuck so now the EMG causes just the index finger and thumb to open and close. An outward flick (lateral direction) will open the fingers and resets the hand. As with many of these systems, in the hands of a talented and practised user, the pause and flick can be made so fast that a casual observer might well not notice it.

The ultimate form of switching is what RFiD tags are aiming at: Context based switching. We know what form of grip we will use in a particular circumstance. We learned this through many thousands of repetitions when we were young: We form the grip instinctively and adapt the hand to the task as it closes on the object. If the prosthesis can detect what sort of object it is and change its shape appropriately, then we need not tell it at all what to do. The tags create this knowledge. This is also the principle of the Southampton Hand (see 10.2.1). Instead of external switches the hand has sensors on its palmar surface. The sensors detect the shape of the object and the hand adapts to the object, forming a grip. For example, sensors on the side of the index finger trigger the lateral grip.

A second way to detect context is something robotics researchers have investigated for years [225–227]. They use a camera to view the object and a computer to determine the correct preshape for the hand to hold the object, much as we do subconsciously. For the longest time this has been the engineer's fantasy. It was a neat idea without any possibility of ever becoming practical in a prosthesis. Who would want to wear a camera on their head and carry a powerful computer on their body? As it turns out, very many people will.

It is now possible to track the gaze of the person; using this a computer could decide when they want to pick up an object, what shape it is and so what grip is best. Alternatively cameras are now so small they can be built into the hand itself and as the hand approaches the object it could change the grip appropriate to the shape. The option of building the camera into the hand would require the most complexity in the product itself, so manufacturers may remain resistant to the adoption of this technology. However, using an existing consumer product (like using smart phones to switch grip shapes), reduces the cost of manufacture of the prosthesis and any inherent reduction in reliability. So the likely route for this solution is the head or body mounted camera triggering the grip.

Of course this will not work for everybody, but there is a surprising number of people who now carry a smart watch who would not be described as gadget orientated. This suggests that this form of grip selection might work for many, *if it is delivered in the correct format*. If head mounted cameras become acceptable to the general public for other reasons, they will also become invisible. It is this set up that will become completely acceptable as a prosthesis control system. Google glasses, developed around 2012 and launched in 2013 would have been the perfect solution for this sort of innovation, but when it was introduced it proved to be a step too far for the general public and it was not accepted, at the time. Perhaps in another few years something like it will become fashionable and so open up the technology for use in prosthetics.

CHAPTER

7 A tale of two elbows

The development of the Boston and Utah Arms

The primary product in both cases is elbows. They are referred to as 'arms' and incorporate the upper and lower arm in the design of the eventual prosthesis. Why the elbow was neglected by the rest of the research and commercial fields is complex, but there are a number of good technical and business reasons. Primarily, the number of wearers who could use an elbow are far fewer than those for a hand alone, making the market much smaller. Added to this is the mass of the additional drives, making any powered elbow a tougher challenge to fit and to wear. To be able to pick up an object in the hand and raise it to the mouth (one important reason for having an elbow), the power requirements are not insignificant, and in the 1960s there were few powerful motors that were light enough and compact enough to be able to perform this task. In addition, the way it would be controlled by the users was a problem that needed to be addressed. Importantly, compact, powerful batteries did not exist. This aspect plagued the Boston Arm from the beginning. Thus the tendency was to fit a body powered elbow and a myoelectric hand. Of course, this does not fulfil the needs of all persons with limb absences, hence the development of the powered elbow. The story of these two elbows highlights the choices that need to be made by designers of prosthetic systems.

7.1 The Boston Arm

The Boston Arm, was created in MIT in the 1960s, and then developed into a commercial product, it remained in production until its replacement was launched in spring of 2019.

7.1.1 The origins of the Boston Arm

The project at MIT aimed to provide an alternative to body powered elbows. Instead of using cables to control the elbow or hand by movement of the shoulder, the first of the devices that became known as the Boston Arm used a computer to interpret EMG signals from biceps and triceps muscles. In turn it sent orders to a motor to flex and extend the elbow, or open and close the hand. The prototype elbow moved at speeds proportional to the intensity of the muscle contraction, imitating the flex and extension of a natural elbow joint. The batteries and electronics were housed in the forearm. It weighed about 1.1 kg and could lift up to 2.27 kg and hold objects over 22 kg in a locked position.

The origins of the Boston Arm are complex and the various players in the story have their own version of events. Engineers tend to be practical and want to see things simply. Historians understand that history has many perspectives. This story of this key piece of engineering has those many perspectives from the people who were associated with the story.

The tale that has come down as modern legend is that in 1962 Norbert Wiener, an MIT professor and the developer of the idea of cybernetics, broke his hip, and while recovering at the Massachusetts General Hospital. There he met Dr. Melvin Glimcher, a resident surgeon, professor of orthopaedic

Making Hands. https://doi.org/10.1016/B978-0-12-820544-0.00016-0

surgery at Harvard Medical School, and an associate of Liberty Mutual insurance company. Wiener was bored lying in bed with little stimulation, and very soon shared his speculations with Melvin Glimcher, that a servomechanism could be made that involved both the central nervous system of a person and of an artificial limb.

Melvin Glimcher was intrigued and recruited MIT mechanical engineering professor, Robert Mann, to work on a small, light, power supply. Mann was chosen because he had worked on the Sparrow missile, which required a light power source. Subsequently, Mann developed a close collaboration with Melvin Glimcher and in 1968 they demonstrated the Boston Elbow, the first artificial elbow that employed myoelectrical signals for command signals. This is the version of events that was repeated in 2014 when Glimcher died [228]. The truth is somewhat more complex.

Norbert Wiener was a mathematician at MIT and a child prodigy. He was born in 1894 in Columbia, Missouri. His education initially consisted of home schooling by his father, a professor of Slavic languages at Harvard University. He graduated from high school at age eleven and received a BA in mathematics from Tufts College when fifteen. In 1913, aged 18, he received his PhD in philosophy from Harvard University. Wiener spent a year in Germany immediately before the war broke out and after the war in 1919 he studied mathematical logic with Bertrand Russell in Cambridge, England. He followed this with teaching jobs at Harvard and the University of Maine, and some journalism. Wiener would remain linked to MIT until his death in 1964. In World War Two he served with the US Army. Here his mathematical skills were used as he performed calculations in a job known then as a 'computer'. For this he undertook to perform the sorts of mathematical tasks that today we would set an electronic device of the same name to perform.

His wide recognition outside the field of mathematics came from his 1948 book *Cybernetics: The science of control and communication in the animal and the machine* [229]. The title of the book is a very succinct definition of cybernetics. Taken from the Greek for 'helmsman', cybernetics studies any system where information on the state of a system is used to control the output of that system. In such a system, an increase in one thing causes a decrease in another, such as excessive speed of a car leads to taking ones foot off the accelerator. This is the servomechanism Wiener mentioned to Glimcher. Wiener's innovation was to include natural systems into the description, such as an ecosystem where too many predator animals cause the number of available prey to go down. This is a *cybernetic system* and a mathematical model is described in 6.1.3.3. In either example, the result of the decline of the second category will mean that the size/number of the first also goes down. So as the foot comes off the throttle, the fuel to the engine is reduced, the engine slows and so does the car. Or as the number of prey is reduced through overpredation, then the number of predators must decline as they starve and die. This is *negative* feedback: The increase in one factor causes the decrease in a second which feeds back to reduce the first. The other side of the coin is *positive* feedback. Here an increase in one factor causes an *increase* in a second and that then boosts the first factor and so on. This most famously occurs in audio systems when the output from an amplifier is picked up by the microphone, and then amplified again and again until the output screams.

Wiener did research during World War II on predicting the future positions of fast flying aeroplanes [230] for targeting anti-aircraft guns. From this he developed his ideas of cybernetics. The troops who aimed the anti-aircraft artillery rode the guns. As the guns swivelled, so did the gunner. The gunner becomes an intimate part of the system. To Weiner, this was much like the way that the helmsman of a small boat moves with the boat, or a cyclist leans with the bike going round a corner. The man and machine become a single unit. Hence he chose cybernetics from the Greek *Kybernetes* meaning helmsman. Underlying cybernetics was the observation that the information processing used to determine the position of enemy fighter planes was the same as that which

> "...lay at the root of all intelligent behaviour... [such as]... light- and heat-seeking movements by plants and primitive creatures; homeostatic processes such as the body's internal mechanisms for regulating appetite and temperature; and virtually every form of higher-order animal behaviour. All those purposeful actions were governed by circular communication processes and guided to their goals by error-correcting negative feedback" [230].

His contribution was that feedback control was not just the domain of man-made systems, but cybernetics could describe anything from the motions of a robot arm to changes in the weather, from the speed of a steam engine, to money transactions in a market.

In his 1948 book Weiner saw the promise to control a prosthesis through sensory feedback:

> "There are two other fields where I ultimately hope to accomplish something practical with the aid of cybernetic ideas, but in which this hope must wait on further developments. One is the matter of prostheses for lost or paralyzed limbs... The loss of a segment of limb implies not only the loss of the purely passive support of the missing segment or its value as mechanical extension of the stump, and the loss cutaneous and kinesthetic sensations originating in it, The first two losses are what the artificial-limb maker now tries to replace" [229].

Important to the story is Melvin Glimcher. Glimcher's first degree was in general science and mechanical engineering at Purdue University, before he went on to read medicine. It is likely that while a medical student he attended a talk Weiner gave to the Harvard Medical School on feedback control. Perhaps this engaged Glimcher's imagination in cybernetics. Later, Glimcher went to MIT to study for a doctorate on the biomechanics of walking. At the University of California he continued to apply engineering principles to human anatomy. In returning to Boston he became associate professor of orthopaedic surgery at Harvard and director of the orthopaedic research laboratories in the rehabilitation clinic of the Liberty Mutual Insurance Company at Massachusetts General Hospital. He was in these roles when the prosthetics work began.

In 1961 Glimcher visited to the Moscow Institute to view a demonstration of the Russian Hand (4.2.1.1). The major problem with a body powered elbow was (and is) that the single body powered cable was used for controlling both hand and elbow. This does not allow for simultaneous control of both axes. So to control the combined arm requires unnatural body movements, switching between the axes. The movements are also unattractive and inefficient. To Glimcher it seemed that myoelectric control was more necessary for users with above elbow losses. For persons with a below elbow absence, they could function well with conventional devices. When picking up an object, the majority reach out with their elbow and open their hand at the same time. The degree of opening is directly related to the proximity of the target: if an object is pulled away while the hand is reaching for it, the hand will close and then reopen as the forward reach recommences [231]. This shows that most natural motions are coordinated actions, in that the hand and elbow work together. In contrast, in a prosthesis with a single control input, these motions have to be performed serially, with the operator opening the hand, then switching to elbow control and reaching for the object. Glimcher felt that this was unacceptable, and that a user should really be able to do both at the same time. This concept has become enshrined in the upper limb prosthetics field as the 'Glimcher Test'. To pass it, an operator must be able to control hand and elbow at the same time, consisting of reaching out and grasping an object in a single smooth motion. Functionally, this still remains a worthwhile goal, and is not always achieved by a modern prosthetic fitting. Before their discharge, new users of prosthetic arms are encouraged to move their elbow around in space and still be able to use the terminal device [232].

During his meeting with the bed bound Weiner, Glimcher obtained suggestions of names of people at MIT who might be able to help him in his quest. One name was Amar Bose. Bose's name is well known today as the innovator of high quality HiFi, especially speakers and noise cancelling headphones. In the 1960s he was already a lecturer in electrical engineering. Bose persuaded his graduate student Ralph Alter, to undertake a PhD thesis studying the possibilities of cybernetic limbs. One aspect of this study was that they tapped into the nerves of volunteers from the Massachusetts General rehabilitation clinic of Liberty Mutual. Under local anaesthetic they succeeded in recording the signals from the severed nerves while the subject attempted to move their missing hand [233].

For his project Alter studied physiology as well as prosthetics [233]. The main project was two simultaneous investigations, one to use nerve signals for prosthesis control, and the other to use a computer for an EMG controlled prosthesis. Alter acknowledged that although the use of the nerve signals to control the prosthesis was the preferred goal, the practical and short-term approach involving surface EMG signals from muscles was a more easily obtained surrogate. It was half a century before this first approach began to be practical for a real prosthesis. The aim for the EMG system was to make control of the prosthesis as much like that in a human arm as possible. To achieve this required significant signal processing from a computer. At the time a computer was a large and immovable system, so that the prosthesis using it could not be portable. Much later, the first microprocessor controlled prosthetic device used in the field, appeared in 1988 [234]. However, the team proved that Weiner's concept of cybernetic control of a prosthetic elbow was possible. As with all prosthetics research, the next stage should be to try the ideas in the field, so it needed a small and compact system, which brought in the engineer Robert Mann and his student Ronald Rothchild.

In the fields of rehabilitation science, engineering and human control, Robert Mann is a well known and revered name. Mann was a student at MIT in the 1950s. His undergraduate, master's and doctoral degrees were all from MIT. After his doctorate in 1957, he joined an MIT laboratory designing the Sparrow and Hawk missiles. Evaluating compact and lightweight energy systems which led to applying these concepts to prosthetics [235].

The requirement for compact power supplies was one of the consequences of the American's initially poor rocket performance compared with the Soviet examples. The Soviet R7 booster has been used in one form or another since the first successful satellite (Sputnik) to the present day. So powerful was the booster that in an evolved form it still lifts the three man Soyuz capsule into orbit to reach the International Space Station [236]. The US contemporary, Redstone could not even put one man into orbit. Every gram on the ground requires many more grams of rocket and thrust to get it into orbit thus the mass of the rocket system is always a driver of technology. Hence USA's significant innovations in power supplies and electronics was a way to get ahead of the Soviet Union and ultimately led to very many major strides in technology including microelectronics.

In 1963, Mann became a professor of mechanical engineering at MIT [228]. In 1974 he helped found a Laboratory for Biomechanics and Human Rehabilitation, and served as its director. According to Mann's account of the project, he had almost no knowledge of Wiener, when in the spring of 1964 he was approached by representatives of Liberty Mutual to develop an artificial elbow [237]. Liberty Mutual ran a clinic in Boston, which fitted prostheses, and had orthopaedic surgeons from the Massachusetts General Hospital who served on the clinic staff. What Liberty Mutual did not have was engineering expertise. Mann said that although *"Wiener's speculations on cybernetic control of prostheses indirectly precipitated what became the Boston Arm"* he and Wiener could never have discussed the project due to Wiener's death in April 1964. Mann said he drew on his experience with power supplies [237].

Weiner's biography tells a different story [230]. In December 1963 an article in *Saturday Review* [238] promoted the originator of the idea as Glimcher with Weiner having only a passing involvement in the process. In response to this article, a letter was published in the issue of *Saturday Review* on 25th January 1964 signed by a vice president of MIT, the dean of Harvard Medical School, the medical director of Liberty Mutual, and others including Wiener, Glimcher and Bose. It expressed appreciation for the magazine's interest in the work, but showed concern that the program was still at an early stage and publication in the magazine should not have preceded completion of the work and publication in a recognised scientific journal, especially if it may lead to false hopes for users. This aspect of journalism and hype is something that remains a concern to those in the field to this day. Irrespective of the details to qualify a statement, a journalist has a need to deliver a good story. This leads to articles which assure us that all the problems are solved and the magic bullet is ready. The reporters are not weighed down with all the details that still need to be fixed. This story plays well to journalists and funding bodies, but is bound to disappoint those who the researcher purports to be helping. It is our duty to be honest and measured in our responses, even when others promise the earth, they will ultimately disappoint the potential recipients.

The article had implied that the Russian Hand was 'crude', but Glimcher's letter countered this assertion. The prosthesis Glimcher saw demonstrated was the product of serious research. This is part of a long running tendency to play down the scientific products of the former Soviet Union. Even today the Soyuz capsule that is being used as a ferry for astronauts to the International Space Station (ISS) is denigrated for its basic form, its origins being from the same era as the Russian Hand. Yet when in 2013, the Canadian commander of the International Space Station, Chris Hadfield, was asked about being nervous of flying to his command in *"Soviet era technology"* he said *"Complexity is the enemy of good enough"*, implying that the 1960s space ship was good enough to get the men to and from the station safely. This dependence on Soyuz was because the USA's far more advanced spacecraft, the Space Shuttle, had been retired due to its dangerous design flaws, leaving the US without any options to get astronauts into orbit for over a decade.

Part of the newspaper story was that the article suggested there was some subterfuge on Glimcher's part to get to see the hand, again this was not the case. The letter corrected the authors saying that Glimcher had not *"bluffed his way into a Moscow clinic to see it in action"* [239]. The visit to the Institute was at the suggestion of Russian scientists and Dr. Glimcher was *"cordially received"* [239].

The threads of the story are hard to untangle. According to the author of Weiner's biography:

> "Wiener laid out a detailed design for the first cybernetic arm... Wiener's vision powered the project. Bose oversaw the venture and guided their team of doctors, electrical engineers, and biomedical technicians through the formative stages. Research facilities were provided by Mass General, MIT, and Harvard Medical School, with additional funding from the Boston based Liberty Mutual Insurance Company..." [230].

Bose reported that after two years of research and development, when the team was readying the first test of the device, a young doctor from Harvard (Glimcher) who had come late to the project attempted to suppress the release of information until it was ready. However, Weiner in an article for the *New Scientist* magazine published in 1964 (as part of a series of essays reflecting on advances in the next 20 years), Wiener said that:

> "The taking of signals in and out of the body higher up in the nervous system than our sense organs and muscles, by the amplification and manipulation of nerve impulses, is well under way in the work conducted jointly by the members of the Massachusetts General Hospital and M.I.T." [240]

making no reference to his own involvement in the process.

T. Walley Williams III, the man most responsible for turning the Boston Arm into a viable product, agrees that the credit for the early days is disproportionate [241]. He said that Mann made sure that everyone noticed his contribution, but no one noticed Amar Bose, despite the fact that he applied Pulse Width Modulation (PWM) to drive the arm. PWM uses transistors to switch the battery voltage on and off quickly to give a range of drive voltages from all off to all on. It is far more efficient with energy than analogue electronics and so extends the life of the battery. Without PWM many modern devices would be bulkier, more expensive and run for less time. Similarly, the Boston Arm would have needed many more batteries. So without Amar Bose, the Boston Arm would not have been practical as early as it was.

Ronald D. Rothchild, completed his master's thesis on the Boston Arm in May 1965. Rothchild designed, built, and tested an artificial elbow controlled using EMGs from electrodes over the biceps and triceps of the subject's residual limb [242]. The work included tests with above elbow users and it demonstrated sufficient natural control of the artificial elbow to convince Liberty Mutual to take the project in-house in 1967. They hired Mann as a consultant and two of his former students to produce a wearable design.

7.1.2 Commercialisation of the Boston Elbow

Liberty Mutual's accident prevention department, formed in 1912, was initially focused on providing insurance policyholders with safety advice. In 1943 they opened a medical rehabilitation clinic in Boston to address the obligations of policy holders and to assist injured workers return to work. They followed this in 1954 by opening a Research Institute for Safety in Hopkinton, Massachusetts (a suburb of Boston). This expanded the capacity to do research and development. It was here that the precommercial design and testing of the Boston Arm occurred. It would become home to Liberating Technologies Incorporated (LTI). The Institute initially focused on investigation of industrial accidents from machinery use and material handling, also industrial hygiene, such as control of dust, vapours and noise. By the early 1950s Liberty Mutual had extended its remit into new product development projects. This included development of collapsible steering columns, arm and headrests, air bags, and seatbelts with researchers at Cornell University. The upshot was that although Liberty Mutual did not have a design engineer like Mann or Bose on staff, they had the cash, facilities and experience in both research and product development, as well as access to users to support the elbow project in its early days.

In 1968, a first version of the Boston Arm was designed and built at Liberty Mutual's Research Institute. Eighteen arms were tested with users and they were universally rejected. They also were deemed unsatisfactory by the National Academy of Sciences. The biggest problem was that the elbow ran on a battery so large and heavy (3.6 kg) it had to be mounted on the wearer's belt (see Fig. 7.1).

FIGURE 7.1

An early Boston Arm, the batteries are external and mounted on the user's belt. (Courtesy MIT Museum.)

This was simply impractical. A second flaw was that it was noisy, according to Williams *"it sounded like a tank"* [241].

The work to design and produce a commercial product was performed by two Liberty Mutual employees, while an MIT mechanical engineering graduate student, Robert Jerrard, used smaller batteries and incorporated them into the prosthetic forearm to make the arm practical. Robert Jerrard had completed his master's degree in mechanical engineering under Mann at MIT from 1969-70. Jerrard built twenty-five prototypes during his three years of employment with the firm, before beginning a tour of duty in Vietnam. Jerrard's master's thesis was a mechanical redesign of Rothchild's proof of concept elbow. The version designed by Rothchild had a ball screw motor, and weighed *"six or seven pounds"* (roughly 3 kg). It had a battery pack of equal weight. Jerrard redesigned it to use a harmonic drive developed by United Shoe Machinery Corporation in Haverhill, north of Boston. The same design of drive was also used in the wheels of the Apollo Lunar Rover. Harmonic drives are much more compact than other gearboxes, but require greater precision to be manufactured, and thus cost more. In this case, the saving in space was justified as it allowed Jerrard to move the motor, brake, tachometer and gearbox away from the joint, making room in the forearm for the battery and electronics.

After his master's Jerrard worked for Liberty Mutual from 1970-73. Jerrard gave much of the credit for the progress on the arm to Mann, although he also said that if Mann had not done it, someone else would have designed a similar system [243]. Asked about the external influences on Mann's work on the project, Jerrard said that efforts to collaborate on a myoelectric arm with the Rehabilitation Institute of Chicago were fruitless. Disappointingly, Mann was not aware of the work in Canada or Europe on myoelectrics [244]. Proximity to hospitals was a minor factor to the success of the project: Jerrard recalled only one trip to the Liberty Mutual Rehabilitation Centre in Boston. However, he said that user tests were a critical part of the development process, in particular the testing with a World War II veteran Hal Paradis.

Jerrard said it was a smooth transition for him to move from MIT to Liberty Mutual, in part because Liberty Mutual was like a university with its research facilities. They had budgets for design engineering projects that did not have to make money [243]. Fittings and testing were done at a facility in Hopkinton, Massachusetts. Jerrard handed-over responsibility for design to T. Walley Williams when Williams joined Liberty Mutual in 1973. Jerrard then left to get married and begin his PhD at the University of Utah under Mann's former student (and creator of the Utah Arm), Stephen Jacobsen. Following his PhD Jerrard took a position at the University of New Hampshire. Jerrard said Williams had much more experience than he did and did a great job with the next redesign of the system.

7.1.2.1 T. Walley Williams III

T. Walley Williams the third was one of the major characters of modern prosthetics. He worked in the prosthteics industry from the 1970s to the 2010s when in he was in his 80s, still working full time for LTI and contributing to the company and any conference he attended. He was always full of ideas and very happy to share them with other delegates and researchers, without any apparent concern over the intellectual property, or where the idea might go. He was free with ideas and open to any suggestion. His personality is much larger than life and he peppers his speech with expressions all his own. This language he calls 'Walleywich'. Walleywich has compact expressions for everything from the way the different systems connect together ('plugology') to the science of making a prosthesis ('socketology'). Even in the interviews for the book some Walleywich creeps into the quotations.

Williams served in the Navy for three years before he went to Columbia university to gain a degree and a master's in physics. He claims not to have *"the mathematical horsepower for a PhD"*, so while his wife was working towards a PhD at Columbia, he obtained a job as the technician looking after the undergraduate physics laboratories and lecture rooms at Columbia. There his creative streak led him to develop a number of new experiments and demonstrations for the labs. In 1965 the Ealing Corporation offered him a job to build physics equipment for them. Ealing produced a range of test and measurement equipment. It was common to see the Ealing logo on equipment in physics laboratories across the world. Williams was with Ealing until 1973, when a recession caused the company to reduce its staff. Typically, they tended to get rid of the higher priced staff workers first, especially their three product managers, which included T. Walley. Faced with a choice of what field he should try and work, Williams tried to predict where things would go in the next 20 years. He decided that the intersection between medicine and electronics was the way things were *"really going to move"*. So he started to go to conferences to meet people. At one such meeting he met Alan Cudworth, who was then in charge of the Liberty Mutual loss prevention lab, and was engaged on development of the Boston Arm [241].

According to Williams, Cudworth had been trying to interest existing prosthetics manufacturers in making a version of the Boston Arm without success, and following the unsuccessful trial, the prosthesis was going nowhere. Williams background was in small run manufacture so when Cudworth mentioned his problems with the arm, Williams response was to tell him what steps he needed to take to fix the problem. The next day he had a job.

That was 1973 and he worked on the Boston Arm for more than 43 years. In his first year he worked on improving the design. He looked at the problems they had been having:

"there is this black dust and we think [it] is the motor brushes. It wasn't the motor brushes! They had aluminum (rubbing) against aluminum, and guess what you get when that rubs? It generates aluminum dust! So there were a lot of things you couldn't just manufacture them the way they'd designed it, you had to go back to square one, kind of" [241].

Even then he was not given a completely free hand. Cudworth insisted the prosthetist should be able to exchange the powered elbow for an unpowered elbow, which meant he had to use the same mechanical interface: A *"turntable from the manufacturer Hosmer for both systems"*. It was *"a terrible decision. It plagued us for years. But that was the decision"*. So he designed a housing that used the harmonic drive, but they had to manufacture the complex shape of the housing, Walley chose to injection mould the housing. The prototype was made used modelling plastic, which could be formed by hand and when set it was used to determine the eventual shape that the moulds they would produce. We used *"plastic goo-ology to make sure it would work"* [241]. Based on the results he got the injection moulds manufactured.

In the end he created an arm which was broadly similar to the product on sale until the new design was released in 2019, it included:

- A brushed motor (brushless later).
- 4.33:1 planetary reduction with 3 planets and a sun.
- (Double) reverse locking clutch.

This last item typifies Williams' attitude. The original clutch only worked in the down direction, so the arm could be pushed up. This meant that the user could not use the forearm to clamp an object onto a table, or any other similar uses. T. Walley decided that this would not be good enough so he contacted the clutch manufacturers to ask for one to work in both directions. *"They said: "You can't do that it's impossible." Oh that's a word that's not in my vocabulary"*, so he designed one and henceforth it was built into the arm [241].

Within a few years he had designed and produced a new version without the flaws of the earlier devices. The first twenty-five working commercial prostheses were fitted to Liberty Mutual patients. The difference was that these devices achieved acceptance from the users as well as third party evaluators and insurance companies and started the Boston Elbow as a genuinely practical prosthesis. By 1976 another one hundred Boston Arms had been designed and produced, with a slimmer forearm and more reliable electronics.

The drive towards the control format of the Boston Arm (and one of its differences from the later Utah Arm) raises the question of which is the most critical aspect; anthropomorphic movement, or ease of use. The force behind the original design goals was Mann and his cocreators. Mann's basic design made the power of the elbow motor proportional to the EMG level. Mann was concerned with what he called 'innateness' to allow them to imitate control of the natural arm, using the same muscles that control a natural elbow. This also meant achieving control of the speed of the elbow movements and independent operation of the elbow, hand or hook. Mann contrasted this with 'access', or how easy it is to understand, maintain and use. In her study of the prosthesis, Sandra Tanenbaum [84] felt that the Boston Elbow's innateness came at the expense of its accessibility, given its technical complexity, training requirements, and need for specialised personnel to maintain the devices. In other words; natural limb movement trumped ease of use.

Williams was the sole director of design activities from 1973 until 1984 when as T. Walley puts it: *"I hired a boss"*. He went to Alan Cudworth and said:

> "This guy Hanson, he doesn't have anything to do right now... Why don't you get Bill Hanson to run the business side of this thing, so I can do what I do best which is keep developing the mechanism, keeping it up to date, keeping the electronics up to date and so on. So that's what we did."

This was a break for Williams as he could run his small team of technicians and also work on other research projects within Liberty Mutual [84].

7.1.2.2 William Hanson

William Hanson went to Wentworth Technical College and gained a certificate in Machine Design. At school he was evaluated as having an *"aptitude for mechanisms"*, this determined the kind of career he sought. He was employed by Barry Controls, an engineering consulting firm. One area they worked on was vibration and noise in machines and the workplace. Hanson worked in this group until a manager left and invited him to join him in Nichols Dynamics. This work centred on jet engine noise investigation during the Vietnam War. Bill then served in the National Guard, before returning to education, to gain a Mechanical Engineering degree from the University of Hartford (with a speciality in acoustics). It was while he was working for United Acoustics Consultants, looking at 'community' noise such as airport noise, that he met Alan Cudworth. Bill was interested in what Liberty did in the textile industry, so he was invited to visit, submitted his resume and was hired.

Textile production was (and still is) a noisy and dangerous profession with many opportunities for injury and deafness. As a result it was a very significant part of Liberty Mutual's interests. Hanson looked at the noise from looms. Liberty Mutual had an anechoic chamber containing a working loom. Hanson studied weaving room noise and other noises from the processes and machines, in order to reduce its impact on the weavers. Although successful, the work stopped because in the 1980s the industry moved overseas.

During this time Hanson also examined the noise emission of the Boston Elbow. He looked for the individual elements that contributed to the overall noise and worked with Walley Williams to reduce the resulting sound generated. Some of these changes remained in the design for decades. At the same time as this work, Hanson took a business and engineering (management) MSc. With this experience, Liberty Mutual asked him what he thought of creating a profit centre from the arm business. In short, Liberty Mutual wished to free itself of the links with the production of the arm and create an independent company. As Hanson was already interested in the idea of such a business, it was not difficult for him to assume the control of the arm production. In 2001 they created an independent company, whose name was sufficiently similar to the old Liberty Mutual brand that it could continue as before without much effort: *Liberating Technologies Incorporated* (LTI) [20]. T. Walley Williams was made its product development director.

When asked why an engineer, who was identified as liking mechanisms as a teenager, would go into management, Hanson said that he: *"liked the business side of it, over the engineering side (most days)"* [20]. The master's degree was aimed at people who were going to manage engineering firms. Hanson has continued to meet CEOs who have little idea about the technology underpinning their products. He thought it would be better to have managers that know something about engineering.

With Hanson being brought into the management of the arm production, the partnership with Williams to run the company was established lasting until the sale of LTI in 2015. For many years,

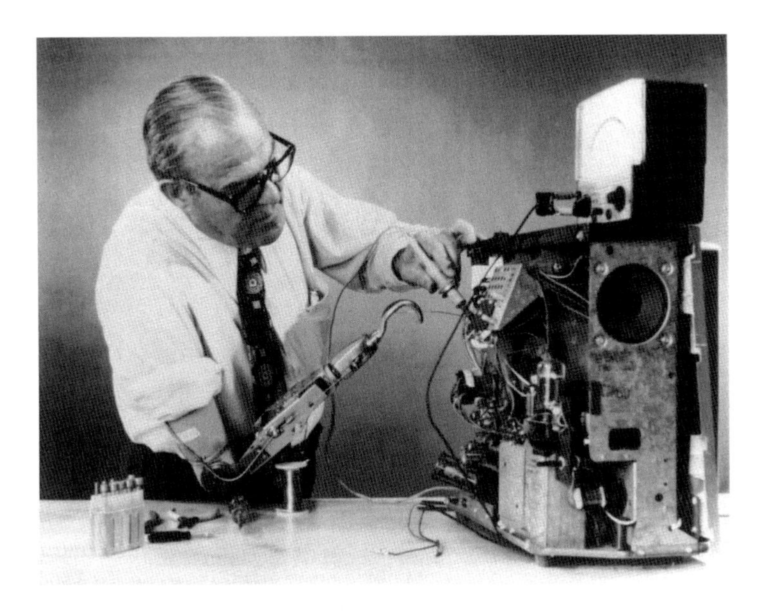

FIGURE 7.2

An early Boston Arm fitting showing the combination of myoelectric elbow and body powered terminal device. (Courtesy MIT Museum.)

to the outside world, Williams and Hanson *were* the Boston Arm. Williams provided his insight and showmanship, and Hanson quietly managed the company and ensuring it thrived from year to year.

Once the original 25 production arms were fitted to Liberty Mutual patients, Walley Williams looked to fitting 25 more to users outside the Liberty user population so that they could begin to recover their costs. However, as this was the first powered elbow, the prosthetists needed to be trained in how to fit and use the arms. One group that signed up for the course were prosthetists from Hugh Steeper, which at the time, was largest supplier and fitter of prosthetic arms in the UK [182]. The representatives from Steeper were impressed by the design and arranged for Steeper to be the UK supplier of Boston Arms. Williams rapidly realised that this was a larger opportunity for both parties. T. Walley was not keen on being part of a one product company. He felt it was a dangerous commercial position to be in. The opportunity arose for Liberty to become the supplier of Steeper's range of products. The first device they imported was the Systemteknik children's hand. This was fitted to the daughter of one of the doctors Walley knew. Other products followed, as well as other collaborations. The company's portfolio grew to include products from Otto Bock, Motion Control and Touch Bionics. The design of the Boston Arm controller and electronics was made to allow interoperability with most prosthetics systems.

The original format of the clinical fitting of the Boston Arm was with a body powered hook and a proportional myoelectric control of the elbow (see Fig. 7.2). This meant that the user could pass the Glimcher test. The upper arm muscles controlled the angle of the elbow, while the shoulder via the cable opened the hand. This continued until 1983, when at an MEC conference in Canada (13.2.3),

FIGURE 7.3

The final version of the Boston Digital Arm. (Copyright: College Park Industries.)

a prosthetist, (Jim Malone) suggested to T. Walley that he wanted an electric terminal device for the Boston Arm. Williams is always ready to assist a colleague, so said he would have a look at it. He had not returned to Boston for more than a few days when Malone was on the telephone asking when it would be ready. Williams said they did not yet have one and yet Malone asked for one immediately. Following this call *"He had it pretty much tomorrow"*, Williams said [241]. This was a one-off to help a colleague, but once that was seen as possible to do, they started to be asked to produce it more often and led to the opening up of a new market.

Liberating Technologies' focus continued to be a supplier of upper limb prosthetic devices for adults and children, it has not been able to follow Liberty Mutual's line of more fundamental research work. LTI continued its engineering design activities to augment or update its existing products. This included making an electronic prosthesis controller for below elbow absence, and a locking shoulder joint for persons with a shoulder disarticulation. It undertook a major redesign of the Boston Arm in 2002 (see Fig. 7.3). They replaced the original electronic controller with a microprocessor based controller for the elbow plus four other prosthetic devices, such as hands, grippers, wrist rotators, and shoulder lock actuators. Bill Hanson said:

> "We were doing some of this ten years ago, but the controllers were less sophisticated. Back then, we might have offered three control strategies and they had to fit all patients. For some, it was not the optimal system" [20].

Williams likes to think ahead a long way [241]. He wrote a specification for the next Boston Arm controller in 1996. He wanted to be able to control the elbow and four other devices. *"No one could figure out why I was asking for one more device, let alone four"*. The new controller was built by 1999, with a Digital Signal Processor chip (Texas Instruments) with most of the rest of the electronics in a single logic array chip. This was at a time when the rivals were not offering anything as sophisticated in their prostheses. Years later when LTI had to build a replacement, using newer technology, they also created a Graphical User Interface (GUI) which allowed clinicians to see multiple input channels at once. William said no other manufacturer did this at the time, and he saw it as very important for providing a better clinical service. The usual interfaces did not allow the clinician to easily observe all the EMG inputs at once:

> "If you only have one channel you won't see that when the patient is trying to generate the signal for motion A then oops, they are generating a lot of cross talk on channel B" [241].

The new microprocessors developed for consumer electronics gave LTI a lot of flexibility allowing them to expand the arm's capabilities considerably. In 2015 Hanson said: *"Now we have about 32 control strategies for patients"* [20]. The types of inputs to the controllers have been broadened to switches and potentiometers to capture rotation or separation of bodily segments, and even the bulging of muscles. *"We can customize the arm to each patient, and if the clinician has done the job right, the patient will learn faster and be more proficient with it"* [20]. With this flexibility LTI could keep up with the introduction of hand prostheses with more than one motor (such as the Touch Bionics and bebionic hands) and powered wrists, the later controllers could use more inputs and outputs and give the prosthetist more choice in control forms.

LTI's size has meant that they could not develop all their own products, and so have worked with universities. They have long working relationships with the Institute of Biomedical Engineering in Canada (see 13.2) and the Rehabilitation Institute of Chicago (see 12.1), but they have not restricted their partnerships to these two institutes. In their last independent decade they successfully sought more federal funding to support their own research. One of Hanson's business advisers observed that other companies who were not any more technologically advanced than LTI were able to gather good levels of federal funding for innovation projects. The change started with the DEKA/DARPA project, with LTI acting as a consultant and commercial partner. More recently LTI have sought to apply their knowledge to new areas. They won a congressional grant to apply EMG control to lower limb prostheses (especially microprocessor-controlled knees). This allowed LTI to expand from five engineers to 13-14 people, with five directly employed on research and development projects. In this way LTI were in a position to explore new ideas and adapt their products to an increasingly sophisticated market.

As a small company with a limited range of products, competitors can move to restrict compatibility with other manufacturers' devices. This was a threat to LTI's ability to prosper. This was one motivation for them to be strong supporters of moves to standardise intercomponent communications (bus) and wrist couplers. Their awareness of other devices and their open approach to competition suggests they would support such moves anyway.

LTI was also closely involved in the testing of the six axis arm with the first Targeted Muscle Reinnervation (TMR) user, Jessie Sullivan. The first full arm was: *"assembled in Walley's basement"* [241]. The arm comprised an electric hand from Kesheng, China (which had hand grasp and wrist flexion built in), an Otto Bock wrist rotator, a Boston elbow and a humeral rotator based on a Boston elbow motor and gearbox (Fig. 7.4). It was topped off with an Edinburgh shoulder (8.2.1). The load was a little too much for the ProDigits shoulder so Walley improvised a set of bungee cords that: *"worked most when they were needed most"*. Initially the shoulder could only raise the arm up ten degrees. The addition of what T. Walley refers to as a *"sine of angle compensator"* meant Sullivan could raise the arm fully. David Gow (the shoulder's designer) maintains the limit was not the shoulder but the amount of current delivered by the LTI controller and batteries.

Talking about the six axis arm Hanson said; *"I think what Dr. Kuiken is trying to do is push the envelope"*, *"It was a lot to ask, [of the Boston Arm] was never designed to do all that, because nobody*

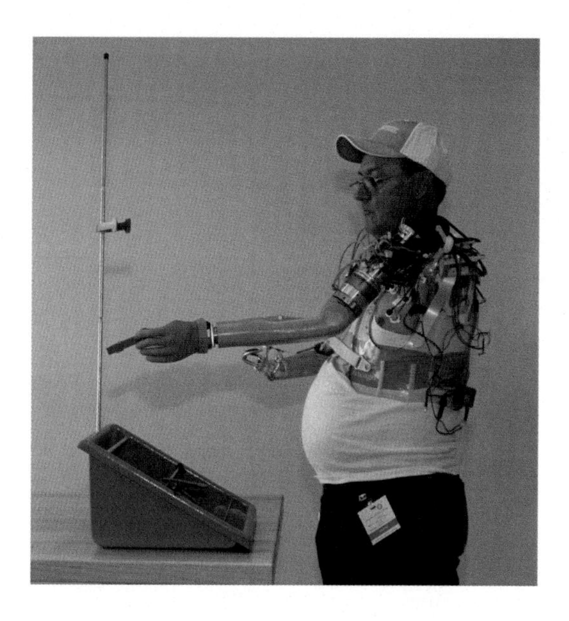

FIGURE 7.4

Jesse Sullivan performing the Clothespeg relocation task with the six axis research arm. He is able to coordinate use of elbow and shoulder with the wrist rotation to take a peg from the horizontal bar and place it on the vertical one [5].

had ever done it before". One critical contributor to the success of that special fitting was that the standard LTI controller could be used to control many axes, which was the very thing that Williams had asked for back in 1996, much to the scepticism of his colleagues at the time.

7.1.3 The new era

The eventual sale of LTI to College Park by Hanson was a move to ensure the company continued into the future. The evolution of College Park Industries followed a similar trajectory to LTI. It was founded in 1988 to exploit the idea of the first anatomically correct, multiaxial prosthetic foot (the *Trustep*). It was their only product for ten years and then in 2001 they released a heavy-duty version. Next, the company moved into a growth phase from 2003 to 2015 with product development acquisitions including LTI. They had no products in upper limb prosthetics before the Boston Arm. College Park also bought the expertise LTI possessed, including their ability to work in government projects and to gain public funding for research and development. In the summer of 2020 in the height of the global pandemic it was announced that Össur had acquired most of College Park, very much for their expertise. The Boston's successor, the Espire Arm, was sold to Steeper, a year later Coapt acquired LTI.

7.1.3.1 The Espire Arm

For the *Espire Arm* College Park engaged industrial designers to create a completely new device. This process was necessary as the Boston was not CE marked and the process of getting an existing arm

FIGURE 7.5

The Espire Arm developed to replace the Boston Arm. (Copyright: College Park Industries.)

qualified would be harder than for a brand-new design. To be sold in the European Union a device or product needs to be CE marked. It indicates that it conforms with health, safety, and environmental standards [245]. It requires documenting every part of the process of design and manufacture of a product, far easier to do with a brand new product than one whose origins are in the 1960s. However, the Espire still has a harmonic drive, and can be used with a free swing. The spirit of T. Walley Williams and LTI are not lost: The design allows for a wide range of controllers and terminal devices (Fig. 7.5). The company aimed to qualify an arm made from components from different manufacturers.

The system has five configurations that get progressively more complex. The simplest for the user is a counterbalanced body powered arm (like the Otto Bock Dynamic Arm 9.2.1). The most sophisticated is a fully integrated arm using a communications network (CAN bus, see 9.2.1.1) and wifi connections with a comprehensive graphical user interface to allow adjustment of the configuration. It also allows auto-calibration of EMG. The user gives the controller no signal and full tension and the system automatically scales the range to use the widest sensitivity.

The arm has been designed with most of the mass close to the elbow so that it is easier to wear. It can lift up to 2.25 kg, and has vibration feedback to tell the user what state the controller is in. The range of colours supplied shows a subtle change in the market from the nominal skin colours. Alongside pink and brown the arm is also available in white, black, silver and grey.

7.2 The Utah Arm

Stephen Jacobsen, studied under Robert Mann at MIT [84] and is the direct connection between the Boston and Utah Arms.

7.2.0.1 Stephen Jacobsen

Stephen Jacobsen was born in Salt Lake City, Utah, on 15th July 1940. After graduating from high school in 1959 he entered the University of Utah's College of Engineering. According to Jacobsen he was not that serious at school [246]. He reportedly set off a widespread explosion in one of the engineering buildings, and rigged vending machines so they would respond only to him. He was asked by the university administration to leave, but was saved by the dean of the engineering college, who invited him to return if he could maintain at least a 'B' average. Jacobsen returned, obtained the appropriate grades, and received a bachelor of science degree at the age of twenty-seven. He stayed at the University of Utah to obtain a master's degree in science in 1970, and then moved to study at MIT, receiving his doctorate in 1973. He spent much of his career creating different high-tech solutions to problems, while keeping an eye on the prosthetics company he founded. He died in April 2016.

According to Jacobsen, the outline of every possible prosthetic hand had been disclosed years before he began his doctoral program, a significant source being the 1919 German publication [2]. This was fine with Jacobsen as his interest was in getting things out into the field, and if prior work helped then so much the better. He participated in a project to directly attach prosthetic devices to the skeleton. He implanted ceramic materials in rabbits and performed tests. Although the implants were strong and effective, the consensus at the time was that it was not a feasible procedure, and on this basis, Jacobsen abandoned the line of research [144].

The University of Utah created a 'Bioengineering Center' in 1974. They chose Willem Kolf, the Dutch inventor of the dialysis machine, to be in charge. They aimed to compete with the bigger universities that could achieve at a range of subjects, by choosing to focus on a smaller number targets and aim to excel at these. Along with Kolf they recruited a team to concentrate on different aspects of advanced bioengineering. Kolf worked with Jacobsen on the Wearable Artificial Kidney (WAK) which was designed to allow home dialysis, or even dialysis while travelling. Unfortunately the government agency that could approve the device, the Food and Drug Administration (FDA), would not allow dialysis outside a clinic, so it could not be paid for by US health insurance [196].

Everyone who knew Steve Jacobsen comments on his brilliance. Harold Sears president of Motion Control, noted that Jacobsen had a broad knowledge of science, mathematics and the history of technology, and he could pull it all together into one consistent picture of the world [86]. Sears was one of Jacobsen's graduate students, he considered that few could see the impact of decisions on their design the way that Jacobsen could [86].

According to Mann, Jacobsen came to MIT to work on fluid mechanics, but there was no one in faculty interested in the work, so he moved to work on myoelectric control of the Boston Arm, using a form of pattern recognition for mid-humeral absences. Jacobsen studied quantitative anatomy, dissected cadavers, and examined the number of degrees of freedom that act in concert to control the position of the hand in space. He showed that the musculature of the shoulder anticipates the intent, preparing for the reaction forces arising from the actions of the more distal musculature. Simply, the back *stiffens* to take the load the hand will pick up, even before the hand has reached the object. Detecting and interpreting the EMGs from the shoulder girdle provided the control information for the prosthesis: People with missing limbs were fitted with EMG monitors, and data was recorded from the shoulder to understand how it could be used to control a shoulder and elbow. While it was possible to devise algorithms to control the hand from this information, Jacobsen determined that the system was too complex for practical use [144].

When in 1973, still finishing his PhD, Jacobsen was offered the job of director of the University of Utah's 'Center for Engineering Design'. He had found the design of the Boston Arm somewhat elderly so when he took the director's job and began to work on a new design of the prosthesis that became known as the *Utah Arm*. Following the appointment Jacobsen realised that money was the driving force for the lab. With grants he could do work, so he went out and pursued money from many different sources: Industry, government, military and charity, to support the work at the Center. Jacobsen vigorously exercised his broad mandate. Projects for the centre included development of photovoltaics, batteries, catheters, stents, drug delivery systems, artificial kidneys, electrical and body powered arms and wrists, microsensors, virtual reality interfaces, and a wide variety of robots. Over the years, the Center developed robots for NASA, Ford, the Carnegie Science Museum, and one hand (see 7.2.1.1), which featured in the movie *Short Circuit*. Jacobsen collected appointments within the University of Utah including in the departments of bioengineering in 1983, surgery in 1989, computer science in 1992, and mechanical engineering in 1996.

Within a year at the University of Utah Jacobsen created the company *Motion Control*, in 1974. Its original mission was to commercialise the range of medical technologies developed at the Center. Over time, plans changed, and it became the commercial vehicle for the development and manufacture of the Utah Arm. The arm was conceived in 1974, and it was developed over the next four years into working prototypes. It first used a computer to control it (1978), and was finally launched as a commercial product in 1981 with Harold Sears at the helm of the company. Jacobsen said: *"I gave Motion Control to Harold"* [144]. In 1983 he established another company, Sarcos and this became the robotics development and commercialisation vehicle for Jacobsen and his team's energies. Jacobsen's contributions to the Utah Arm was reduced because of commitments to this new company. Sarcos went on to apply robotics to diverse themes such as exo-skeletal systems and animatronic dummies for film studios' theme parks. While there was little interest for the exo-skeleton as a powered orthosis, he managed to obtain a grant from the US military research agency, DARPA, to develop a military version of the system. This led to prototypes in 2006 and to the company Raytheon's purchase of Sarcos in 2007.

To develop the Utah Arm, Jacobsen's team sought a technical solution closer to the natural motions of the human arm, which they felt had not been sufficiently approached by existing prosthetics technology. However, they found these prior works had not achieved overwhelming levels of success largely because while most performed well in the laboratory, the devices could not withstand the rigours of continual daily use [247]. The mechanical systems of the past, they wrote in their 1982 article, were poorly received by users, and more recent developments were plagued by problems. While they considered priory work as promising especially from Liberty Mutual, Scott from UNB and Simpson from Edinburgh, they took a pragmatic line to produce an arm that would be reliable in the short term, while being compatible with existing prosthetics systems [247]. Post development activities were mapped during the basic design phase, such as the cost and performance of fittings, maintenance, repairs, government regulatory affairs, and the funding and purchasing of prostheses for clients. They conceived it as a multidisciplinary project, which required:

> "simultaneous and coordinated work in a number of normally unrelated areas, such as physiology, biomechanics, prosthetics, biological signal acquisition, information processing, control system design, materials, mechanisms, structural design, electronics, rehabilitation, psychology, etc, etc" [247].

While they acknowledged they existed in a research environment, they did not think of themselves in this way, but as an engineering design and production group oriented to market considerations.

According to the papers published at the time, the arm project progressed in three phases: (i) development; (ii) field trials; and (iii) third party evaluation. Phase one involved continuous design, development, and evaluation over a period of six years. Ten artificial arms were developed, beginning with a purely laboratory prosthesis and building to a self-contained system. Testing was performed primarily in the laboratory (Fig. 7.6), with some limited experience in the field. Evaluation protocols and user questionnaires were designed for use in later phases. The next stage involved extensive field trials of the prosthesis. The goal was to more clearly understand the needs of the user, when employing

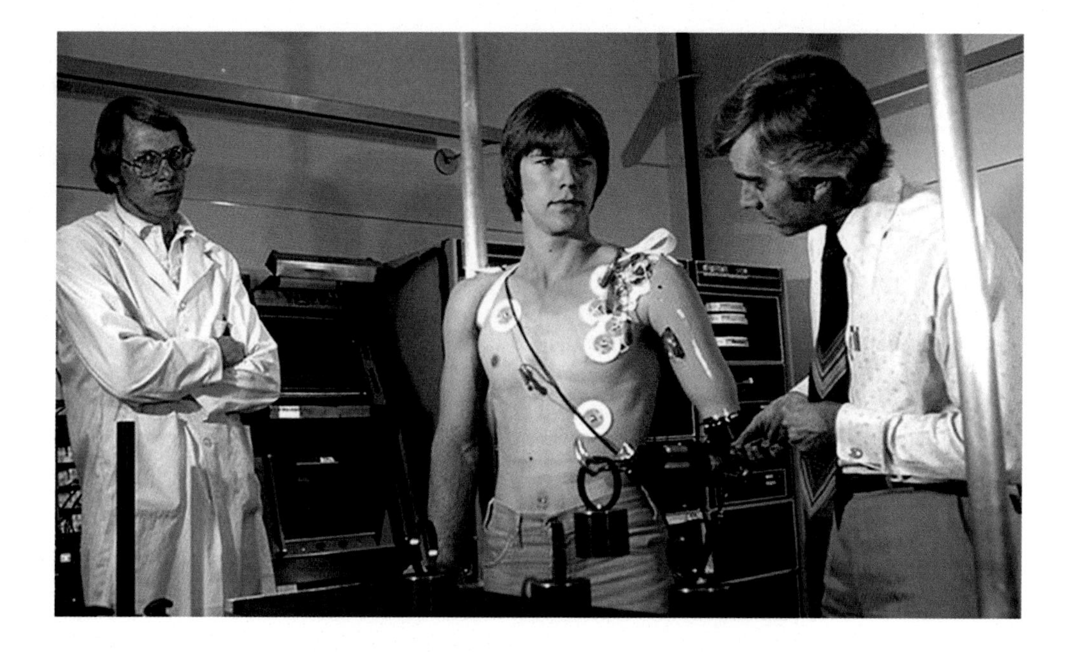

FIGURE 7.6

Harold Sears (left) and Steve Jacobsen (right) with volunteer using early Utah Arm. On the back wall is one single computer that was used to process data. (Courtesy: Fillauer Companies.)

a myoelectric prosthesis in activities of daily living. Twenty-two engineers, prosthetists, physicians, and users were involved. Seven prostheses were taken home by users and operated in the field. Electromechanical problems were identified and suitable modifications made. The final phase compared the Motion Control prosthesis to competing systems and against available standards. To remove bias, the evaluation was conducted by external organisations [247]. Harold Sears recollected that the trials were not as formal as this implies. The development followed the same lines as much of the Motion Control projects; the devices were placed in the field with paying customers and user experience drove the improvements [196].

The original version of the Utah Arm (without terminal device) weighed 1.34 kg, which is lighter than a natural arm (2.6 to 3.6 kg), and was powered by a twelve-volt rechargeable battery pack with a (now quite modest) 450 mAh capacity (Fig. 7.7). The location of control circuits in the upper arm allowed its shape to resemble the natural arm. The aim was that the arm should look and feel more natural than the Boton Arm [196]. The placement of the battery in the most distal part of the humeral section (not the forearm) allowed it to be exchanged one handed by the wearer. If it could be exchanged quickly and easily the battery mass could be reduced and the centre of gravity kept closer to the body. The EMG controller was arranged so that the elbow angle reflected the muscle tension, so when the wearer relaxed their muscles the hand fell freely to allow the operator to lower their forearms with minimal effort and trigger a free swing of the elbow.

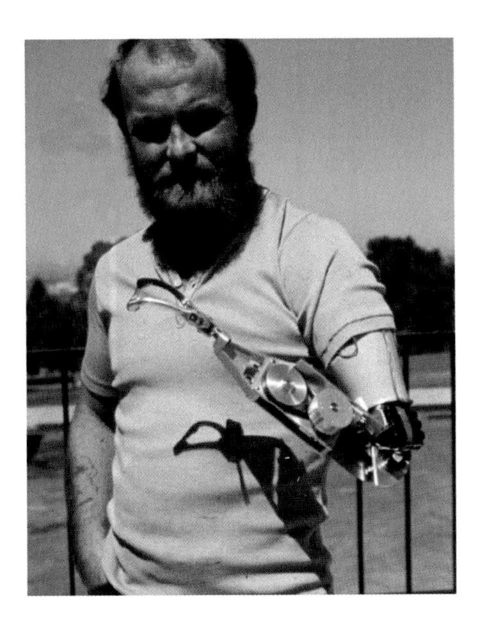

FIGURE 7.7

The first user of the Utah Arm showing the location of the batteries at the elbow, as proximal as possible to reduce the lever created by a more distal mass. Ultimately the design also allowed the user to replace the battery easily with his other hand. (Courtesy: Fillauer Companies.)

While the short-term goals were achieved, the team has continued to evolve and innovate with the same basic design of arm. Its initial commercial launch was in 1981. The Utah Arm 2, released in 1997, was made more rugged and dependable, as a result of improvements in electronics, battery and motor technologies. A major change came with the introduction of microprocessor control, giving more flexibility in control. The second generation of its microprocessor controller was released in 1999. Improvements to the Utah Arm 3 (released in 2004) allowed then to incorporate advances in electronics meaning: Reduced current consumption, increased bandwidth, and decreased size, as well as finally the ability to pass the Glimcher test.

Motion Control saw a market for water and dirt resistant devices. Vice president and principle engineer, Ed Iversen, was told the story of a girl in England who was so relaxed with her prosthesis that she went swimming in the English Channel still wearing her hand, irreparably damaging it [174]. He felt that a prosthesis should not keep users from having a full life because their devices are too fragile. This provided a strong motivation for the development of entirely new product range. It was only after the release of a totally new design of arm in 2019 that this goal was fully realised.

In 2002 Motion Control expanded its product offering to include their own 'Motion Hand' to be used instead of an Otto Bock single axis hand. In 2005 the Electric Terminal Device (ETD) hook device (Fig. 7.8) were launched. Both terminal devices were designed to accommodate a built-in wrist flexor without increasing length. The hand was aimed to be the same overall size as the Otto Bock hand and to be as strong and light. Even if the very first versions may have fallen short, the later models have

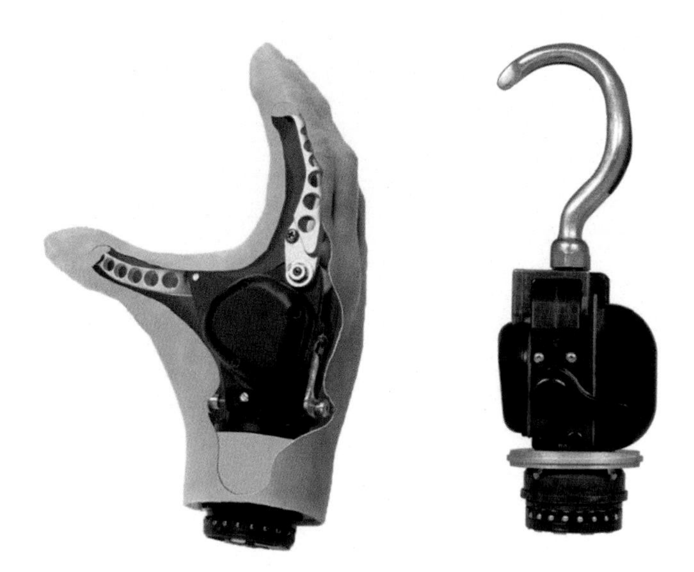

FIGURE 7.8

The Motion Hand and the ETD, terminal devices developed by Motion Control. (Courtesy: Fillauer Companies.)

achieved this goal [174]. The ETD is one of the few devices that compete with Otto Bock's Greifer. It uses fairly standard hook tines and has a water-resistant body. The tines can be replaced, if damaged.

Motion Control has continued to incorporate innovations when practical. Ed Iversen said the hand was designed with the goal of saving space in the envelope of the hand for a flexion wrist [248]. Motion Control was also the first commercial powered device to incorporate separate wrist flexion within its design. While the hand from the Swedish company Centri had a combined wrist flexion with the closure of the hand, and the Otto Bock Greifer could be pushed into a form of wrist flexion, the Motion Control wrist was the first design with separate flexion that allowed the user to select one of three positions and fix the hand at that level. A second advantage was that the hand mechanism was designed to be sufficiently short such that the wrist could fit within the envelope of the hand and not sacrifice any length. This was the first time this had been achieved in a commercial hand. A mark of its success is that it was not long before Otto Bock followed suit, making a similar wrist (with five positions) for their transcarpal hand (which unlike the Motion Control hand was only capable of a lighter duty than their conventional single axis hand).

For the longest time the product range from Motion Control was an elbow and interface systems to use other manufacturers' products for input signals and the terminal devices, such as a hand or powered hook. The hand of choice was the Otto Bock single axis hand. It used a 6 V motor and it opened and closed a little slowly for the users' tastes, but it had a significant grip force (>100 N). To get sufficient power at the elbow, the motor has to be run at a higher voltage. In the Utah Arm application they tended to supply 12 V to the elbow motor so the same voltage drove the Otto Bock hand. This allowed the hand to move much faster, something the users appreciated, while Otto Bock did not. Running a motor at a higher voltage than its specification is not generally a good idea, as it will tend to shorten its life. It is a measure of the Otto Bock design process that the motor and drive train were capable of significant overloads. The practice of over powering the hands did not have any appreciable negative effect on those hands used with the Utah Elbow. When Otto Bock released their own 'SensorHand Speed', it appeared to be the same mechanism, run at 6 V, but capable of a considerably higher flexion speed, showing the capacity the design had for enhancement.

Meanwhile Motion Control began adding to their range. Along with creating their own hand designs that allowed for new wrists, a second change was to create a body powered version of the elbow, using the same basic structure. This was one of the long-term goals for Sears. It was balanced against gravity like an Anglepoise lamp or the Otto Bock body powered elbo (ErgoArm). The user can 'throw' the arm up with a shrug and catch it with the brake at the top of its swing. More recently they have also produced a shorter and more compact ETD2 prehensor.

7.2.0.2 Harold Sears

Harold Sears' original inspiration for a career choice was the 1968 film *2001: A Space Odyssey*. This created in him an interest in aeronautics, but when he got to Stanford to take his degree, he found the subject daunting. One of the lecturers (Max Anliker) was doing research for NASA. This was the era of the late-Apollo space station project, Skylab. At the time this project was seen as part of NASA's path back to the moon and on to the planets. Space stations in Earth orbit were going to be an intermediate step to the stars, rather than the eventual goal of the work (as it became for the International Space Station). NASA was beginning to look at long stays in space. Following the failure of their own moon rocket program, the Soviets had already changed their focus from the moon to near Earth orbit with the Salyut stations, but Skylab set a new endurance record that it took a while for the Soviets to comprehensively beat [249].

The impact of low gravity environments includes significant bone loss and so the space agency was interested in measurements such as detecting bone density noninvasively using ultrasound. Anliker had demonstrated this. Sears found this entire field engaging, and so he followed this trail. Stanford allowed students to make up the content of their own degree, so Harold was allowed to change from aeronautics and study physiology at the medical school. There he met physicians and realised they found the material they were studying no easier to understand than Sears did. This was stimulating. At the end he emerged with a degree with the title of *Engineering Science* because there was no title such as *Biomedical Engineering* at the time [86].

After Stanford, Sears was not drafted into the army because he had problems with his back and feet. Instead he joined the Peace Corps, and travelled to Asia. There he was assigned as a school teacher in Nepal. When he got back to the USA in 1974 he returned home to Utah. He was looking to get into biomedical engineering so he wrote to Kolf at the Utah Biomedical Engineering Centre, but at the time Kolf had no use for mechanical engineers so he passed Sears on to Steve Jacobsen. When they met, they hit it off. While Jacobsen did not have any money at the time, he promised Sears a job when there was funding and Sears started his master's at Utah. Less than a year later he began working with Jacobsen and the master's evolved into his PhD project.

Sears' dissertation was on a body powered hook [250]. The design of the hook came from an earlier student's master's project. Prototypes had been built, and it was up to Sears to test it. *"I got the mission to do a field trial"* [14]. For Sears it was a wake-up call. While he considered his mentor brilliant, he rapidly realised they were still at the very beginning. It was; *"A huge lesson... we didn't know anything about the wearers"* [14]. He recruited ten unilateral transradial hook wearers. The subjects were heavy users and they employed their hooks for up to sixteen hours a day, so they did not want to lose any function. They universally hated the new design. Sears' work turned more into a user needs study, and by doing this he also began to define the method of user focused development that later characterised Motion Control's approach.

A new hook was evolved based on what the users demonstrated that they needed. In its third form, eight out of the ten liked it. It was not made commercially at the time. According to Sears; under the reimbursement structure in the USA there was no way to charge enough for the hook to cover manufacturing, let alone the development costs [14]. In the States, the *L codes* set the levels of reimbursement to pay for the supply and fitting of established products. While it does not apply to the entire market, the L Codes tend to influence the rest of the field. The level set for hooks was too low to pay for the development of a new one. Only those who could self-pay or had funds from workers compensation might pay more for a new and better hook. The Motion Control ETD used APRL hook fingers, it was cheaper and simpler to use the existing tines in their product. However, his earlier hook design influenced the tines in the ETD2 that was developed between 2012 and 2014. The ETD2 is a shorter design than the previous device [14].

Sears considers one of the papers he is most proud of to be also the simplest; it was on prescription guidelines. He developed an approach to drive the prescription process with peoples' needs [251]. He now realises that the needs can change by the minute, making any choices very complex. However, he adopted a very practical line. For the right prescription, all the factors need to be considered, from the physical to the social, as well as the psychological. Before seeing the device in use many of these are guesswork. Clinicians are left with only one practical option, to undertake a trial fitting [86].

Even now, many years after moving into the biomedical field, Sears tends to think in aeronautical terms. He likens the history of prosthetic arm development to the time before successful powered flight was a reality. Back then, there were many people trying many different ideas for aeroplanes, but many failed and the inventors were *"crashing into the hillsides"*. He believes that many crashes were necessary to explore what was not possible, so none were futile attempts. Eventually as the knowledge grew and inventors learned from their mistakes, flight became more stable and predictable, people stopped crashing into the hillside as much. In the same way, all over the world people, are pursuing what seems to be the most obvious forms of design and control for a prosthetic arm. Designers of prostheses will inevitably *"crash and burn in many directions"*. With hindsight, he reflects, even those with established products might not yet be flying straight and level. For example; surgical techniques or pattern recognition, these may be blind alleys. *"We cannot say what arms will look like in 10 or 20 years"* but he hopes many of the problems will be solved. He suspects that by then it will be revealed that *"we were all some of those people who crashed into the hillside"* [14].

In 1981 Harold was hired to work for Motion Control, but he did not finish his PhD until 1983 [250]. The job was his first commercial position, and the beginning of a transition from the culture of academia to commerce. He also supplemented his knowledge gained at Motion Control with a technology management course taken from the University of North Carolina in the late 1980s. Ironically, it is the Motion Control president, Sears, who said that the entrepreneur was Jacobsen, the university professor. Sears' initial work was transferring the Utah Arm technology, and performing fittings for the first few paying customers. Sears said:

> "I just happened to be finishing my degree at the time and Motion Control lost its product manager. I suspect he was scared to death of this new technology with high technical risk and not a very long track record" [86].

He had good reason. Translating successful laboratory systems into commercial products is always a difficult step, but Sears feels that Jacobsen's previous observations of the Boston Arm during his own PhD studies was a help. Sears also had his concerns, but they were overcome by his confidence in Jacobsen. Sears said:

> "I had eight years of graduate school, but we had this fearless guy, Stephen Jacobsen, with almost total mastery of technology, as far as we knew. He understood mechanical design and actuator design, plastics, etc." [86].

Sears thought this amazing, to go from idea to product, it seemed like a big job, but Sears had faith in Jacobsen and so he agreed, he *"just wanted to hang on for the ride"* [14]. Forty years later and Harold Sears was still on the horse and riding. He had become one of the grandees of the upper limb prosthetics community. He brought his philosophy and personality to every meeting he attended. Less widely promoted is the fact that through Sears' generosity of spirit Motion Control encouraged others to undertake research and development through the easy access to their products. In the late 2010s he took retirement but retained links with the company.

7.2.1 Origins of the Utah Arm

The start of the Utah Arm was one of the successful moves by Jacobsen to generate money. In 1980, the Seattle City Lights Company wanted a pair of electric arms for one of their workers who had lost both arms in an accident. The injuries meant that it was a very complex fitting. The potential user had little anatomy to hang the arms on and nothing to control them with. Sears and another Jacobsen student, Sanford Meek had to make it work with the support of their supervisor and the team of machinists, electrical and mechanical engineers, and prosthetists.

The subject had a bilateral forequarter amputation, with no scapulae and only a half to a quarter of a clavicle left on each side. They fitted him with two prototype Utah Arms. On the right, he had a modified Hosmer body powered hook with an electric winch to pull the cable, placed proximally to keep the mass closer to the body. On the other side he had a humeral rotator, an Otto Bock wrist rotator and another powered hook terminal device.

They originally tried to get EMG signals from his torso, but they noticed that every time he generated EMGs, the skin moved. It seemed that post surgery there had been some nerve or muscle growth into the skin of his back. They put a pattern of marks on the user's sides so they could observe the motion of his skin as he made voluntary movements. This allowed them to identify where the motion was biggest, which could then be used as a control input. The area they found was along his sides. He could move in two directions. So they rested two potentiometers on his skin, one for each axis, creating a joystick. Additionally, he could also move his side away from them and so 'unplug' himself. As a result there was no risk of unintentional movements, a very useful feature. They then scaled the joystick inputs to the level of perceived effort by the user, so that while forward and back movements on the skin were not necessarily the same displacement for the same effort, the resulting prosthesis movement could be made to be equal movement for equal effort. It felt to the operator that similar effort in both directions led to similar movements of the arms [14,173].

The team used these inputs to control the arms instead of the EMGs. The movement was tied to the flexion of the joints (position control). On the left side, this control was for the terminal device and elbow using the elbow lock to transfer control between the devices (just as the production version of the

Utah Arm would do). On the right, the user could switch between elbow/humeral rotation to wrist/terminal device in a similar manner. He could also switch between wrist and TD with a rapid motion of the joystick, like the short cocontraction is used to switch between hand and wrist in myoelectric systems. The user had two axis shoulders (unpowered) which were set to normally free swinging, but could lock into position with a chin switch. To operate the prostheses, the wearer could put the arms into position, lock them, and reach with the elbows. The user was reasonably successful. In the training sessions, the user could eat, dress (mostly), and hold his cigarette. Importantly for personal independence, he could don and doff the arms himself. The set-up had two shoulder caps linked together, so if it was hung on the wall (or laid on a bed) he could get underneath them and then shrug it on. The entire system was then pulled tight on to his torso with a chest latch similar to that for ski boots. The sockets were open frame to keep him cool, but sadly, the 66 kg total mass was too heavy for the lightweight user to tolerate for more than a few minutes at a time [196].

Sears said:

> "As a learning experience, it was a little (or a lot) like jumping off the pier to learn how to swim. As a team-building exercise, it was like akin [sic] to the entire team crawling under barbed wire with live machine gun fire over their heads. As a clinical effort: it was both 'great', and 'terrible'. It was a great example of a patient-driven effort – the client was involved with every aspect, and could accept or reject every design decision. And he definitely knew that it was unlikely that any other laboratory in the world (at that time) could have assembled a prosthesis with the capabilities of those arms. But until the final prototypes were assembled, neither the total weight of the final system, nor the wearer's tolerance of that weight could be assessed properly" [196].

This was quite an advanced fitting for the time, but Jacobsen chose not to publicise it. He was not interested in publishing incomplete or unsuccessful results, and publication was not encouraged as it is today, *"so a lot of really cool stuff [that] never got published"* [14]. This makes it hard for researchers today to know if a solution has been tried before, and if it was ever a useful approach. Sears and Meek did get some good public relations out of it, Meek said: *"We actually had a picture in National Geographic for that, Harold and I, [our] mothers were very proud of that"* [173] (Fig. 7.9).

Despite their common origin, the Boston and Utah Arms diverge in terms of their execution. This very much due to the differences in the aims of the two men driving the projects forward. Importantly, while Glimcher had a fixed goal of separate control of the hand and elbow, the basic Utah Arm was not originally designed to be controlled this way. While it is the primary format of the Boston Arm, it is only a secondary option in the Utah. The Utah was an arm for light duties that could move quickly. As the majority of prostheses are unilateral fittings and are used in a support role, this is a reasonable compromise; *"get the hand to the right place as quickly as practical"*.

The important consideration for Jacobsen was the aesthetics, and according to Sears its appearance comes completely from Jacobsen [14]. The Boston Elbow, in contrast, had a boxy forearm. It also weighed more than the Utah Arm, but it could lift significantly more weight [84]. This is less significant as a prosthesis acting as a nondominant arm is typically used in a locked position [196]. The difference in the two movement options may be part of the reason why no intellectual property issues ever arose between the owners of the two systems. Functionally, the Utah Arm had a completely free swing mode so that when a person is walking with the arm it appears more natural than the Boston Elbow, which has only thirty degrees of free swing. The value of free swing to users is in making the arm feel more natural. This cannot be underestimated. Campbell Aird, the golden user of the Edinburgh arm (see

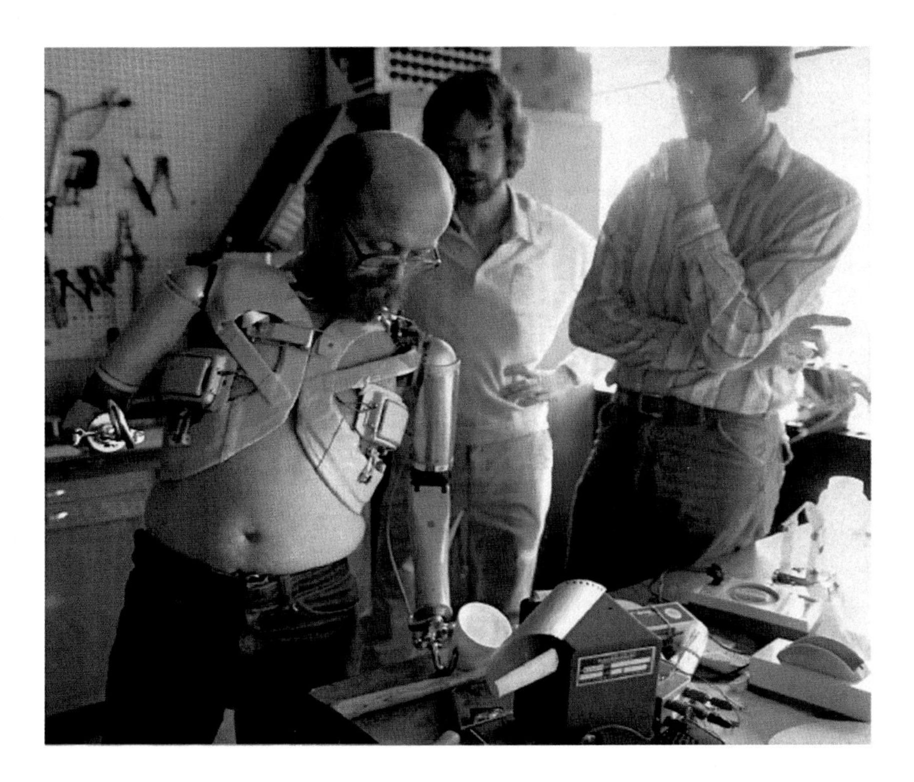

FIGURE 7.9

The first bilateral fitting of the Utah Arm. The user operated the device using a joystick driven by skin movement on his sides, their mountings are visible. With him is Sandford Meek and Harold Sears. (Copyright: Chuck O'Rear.)

Chapter 8), reportedly practiced driving his arm forward and back as he walked in order to create the appearance of free swing, despite using the completely rigid Edinburgh shoulder. When Gow learned he was doing this he modified the arm to allow the shoulder some movement.

The Utah Arm is an exo-skeletal structure, with the structure and cosmesis as the same elements. Sears said they were advised to make it more endo-skeletal, but that would have meant making the overall shape from a foam cover, which would have rapidly become damaged and unsightly [14]. The Utah's profile is clean and simple (at least for a few months before the active users scuff it up). The attractive lines also were achieved through the use of a four bar linkage for the elbow [252]. This allows the lower arm to flex around the humeral section. This is unlike the Ergo or Boston Arms that have an anatomical elbow but a nonanthropomorphic forearm, they both have a section of the upper surface of forearm just below the elbow scooped out so that the elbow can flex further.

Jacobsen saw that in the future there would be innovations in the types of plastics available to make the shells to create the shape of the arm. So he designed the arm with moulded shells assuming advancing technology would create newer materials that could be incorporated into the designs. This process required moulds to be made. This is an expensive process, it needs careful design of the way the molten plastic flows into the mould replacing the air that escapes from it. The moulds were made in 1981, and the significant investment has been worth it. Over the years, the same moulds have been used for four generations of plastic, each one making the arm a little stronger or lighter. Sears said they could not get an outside company to make the moulds for less than many thousands of dollars [14]. This was clearly was too expensive for a product that they thought would be able only to sell in the hundreds. Instead the moulds were made in house by a skilled engineer who: *"Could build incredible things, over night"* [173]. Ultimately Motion Control has kept the overall shape of the arm (and so the moulds) as they have improved the mechanism inside with three major revisions. The moulds did not wear out, and because of the great cost to produce them they were treated very carefully, they were *"nurtured and pampered"* [34]. For Motion Control's Founder's Day in 2013, they made a Utah Arm for Jacobsen from clear plastic. It had been Jacobsen's long standing wish to have a transparent arm, but the clear plastic (Lexan) is hard to control and trying to produce items from the plastic could potentially have broken their valuable moulds, so the company only decided to attempt it when they had control over the plastic [14].

The process of creating a reliable product is not a simple one, but the challenges are rarely large conceptual ones. As Williams knew when he created the viable version of the Boston Elbow, it is often the simple details that need resolving to make the device practical. When the Utah Arm was being created, existing electrical connectors between the various components of the hand-arm system were unreliable. Better quality connectors were not available, and the existing Otto Bock connectors were not the right shape to fit within the arm. In the end Motion Control designed and built their own connectors [86], another expensive process.

As with the Boston Arm, the other requirements of increased battery power, longevity and reduced recharging times were still a problem. When the Utah Arm was released, it was still quite novel to fit powered elbows. Moreover, the use of the arm was intended to be different to that of the Boston Arm. So it was necessary to develop new techniques for fitting users with very different needs, especially persons with a transhumeral absence, who Sears said were the most difficult to fit [86]. Sears' responsibility included figuring out how to train prosthetists to fit wearers. With only one other electric elbow on the market at the time, and a small number of overall sales, there was little experience among prosthetists with these devices. This meant a step-up in training for prosthetists. At the beginning Sears also had no experience doing professional training. He learned to train by trial and error while travelling and doing on-site training and follow-up visits. Training visits had an unanticipated benefit They allowed Sears to recruit users and prosthetists to participate in product development. Local prosthetists identified potential users who would use the device and feed information back to Sears [86].

Motion Control maintains standards of fitting by controlling the arm's purchase. No one can buy a Utah Arm if they have not been on a training course. Here they learn the appropriate casting technique, make the temporary thermoplastic arm sockets, mount electrodes, make a harness and attach the arm to the socket through a simple plastic lamination collar. *"You can put an arm together and the patient is wearing it at the end of the day"* [34]. They enforce this rule exclusively and *"self righteously"* [34].

The development of new products for Motion Control remains in this form today. Any public launch of a new product to the field will be early, and will consist of the salesperson promoting the value of the product through illustrations of their effectiveness from their volunteer panel, who have been using the devices for more than a year beforehand.

7.2.1.1 Ed Iversen

The other critical person behind the continued evolution of the Utah Arm was Edwin Iversen. Ed obtained his master's degree in mechanical engineering from Brigham Young University. He went to work for Jacobsen and was cofounder of Sarcos and general manager from 1983 to 1992, when he then joined Motion Control as its vice president for research and development in 1992. In this role Iversen sees himself primarily an electromechanical engineer and designer of systems, collaborating within a team of engineers [248].

Iversen's interest in biomedical engineering goes back to when as a teenager. He rang up a local company in Seattle *(Prosthetic Research Limited)* and volunteered for a summer job. Prosthetic Research was run by the Orthopaedic surgeon Dr Ernest Burgess who developed the Seattle foot.

It was here that Ed produced his first device that was used clinically. Bone responds to the loads placed upon it, so in zero gravity, bones lose their strength. Similarly, when a fracture in a bone is healing there is a sweet spot in the amount of loading the fracture can tolerate: Too little and the healing is slowed, too much and the healing fails. Enough loading in the correct zone will encourage faster repair. For Iversen's project, a sensor was placed on the bottom of the cast on the foot of patients with a leg fracture. This detected the load the patient was putting on the injured leg and produced a tone if the load for optimal healing was exceeded. This meant that the patient could load the leg (and so encourage healing) but not exceed a set force which would impede repair, thus accelerating recovery. This gave him a taste of orthopaedics.

Ed's first job after graduation was in mining. This was not satisfying, so he pursued biomedical engineering through a master's degree and then a job in Salt Lake City in biomedical engineering. Too soon the company moved away, but Iversen stayed in town and sought out the appropriate people to work with. He contacted the University of Utah asked if there was a professor in biomedical engineering. Like Sears, he was directed to Steve Jacobsen, and similarly, this was a pivotal decision. Jacobsen became his boss and mentor. Between them they set up the company Sarcos and soon Iversen was in charge of numerous projects from underwater vehicles to animatronics displays. At this time the Utah Arm was beginning to be tested, so while he had no direct contact with the arm for many years he was able to observe its progress.

Iversen sees all of the project at Motion Control as teamwork. Early on the Sarcos team included engineers who had worked on the Utah prosthesis, so they brought that knowledge to the MIT/Utah dextrous hand.

This device was a sophisticated four fingered robot hand that allowed experimentation in aspects of grasping and control. The robot hand used tendons as flexible links to the drives in the forearm. The tendons were made from Kevlar and the team had to learn how to weave it to create the ribbons that made the tendons. The tendons often failed as they passed over pulleys, with a small radius of curvature which caused the Kevlar to crack and break. The team had to work out the physics on how the fibres cut themselves during operation, in order to route them more effectively. They then discovered *Spectra cable*, which was better than anything they made themselves [14]. Spectra is the flexible link most often chosen by modern citizen engineers who aim to build cheaper prostheses. Sarcos made a dozen of these hands and they were sold to labs who were doing research in robot manipulation. These days firms such as the British company Shadow and the German DLR provide anthropomorphic hands for research robots in the same manner.

Iversen stresses that there are big differences between the two fields. It is common to see publications on aspects of robotics that say the idea can be applied to prosthetics, when the truth is they can't be applied easily. Iversen says:

> "It is the same in any field; if I am in robotics and think I have something that can be used in prosthetics, or if I am in prosthetics and think I have something that can be used in robotics, I am probably wrong... It is a lot easier to design something in a field that you are not working in, because you don't understand all the constraints" [174].

Iversen believes his own contributions to the Utah Arm are the incremental improvements in the push for ever more rugged prostheses. But he also added wrist function to the arm, which has altered the upper limb market in general.

As well as adding the flexion wrist to the Motion Hand, the second wrist Iversen conceived was one that had two passive axes of motion: Flexion/extension and radial/ulnar deviation. It is sprung back to the neutral position and can be used locked or unlocked. This gives the hand greater compliance when gasping objects. Not only can it be used to perform activities that need some passive motion, such as rowing or biking, but it also allows the hand's approach towards the object to be less restricted. In the natural hand the fingers can adapt to an object by changing their precise alignment to the target. With a simple hand this adaptation to the object cannot happen. A single axis hand must approach an object along the same line every time [253]. By adding the compliance to the mechanism the use of the hand is easier and gives it a greater functional range [254]. As it grasps the object the wrist allows the hand to move a little to align it more with the object. The market has noticed this advantage and Otto Bock have added a compliant wrist to the Michelangelo Hand.

The idea for the wrist came not from the market, but according to Iverson, from the field of robotics, and watching what people had to do to eat [248]. It is known that for a robot arm to be able to pick something up from one place and put it down at another, it needs six separate degrees of freedom. While the intact human arm possesses more than six, and so can get the hand in most positions needed, the prosthesis bearing arm generally lacks one or two motions, and so the user has to compensate with movements of their trunk and legs. By creating this wrist Iversen gave back some flexibility.

7.2.2 The future for Motion Control

The range of Utah Arm 3 prostheses with the choice of colours and terminal devices are shown in Fig. 7.10. It shows the adaptability of the original design and the partnerships developed to create choices for the users and prosthetists.

The collective aims of Iverson, Sears and the company are for more robust devices. Their response to the current trend towards multifunction hands was to look for a more functional hand that can be used in the workshop rather than the office. Put one of the multifunction hands to use in the garden or workshop and generally it will not survive as long as simpler hands and grippers. From 2014, Motion Control began the process of creating an entirely new arm system incorporating many of their ideas. First, it had some more compliance in its joints. This is important to Iversen: If a device is rigidly jointed, the forces that go through an object are high frequency and will need fast actuators to control it. The human arm has a series of elastic elements as part of its drive and our muscles are quite slow moving. As a result we do not feel all the high frequencies that our arm receives [174].

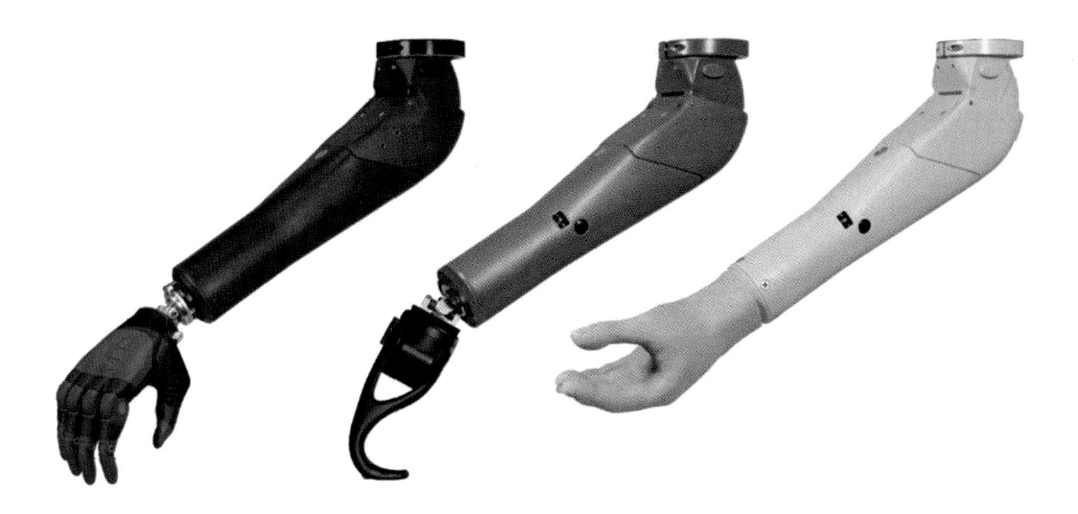

FIGURE 7.10

The range of the Utah Arm 3 with terminal devices (Left to right) – Taska hand, ETD2, Motion Hand. (Courtesy: Fillauer Companies.)

A second aim is ease of use. The correct positioning of a terminal device to grasp an object remains an important goal. The next development included a modular three degree of freedom wrist, which can be made up from three separate drives, so that the wrist can match the needs, anatomy and capabilities of the potential user. The wrist does not use the nascent wrist coupler standards to join the parts together, but retains the Otto Bock quick disconnect wrist standard.

The third innovation builds on the collective aim to make the devices waterproof. The entire arm is waterproof to the IPX7 Standard [34,255]. The company took a pragmatic approach and did not apply for separate qualifications for dirt and water resistance. Each test has to be conducted by an external company and costs at least $10,000 per test. They reasoned that if the device resists water then it is likely to not allow dirt in either. This standard is described as 'water resistant', but the test is not for a device to be splashed a little with water, but to be placed in one metre of water for half an hour and for this to have no ill effects. To be described as fully waterproof means the device must be usable in diving. This is really a technical nicety. In practice the arm should work in most domestic situations, freeing the user from worrying about getting the arm wet [34]. The girl Iversen heard of would probably be able to swim with the new arm. This is a significant improvement in the performance of powered arm prostheses. The Otto Bock Genium leg was water resistant from 2015, but all previous arms and hands relied on the gloves and were not sealed. It is why the representatives of Motion Control were receptive to Mat Jury of Taska Prosthetics when they met in 2017 (see 14.1.3). Taska was the only other manufacturer who was pushing towards water resistance at the time, but Jury did not think this advantage will last for ever: We *"...took on the monster of waterproofing, which I am confident everyone else is working on now"* [118]. The Taska hand does not match IPX7 standard yet, but greater reliability remains a goal. The result is that the entire Utah Arm is waterproof, up to the socket. With a modification to the socket design to create a sealing lip, even the socket resists the ingress of water.

The final touch is that the arm can be dropped from a metre height in any orientation and survive the fall [252]. Combining this with the joint compliance and the water resistance, the arm is more prepared to be used in the outside world. Iversen does not see any revolutions coming to the field in a while, instead he sees steady progress towards better devices.

Motion Control relies on external funding to make these projects possible, and they continue the Jacobsen tradition of being very effective at obtaining this money. They won funding from CDMRP (Congressionally Directed Medical Research Programs). These are distinct from the DARPA program, CDMRP funds medical research with less of the complexity of the Defense Advanced Research Projects Agency that styles itself on funding 'hard' projects. So far, these DARPA projects have been more expensive and less successful (4.2.2.3). The CDMRP grant was the biggest grant Motion Control had received, $2.3million over four years. Out of it came Electric Terminal Device (ETD2), Force Limiting Auto Grasp (FLAG), powered wrist flexion and the waterproof Quick Disconnect Wrist (QDW) [34]. The first of these was a new ETD2 is shorter and more attractive than the existing device, which Sears describes as *"ugly, but effective"* [14]. FLAG is the latest version of sensorised Motion Control prehensors. While Sears moved into commerce with Motion Control, Meek stayed in the university and has worked on different aspects of sensorised grasping [256]. FLAG incorporates force detection into the Motion Control hands. Sears claims its the 'first practical force sensor' [34]. He compares it only with the Otto Bock Sensor Hand. The latter uses shear forces to detect objects slipping from the hand's grasp. The FLAG feature uses sensors at the base of the finger to detect light contact with the object. A buzz from the motor tells the user and they are then able to take over control of closing the hand as much as they wish.

Both techniques give the user greater control of the grip force of their hand. With the addition of a sensor to detect contact with the object, the relationship between the EMG level and force imparted on the hand can be made more progressive. Rapid opening and closing can be related to the EMG when the hand is not closed on an object. When the hand is holding an object the EMG can then control the grip force more subtly. All these features together form part of the control for the Southampton Hand (see 10.2.1) used experimentally in the field since the late 1980s.

For the FLAG, the increases in grip force are the result of pulses from the command muscle. This gives the user greater feel of control over the grip force of the hand. In contrast, the Sensor Hand Speed form Otto Bock has a very fast motion, which is useful, but the penalty is that the initial grip force for the hand is considerable. The user tends to want to dial the speed down to increase the feeling of control of the hand. This feeling of control in the FLAG hands should continue as users can turn the feature off by holding the hand open for a couple of seconds, two buzzes are the acknowledgement from the controller. Sears maintains that users tend to want to turn the grip reflex on the Otto Bock hand off, while they keep it on in the Motion Control devices [34]. The FLAG feature can be built into the entire range of terminal devices.

7.2.2.1 Innovation in the prosthetics market

The new Motion Control systems are more robust than the old models and incorporate many developments in technology that required more than simply tweaking the original designs. To completely replace an existing and successful design with a new one, is a bold step, no matter how many good reasons for it can be made. The old design represents the majority of Sears' professional life's work, so it is even more remarkable that Sears eagerly awaited the launch of the new arm.

It is one of the curious aspects of the prosthetics industry that while there were three or four different companies in Europe that produced the original simple electric hands, and the first four multifiunction hands also emerged in Europe, the first two powered elbows came from the United States.

When Harold Sears is asked the question: *Why are hands built in Europe and elbows in America?* He suggests a couple of influences. He says with a small market there is little reason to build a hand to compete with the Otto Bock device. In the past there had been some attempts, but they were not that successful. In the end Motion Control needed to build their own, to make hands that were compact and robust enough for their aims. Sears thinks that a second reason is that the market in North America is smaller than Europe [34]. Europeans, he says, want hands, they don't want hooks. The Greifer was a product for North America. Motion Control has always sold many more Greifers in North America than Otto Bock sold in Europe [34]. *"In Europe you must have a hand"*. In the new era of multifunction hands and small start ups there are some American companies beginning to build hands, so it will be interesting to observe if this bias influences their futures.

The story of the two elbows has been united by origins and circumstance throughout their existences. Created in university programs to explore the control aspects of arms, they went on to become successful products with long lifespans and many users. They continued this parallel relationship as they were replaced by their successors at about the same time. Again the resulting arms are different with the Utah Arms perhaps being the most revolutionary. However, the cost of either of the prostheses might be the ultimate limit on their use in the field.

CHAPTER

The Edinburgh arm and the i-Limb hand

8

On the outskirts of Edinburgh was the Princess Margaret Rose Orthopaedic Hospital (PMR). Facing away from the city, from its canteen you could look out across fields to the ancient hills beyond. It was founded as the Edinburgh Hospital for Crippled Children in June 1932, it changed its name to Princess Margaret Rose Hospital for Crippled Children shortly after. It became the Princess Margaret Rose Orthopaedic Hospital around 1957. It closed in 2002 and the site was redeveloped for housing. The PMR had the leading research centre in upper limb prosthetics in the UK. In its connection with advanced prosthetic limb development, it has many similarities to the stories of UNB's IBME and the Bloorview Research Institute in Canada. They each formed a bioengineering centre to conduct research and development into powered prostheses as a result of the needs of the Thalidomide affected children. Similar to the IBME and the Bloorview Research Institute, founding director David Simpson developed an international reputation for his centre at the PMR. Like Robert Scott in Canada, David Simpson was not a physician, rather he was a medical physicist and therefore at home in orthopaedic hospitals. Both also worked closely with users when designing and fitting upper limb prosthetic devices. Both men shared a patient-oriented motivation, with Simpson having a particular insight into the user experience arising from his lengthy hospitalisation due to tuberculosis. Simpson's proudest professional moment occurred when he saw a child feed herself with the use of an arm system he had designed and fitted [257]. The Edinburgh approach was generally different from other institutions. This was evident in with their original arms which used carbon dioxide as the power source in the 1960s and 1970s, not batteries and myoelectric systems. Simpson famously did not regard myoelectric signals as a useful control input. This attitude continued with David Gow, whose independence of mind, led to a revolution in prosthetic arms.

8.1 David Simpson

Germany was to have a profound influence on the life of David Simpson. Simpson was born in 1920, and he originally trained to be an accountant. These plans were changed with the outbreak of the Second World War in 1939. From the summer of 1944 to the spring of 1945 Simpson fought with the Highland Light Infantry in Belgium and the Netherlands. Only weeks before the war ended, shrapnel cut into his right shoulder and neck, severely damaging the network of nerves responsible for control of most of his right arm.

Injured and traumatised by the war, and without the use of his right arm, the right-handed Simpson nevertheless entered the University of Edinburgh in 1945, obtaining a degree in science. It took a series of operations for Simpson to regain most of the use of his right arm, only after finishing his first degree.

Making Hands. https://doi.org/10.1016/B978-0-12-820544-0.00017-2

At Edinburgh, he studied physics under Max Born, the father of a childhood friend he had first met in Germany in 1938. He obtained a PhD in medical physics in 1952 and then stayed on to design monitoring equipment for use in transplant surgery and a foetal heart monitor.

The conditions in the United Kingdom after the Second World War were significantly different to those in North America. The population was exhausted after more than five years of war. The infrastructure was badly damaged from bombing. The country then owed the US for the materiel help given to them prior to USA's own entry into the war following Pearl Harbour. As a result the level of progress of reconstruction was slow during the 1950s. Food and clothes rationing worsened after the end of the war and many items stayed rationed into the 1950s. The debt itself was not fully paid off until 2006 [258]. In the UK the image of affluence and excess that North America possessed persisted long after the gap had actually been closed. Against this background research and healthcare had limited funding and any technological progress had to be affordable.

Common with many hospitals of its era, the PMR was a collection of low buildings housing the departments which had accreted onto the main buildings over time, as the requirements of the hospital changed. In the late 1990s the orthopaedic aspect of the PMR was relocated into the main hospital site in the centre of town. Afterwards, rehabilitation (prosthetics, orthotics, wheelchairs, seating) moved twice, finally becoming the South East Mobility and Rehabilitation Technology Centre (SMART Centre) housed in a brand-new purpose built building in the Astley Ainslie hospital site, closer to the centre of Edinburgh than the original PMR. As head of SMART, David Gow was responsible for the rehabilitative services it provided, including wheelchair, seating, prosthetics, orthotics, driving assessment, occupational therapy and physiotherapy. The move required the intermediate step of five years in a different, older, hospital on the east side of town (the Eastern General Hospital from Easter 2002 to March 2007). It was during this period that Gow also set up the company, Touch Bionics.

Edinburgh itself is the ancient capital of the region. Its set piece landmarks of; the castle, the park, and the rock promontory at one end of the city, have been added to by the building that holds the devolved parliament of Scotland. For centuries Scotland has been a world leader in social, intellectual, medical and scientific developments [259] and by the late 1990s it was beginning to set a new agenda in prosthetics.

As in Canada, it was 1963 when major institutional funding programs began as the result of Thalidomide. In Scotland, the National Health Service (NHS) created and funded a new Powered Prosthetic Unit (PPU) at the Princess Margaret Rose. The technical and research organisation concepts came from West Germany. In May 1963 Simpson visited the University of Heidelberg's Orthopaedic Clinic [257]. The clinic, originally established in the 1890s, had become internationally recognised as one of the world's foremost orthopaedic centres. Simpson spent a few days with the University of Heidelberg's Ernst Marquardt, who had created surgical techniques for improving prosthetic suspension [38]. In the 1950s Marquardt developed the Heidelberg arm, powered by carbon dioxide, which proved influential. Variations of it were used in a number of centres. By 1963, when Heidelberg was confronted by greater numbers of children affected by Thalidomide than anywhere else, the arm was ready for application. Although the arm was still at an early stage of development, Simpson was sufficiently impressed to accept the generous offer from Marquardt to take some of the prostheses back to Edinburgh for fittings with Scottish wearers.

The other significant idea brought back from Heidelberg was locating the clinic at a hospital. It was conventional for Marquardt to work in a hospital; he was a physician as well as a researcher. Simpson adopted Marquardt's approach of a group of physicists and engineers, not in a university electrical engineering department, but in a hospital, close to the population centre with other professionals, such as occupational therapists, prosthetists, and engineers.

The Edinburgh PPU grew incrementally as patient relationships became more and more central to the work of the staff. The first home for the PPU (from 1963 to 1965) was in the basement of a now demolished house which was part of the University of Edinburgh, near the city centre, five miles from the hospital. In 1965, Simpson and four technicians moved into a wooden hut on the Princess Margaret Rose site. Until 1969, the hut was the home of three workshops: mechanical, plastics, and electronics.

Growth in the PPU's patient services workload and staff meant that more space was needed. The next move coincided with a name change, from Powered Prosthetics Unit to the Orthopaedic BioEngineering Unit (OBU, also known as *'the Bio'*), indicating an expanded mandate. It now included the design of beds, wheelchairs, and other biomedical devices. It also reflected a change in priorities for the governing groups, Thalidomide was fading from the public consciousness and other physical impairments were making themselves seen. Funding from the Medical Research Council and the Scottish Home and Health Department was provided to build and equip new facilities in the Princess Margaret Rose to accommodate the now approximately twenty-five staff in the OBU. There was also a change in relationships to patients. Originally, the PPU had served other hospital departments, but increasingly the OBU provided services directly to patients. What this meant was that physicists, engineers, and technicians dealt with the users of their products on a daily basis. They had to learn how to combine the performance of research and development at the same time as staying ahead of the needs of their patients. In 1976 Simpson stood down as director of the OBU to accept the post of assistant dean of the University of Edinburgh's Faculty of Medicine.

Simpson's work in the 1960s was directed towards the production of arms to help the children affected by Thalidomide to perform activities on their own. A principle target was that of feeding, bringing a fork or spoon from the table to their mouths. One of the ideas that emerged from Simpson's work on arms was what became known as *Extended Physiological Proprioception* (EPP) [260].

EPP emerged as an idea because controlling an arm with multiple joints is hard to do well without feedback. If someone is operating a hand alone, it is possible for them to see what the hand is doing. In fact, users of prosthetic hands perform most of their actions with their prostheses within their direct gaze on the hand or the object being manipulated. Persons with natural hands tend to keep their hands within their peripheral vision and are more concerned with looking ahead to the next part of the task. When people operate their own arms they are less concerned with what the joints are doing, and more interested in where the hands are in space. They do not focus on the elbow or wrist. Thus guiding a multijoint prosthetic arm, such as those that were being developed by Simpson, is much harder without additional feedback as the user will have to look at each joint in turn. The feedback we use is *proprioception*, which is the sense we have that we know where our limbs (or any other part of the body) are in space without looking. Without proprioceptive feedback it would be almost impossible to touch one's nose in the dark.

A tiny number of people do lose their proprioception, generally through an infection. The result is they cannot control their limbs without looking at them at all times, so that the act of standing is an incredible feat of concentration. Indeed most of the handful of suffers of this condition spend their lives in bed or a wheelchair. One man (Ian Waterman) through sheer persistence and long practice and concentration has managed to live with the condition and have a career. The effort required is so enormous that reading his biography leaves one awed by his abilities [261]. The usual operation of a

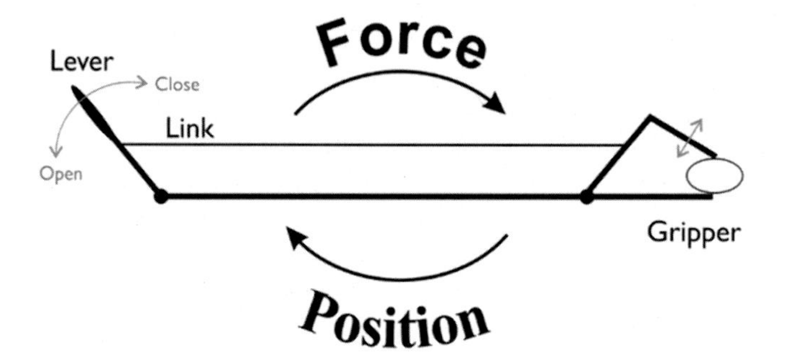

FIGURE 8.1

Extended Physiological Proprioception. The gripper is opened and closed by a link between the lever and the gripper. As the gripper closes on the object it stops moving and the operator of the lever feels it stop. Greater pressure on the lever will cause the gripper to close further and the operator feels the link move.

conventional prosthetic limb is similar. We can see it but we cannot feel what is happening and this must get close to the feeling that Waterman must have, but for one part of the body.

EPP attempts to get round the problem by feeding back the position of the limb to the operator. The important concept in EPP is that the feedback is appropriate in form and is fed back to the point of application. Many other forms of feedback have been tried [262], but they have not been used routinely. David Gow suggests this is because the feedbacks are not appropriate to the task and are fed to a point remote from that of application [257]. For example; force feedback has been tried by Scott's team in New Brunswick feeding an electrical stimulus back to the muscle that commanded the prosthesis [189], but it was generally not used or appreciated. While the stimulation was in proportion to the force the prosthesis encountered on touching the object, it did not relate the action to the stimulus like EPP. It has never proven to increase the functional performance of the user fitted with it.

The simplest illustration of how EPP works is to think of a device like the mechanical hands mounted on a handle one can get in joke and novelty shops. Imagine that there is a rigid link between the lever in the handle and the hand at the other end (Fig. 8.1). If one pushes/pulls on the lever, the hand will close or open. While we use the force of our muscles to move the lever, what we are aware of is that the lever is moving. If the hand closing is obstructed with an object, we feel this as the lever stops moving. If the object is compressible, then a greater force on the lever will compress the object, but we will know this has happened because the lever moves again, just more slowly.

Simply put, EPP is *force input*, but *position return*. We pull on the lever (force), but it is the movement of the lever we feel (position). This is not dissimilar to the way our muscles work; 'effort' is the force the muscle puts out, but it is the Golgoi tendon organs in the muscles that tell us the muscle has contracted as the hand/arm has moved, so what we sense is the arm's position. *Extended* Physiological Proprioception is when we can extend this sense to beyond our skin. If we are skilled at use of a device, such as a tennis racket, when we pick it up our body extends to include the racket and it becomes part of our idea of our self. Once integrated into our bodies we can judge where to put the racket with great precision. This only happens when we have much practice with the racket [260].

To control his pneumatic arms, Simpson found it straightforward to link the lever on the valve that released the gas. This in turn initiated the movement of the arm. He used cords running through pulleys, with the cord attaching to the lever further down than the pivot [263]. As the arm moved the lever changed angle and so more effort was needed to move the arm further. It is possible to create a similar format with an electric motor and servo systems for an electric arm, but the power required for effective performance is demanding. Additionally, the feedback has to be fast enough to return the signal immediately and crisply to the operator. A long enough delay in the response will destroy the feeling.

EPP is one of the very few feedback mechanisms that have worked successfully with prostheses. With EPP users can confidently employ their devices, and so it should be possible to successfully touch one's nose when blindfolded. EPP has also been adopted by other researchers, most notably Dudley Childress [74,264]. The problem for using EPP in the field is still the difficulty of making feedback devices that are compact enough and powerful enough to be useful. Despite this, the industry took on the idea that EPP is a desirable feature. So much so that when, in 2002 a salesman wanted to suggest his new EMG amplifiers were the best, he invoked the idea that they provided EPP. They do not, he had completely misunderstood what EPP is, all the user of the amplifier will feel is the length of her muscle and not the position of the device no matter how well the device is controlled. However, he felt that EPP was a quality that the product should be (erroneously) associated with.

Simpson had a pragmatic approach to prosthetics. He recognised that, despite any hype at the time (or since), that the devices they were producing aimed to alleviate the challenges of missing limbs. They were not there to replace them and they would always be inferior. Many years later David Gow, mimicking Asimov's *Laws of Robotics*, coined *Simpson's Laws of Prosthetics*, the first of which is: *You have failed* [257]. This is not a negative judgement. It is a simple statement of the level of humility an engineer should have. In this law he encapsulated the fact that as engineers we need to be humble and admit that no matter how sophisticated and advanced the devices we make, they will never be as good as the one natural selection created over 365 million years [11]. This was true in the 1960s and, despite assertions from people since about 'bionic prosthetics' making people 'better' than they were, it is still true today.

Gow's creation of subsequent Simpson's laws are based around the practicalities: *It's too heavy. It's too big. It is hard to use* etcetera.

In the same vein, Simpson did not think that electromyography was a good way to control a prosthesis. To Simpson EMG is not directly related to what the arm is doing so generally it will not give meaningful feedback to the operator. All they will feel is that they have tensioned their muscles. Enthusiasts who use body power would agree [4].

Simpson produced gas powered arms that had multiple joints, and were based on the geometry of the Michigan feeder arm [265] (see 4.4), using EPP to help control the end of the arm (Fig. 8.2). The arm was aimed at a person with a high-level bilateral loss, for the dominant side of the wearer. The arm on the nondominant side contained the gas bottle and had more limited movement. One of the reasons that attempts at marketing the next generation of full arm prosthesis failed, according to Gow, may have been the restrictions in transporting bottled gas in the USA [257]. Unfortunately gas power is not as convenient as electric motors. The principle problem is that recharging an electric battery is a simple process that can be done easily, while gas bottles need a compressor to replenish them. Also the Edinburgh Arm required two hands to replace the bottles. This does not make it a tool of independence.

FIGURE 8.2

A gas powered feeder arm developed in Edinburgh in the 1960s. The right arm brings the terminal device up to the mouth. The left arm is passive and contains the gas bottle. (Permission: David Gow.)

8.2 David Gow

After David Simpson moved on, research continued in powered arms in Edinburgh and David Gow became involved in the early 1980s. Gow was born in 1957. He studied mechanical engineering at the University of Edinburgh, graduating with an honours degree in engineering science in 1979. After completing his degree, similar to Simpson, David was appointed as a research associate at the University of Edinburgh working on control systems for artificial limbs. He worked at the University from 1981 to 1984, when he left to join the Princess Margaret Rose Hospital and then on to the Bioengineering Unit in 1986. Like Robert Scott of the IBME, Gow identified himself as an engineer first, although one who did research and development [266]. Gow began his prosthetics career in the academic world but soon realised that his real strengths lay in a more practical sphere within the NHS, which would allow for more extensive interaction with patients and open up opportunities for direct practical research and development in the field. The possibility of a doctorate was still open to him at that time. However, David decided to follow the NHS path full-time alongside continuing professional research. The decision to work within the NHS and not to pursue a doctorate worked out well as it paved the way for the Edinburgh Arm and the i-Limb [257].

Gow has a retained a healthy cynicism about the usefulness of university research to the field (perhaps based on his own experience) and the impact of the press. He is master of the pithy one liner. In the 1980s Japanese robotics researchers conceived of the idea of a robotic guide dog for the blind. This was promoted throughout the rehabilitation world, with a picture of a smartly dressed woman in dark glasses being led by a low box on wheels with a laser range finder on the top [267]. The box was the robot 'MEL Dog' which would know its way around because the person would be living in a specially prepared city, set out for the needs of the disabled population; the *"city of the blind"*. This was a genuine aim to create a space that was perfect for the specific subgroup, as opposed to the alternative idea of making all cities accessible. Gow's response was simple: *"A muggers' paradise"*.

It was David that coined the idea of Simpson's laws of prosthetics as a simple way to encapsulate the knowledge and dispel myths of how advanced the prostheses really were. In a similar vein he chose to refer to the era of prosthetic technology at the start of the 21st century, not as advanced, but technologically more like the Stone Age (the Neolithic era). So we were not in the *Bionic* era but the *Biolithic* age. When the press wished to suggest a standard single axis myoelectric hand that had been on the market for two decades was 'bionic' (as they often did), exasperated Gow exclaimed *"When I hear the word 'bionic' I reach for my gun"* On a cultural note; this is ironic on many levels as no one outside a very small minority would even consider owning a gun in the UK. It is doubly ironic therefore that it was his start up company that made the word bionic de rigeur in the prosthetics industry. However, it was Stuart Mead, CEO, who takes the credit/blame for the name change from Gow's 'Touch EMAS' to 'Touch Bionics'.

When Gow became responsible for the service and research areas of the OBU, his initial work was funded by government research grants and was supposed to be focused on development of more cosmetic prostheses. The actual work he did was more varied. What was obvious to Gow was that improving the aesthetics of the prosthetic hand was not just about the cosmetics, but the overall design [266]. It was about creating something that looked and functioned more like a human hand.

The focus on practical prostheses in the clinical service at PMR was encouraged. It ensured that Gow looked at devices that would fill gaps in the current provision, persons with a shoulder loss and children with partial hand losses. In 1988, Gow started work on electronic shoulders, wrists and hands. At this time there were few groups looking at prosthetics research at all. There were only two groups in the UK undertaking research with any clinical output, in Edinburgh [268] and Southampton [269]. Otto Bock, LTI, Motion Control, and VASI had all manufactured elbows and hands, but none had integrated it with a powered shoulder. Moreover, none of the other research prostheses with a powered shoulder had made the transition into the field. The Edinburgh Arm did [270]. The first device Gow worked on took the existing structure of the gas powered, whole arm prostheses and addressed one problem that had afflicted them, their power source. Since gas was inconvenient and expensive the choice was to look at electric power. The mechanical structures for the gas arms had proven robust, so David sought ways to generate the same linear motion that the gas cylinders produced (Fig. 8.3).

One way to create linear movement is to place a nut on a threaded shaft. Turn the shaft and stop the nut turning and the nut rides up and down the threads (Fig. 8.4). If a motor and gearbox are attached to the end of the shaft to rotate it, the nut has straight-line motion, rather than the rotation of the motor. A disadvantage in these devices is that there are large mechanical losses in the nut running along the threads. The power requirements for an elbow or shoulder are already large without losing more in the

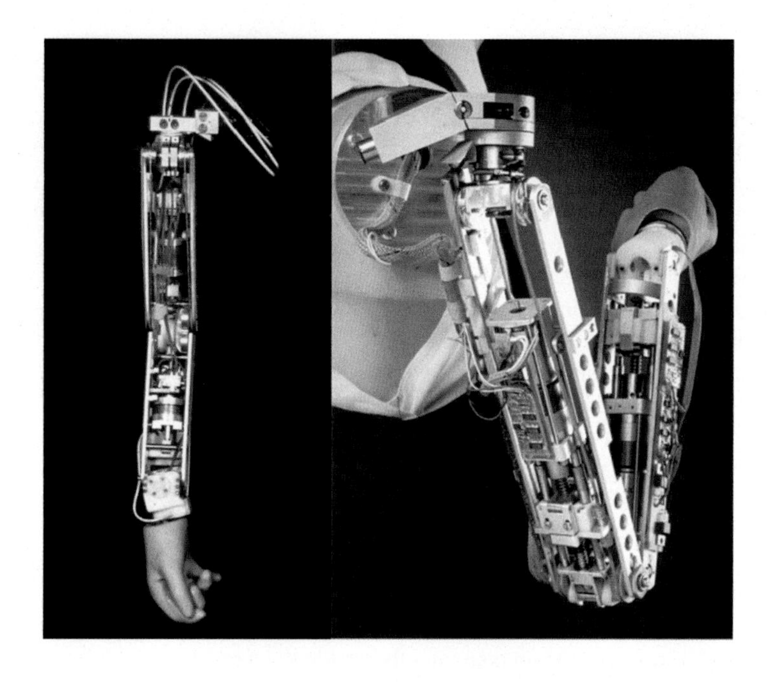

FIGURE 8.3

The Edinburgh gas arm and the electrical version made by David Gow using linear ball screws. (Permission: David Gow.)

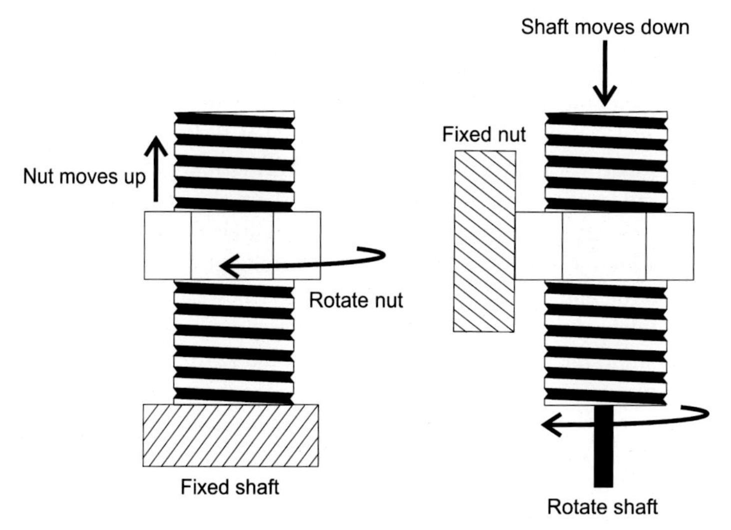

FIGURE 8.4

Operation of a leadscrew to give linear motion from a rotary motor. When a nut rotates on a thread it moves up. If the nut is fixed and the shaft is rotated by a motor, the thread moves down.

actuator. Instead Gow opted for a device known as a *linear ball screw*. These are mechanisms, devices that fit round a plain shaft. They contain a loop of ball bearings in a track in a spiral form. When the shaft is turned, the ball screw moves up (or down) the shaft. The ball bearings on the shaft can only grip the shaft below a fixed load, pushed too hard the device will slip. This gives the designer an opportunity to protect the mechanism from overloading. If the designer wants increased load capacity, then two units will double the holding load. The ball screws are more efficient than the nut on threads. This simple structure of motor gearbox, shaft and ball screw dropped very neatly into the arm frames resulting in a new Edinburgh Arm with minimal redesign. The arm was then tested in the lab by a user; Campbell Aird. A man who would prove pivotal to the story of the Edinburgh prosthetics systems.

8.2.1 The development of ProDigits

The second gap David investigated in 1993 was a partial hand system. There were no powered solutions for partial hands at this time. A person with the loss of only part of their hand (say all the fingers distal of the knuckles) would not be able to wear a conventional prosthesis without making their arm much longer. What was required was a shorter hand, or powered fingers. The occasional attempt in research projects to fill this niche resulted in huge and impractical devices. Gow's design work initially focused on a partial hand for children as young as nine-years. This was different from the conventional approach of first designing an adult device and then scaling it down for a child.

Gow patented the design and coined the name: ProDigits. This was novel and not one of the mechanical solutions already listed in Schlesinger [2]. The finger was based around worm and wheel gears (Fig. 8.5). Conventionally, the motor turns the worm which drives the wheel at a reduced speed. Gow's insight was to reverse the roles: To fix the wheel and let the worm rotate around the wheel. In ProDigits, the wheel (gear) is placed where the joint should be and is fixed. When the motor turns, it rotates the worm so it runs around the fixed wheel. If the motor driving the worm is in the finger, and the wheel is at the knuckle, a powered finger is created. The powered digit was the basic module, which was then adapted for the specific finger or thumb. The initial prototype design, building, and testing of the first individual digits occurred over a period of two years, from 1996 to 1998. OBU staff did most of the work as part of the twenty percent of their time that was permitted for projects outside the strictly clinical service. Gow was the project manager, applying what he called a *"benign dictatorship"* management style, keeping in touch on a daily basis with work being performed by the team [266]. The overall costs from initial design to filing of the first patent were incredibly modest for this sort of venture, about fifty thousand pounds for materials and supplies to produce a prototype [266].

Once the finger was produced, it occurred to Gow that size was not a limitation for this device. Five fingers could be placed together to make a complete hand, with space in the palm for the electronics and perhaps the wrist flexor. If he scaled up sufficiently, the motor and wheel could drive a wrist, an elbow or a shoulder. Indeed, the size would be defined by the user, so a child's shoulder could be the same size as an adult elbow. Additionally, the elbow did not take up much space. As a result it could be fitted to people with long remnant upper arms, with the motor below the joint as with the other elbow prostheses. In contrast, if the humeral remnant was small and there was space for the motor in the upper arm, the orientation of the motor could be rotated to lie in the humeral section above the joint. With this change, the motor was not moving itself and the rest of the prosthesis, making the moving mass of the forearm much smaller, as well as closer to the body. The arm is less cumbersome (Fig. 8.6). This was the concept of the *Edinburgh Modular Arm System* (EMAS) [270].

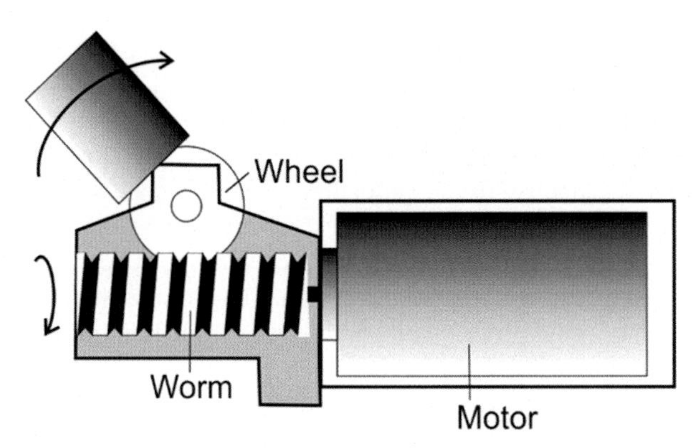

(a) The design of ProDigits. A motor and gearbox is connected to a worm and wheel. The motor turns the worm which rotates the wheel. If the wheel is fixed then the motor rotates about the wheel, this can make a powered digit.

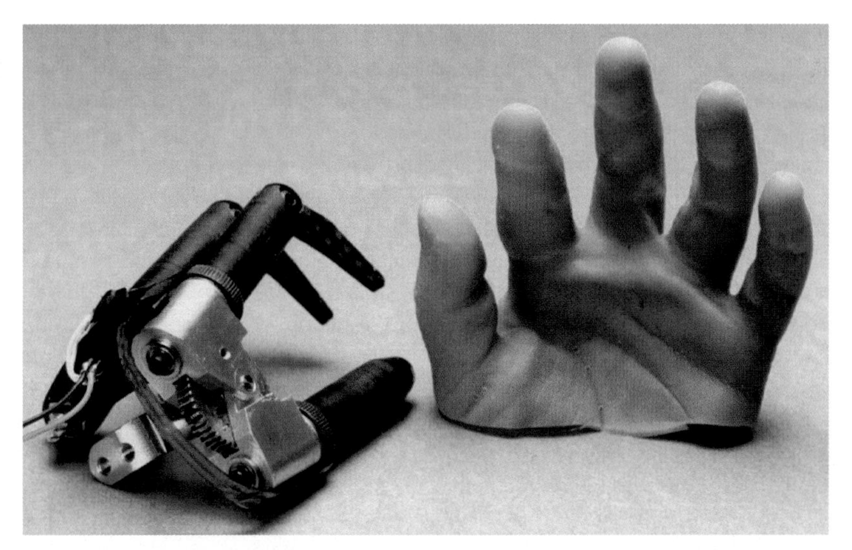

(b) ProDigits as prosthetic fingers - if the wheel is mounted at the knuckle powered fingers are created. (Permission: David Gow)

FIGURE 8.5

The design and use of ProDigits to create fingers for use with a partial hand loss.

In 1998 newspaper stories announced an EMAS fitting of the local hotelier, Campbell Aird, whose right arm was amputated after he was diagnosed with cancer of the muscle [271]. Aird turned out to be a major force in the initial development of both the EMAS and ProDigit technologies, largely due to his persistence in seeking improved prostheses from the NHS, and his skills for generating press interest. Aird had already been fitted with the electric version of the earlier gas arms, so when Gow wanted to

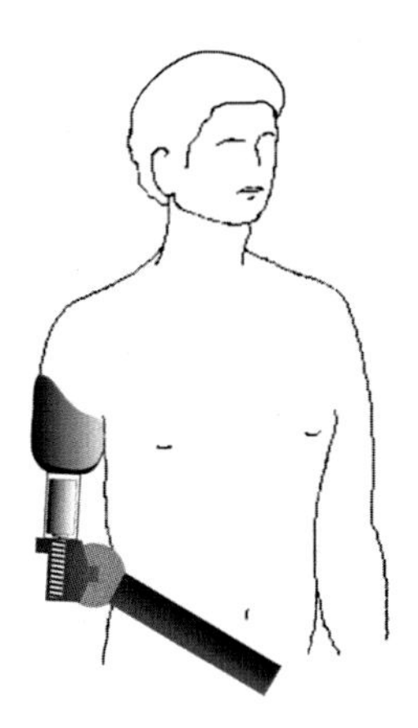
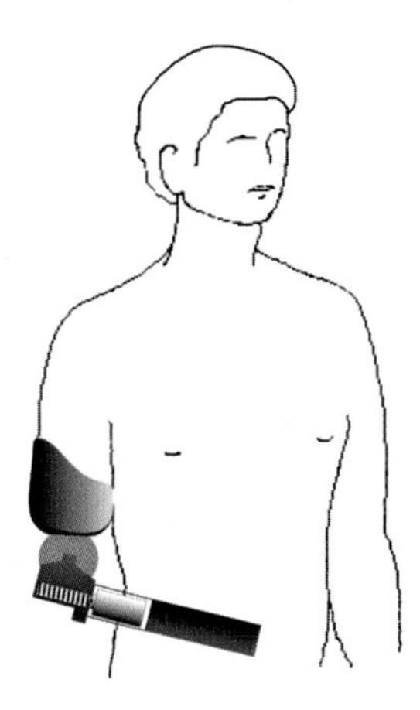

FIGURE 8.6

The design of ProDigits based arm components allow the position of the motor to be above the joint if the remnant limb is short enough, but with a longer humerus the joint is rotated and the motor is below the joint.

try out the EMAS arm he turned to Aird. According to Gow, during the development process, Aird went so far as to announce dates for the fitting of his arm and hand system, before even consulting fully with Gow. In this way, Aird's drive and enthusiasm helped push the project forward and David, sharing Aird's ambition, agreed to aim for his challenging deadlines. It was the resulting international press in 1998 that marked the turning point for Gow, as it was then that investors became interested in his ideas.

To watch Aird in front of the press was a privilege. He was relaxed and a willing talker. He was happy to show off the arm's capabilities and thought of ways to use it to make good demonstrations. Originally the arm did not posses a passive swing for walking, so it moved rigidly when he walked. To make it move more fluidly Aird practiced driving it in synch with his stride to match his other side. When Gow discovered this he altered the design to allow a little 'free swing' at the shoulder. I saw Aird at a large press gathering at the World Congress of the International Society of Prosthetics and Orthotics in Glasgow in 2001 (Fig. 8.7). Aird readily showed off the arm that the PMR team had spent the preceding weeks making ready for the demonstration. The arm was all in black carbon fibre with silver aluminium fittings. Aird was dressed in a black tee-shirt and trousers with a shock of white, well-groomed, hair, matching the arm perfectly. Despite the team's best efforts, the night before, the lower arm (the wrist and hand) stubbornly refused to work. Aird knew this and worked around it. It is unlikely that at the time any of the press even noticed. It even took me, an experienced observer, a few minutes to realise the fact. Sadly, in 2008 Aird's cancer returned and he succumbed, leaving behind a legacy that continues to grow.

Some funding for the development of the partial hands had come from the charity *Reach*, which was formed by parents to help children who were born with arm absences (10.1.1.1). This gave the project a focus on children's hands. In 2000 Steeper offered to built and clinically test some hands with an interest in seeing if they could be a viable product. In order to create hands that fitted the interests of Reach, and also to push the technology further. Gow insisted that the hands built for this trial would

FIGURE 8.7

Campbell Aird was the demonstrator of the Edinburgh Modular Arm System at the ISPO World Congress in 2001.

be small children's hands, and not the easier adult prostheses. The result of announcing the trial was a press frenzy. The car park for the hospital in Nottingham where the fittings were being made was filled with TV vans upcasting their stories to many nations.

During this time Gow was also involved in the ToMPAW project with groups from Oxford and Göteborg in Sweden (10.2.1). While the eventual aim of the project was to produce a commercial product (that never appeared), the first-generation arm was composed of new electronics for older prosthesis systems, an EMAS elbow and wrist and a Snave Hand [272]. Therefore two prototype ToMPAW Arms were produced (one in Göteborg one in Oxford) and were fitted to users. This was the first prosthetic arm that used a local area electronic communications network and it was also the first to be clinically fitted and used at home. A second version of the arm is still in use at the time of writing (Fig. 8.8).

One important part of the design of the hand is a glove that is flexible and strong enough to protect the hand and look good. While PVC had been used for many years, it stains easily and looks less lifelike

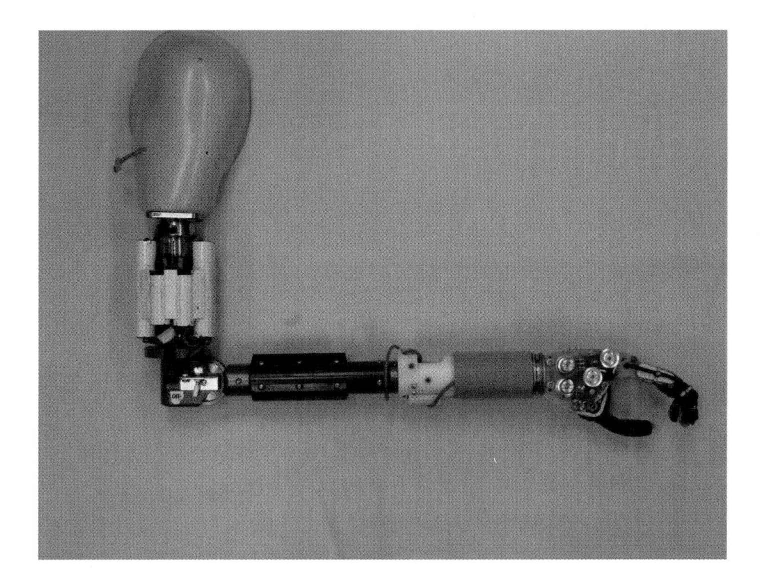

FIGURE 8.8

The second generation ToMPAW Arm was derived from a ProDigits elbow and wrist, combined with a Snave Hand from Oxford. ToMPAW employed the first intercomponent communications bus to be used in a prosthesis system.

than silicone. Silicone is a better material for appearance and flexibility, but it is more easily torn. Even so, during the 1990s, there was an increasing interest in silicone for visually appealing prostheses. Like technologists did at other hospitals, Gow began producing silicone prostheses for wearers and prostheses. He set up a small manufacturing facility in the engineering unit to make custom cosmetic terminal devices for people attending the clinic. By their nature, each device was unique and required close attention to the details. Today, manufacture of these custom silicone parts has been moved to suppliers and companies like Otto Bock who offer a service to the clinics. Prosthetists make casts of what they want to replace or restore and then they send the casts to the manufacturers with a colour match to the person's skin tone.

The facility in the PMR allowed David to begin the long process of investigating different ways to produce more durable silicone gloves for his hands. At the time, the small band of makers of silicone prostheses created their own secret society. It could be observed that when members of the society (prosthetists, engineers, therapists) met at conferences they would conspiratorially pass to each other samples of the latest silicones they had, and mutter the name (usually a code of letters and numbers) like some secret password. The fellow society member would then show their membership of the band, by assaulting the sample to test its mechanical properties; shearing it, bending it and feeling its performance. They would then nod sagely, agreeing it was better/worse than anything they themselves had to offer. Over the next few years, David Gow continued to try different ways to reinforce the silicone to make it tougher against rips and tears, or more flexible to work over the hand mechanism more reliably, before his company was formed. Once Touch EMAS existed, he continued to investigate it at weekends for a number of years thereafter. He finally had an opportunity to hand over this responsibility when Touch Bionics first partnered with, and then bought, the US company *Livingskin*, which had been created to make high-definition silicone prostheses.

8.2.2 Touch Bionics

In early 2003, Gow led in the creation of a spin-off company, Touch EMAS from the NHS. Scottish Health Innovations Ltd (SHIL) played a role in the transaction, as did Scottish Enterprise, which provided partial finance to get the company started. Scottish Enterprise was Scotland's main economic development agency and is a nondepartmental public body of the Scottish Government. It worked with partners in the public and private sectors to identify and exploit opportunities to deliver a significant, lasting benefit to the Scottish economy. Subsequently Archangel Informal Investments, TriCAP, Clydesdale Bank and the Scottish Co-investment Fund invested in the company. In 2005 the company was rebranded Touch Bionics, dropping reference to the Edinburgh Modular Arm System. In 2007, the i-Limb hand was launched, followed by launch of i-Limb digits in 2009. The grip force and mechanical strength of the hand was increased and in 2013 the powered thumb rotation into the lateral grip was added. This last addition filled a significant gap in the hand design. In 2016 the Icelandic prosthetics company, Össur, bought Touch Bionics.

The ProDigits hand was an innovation that used components in a way that had not been seen before and it pioneered a new market. It was the first commercial hand with separately moving powered fingers to be produced and marketed in numbers (Fig. 8.9). The separate fingers meant that the hand could assume a wide range of different grasp types. The original commercial hand had four fingers that curled at two places and a thumb that was fixed at the distal joint. The choice to fix the thumb's joint meant that the tips of a finger and the thumb could come together and form a grip.

A patent application was filed for the prototype prosthetic finger in 1995, naming Gow as the sole inventor [273]. Its novelty was the motor and gear-box inside the prosthetic finger, and the ability of the finger joint to rotate at the knuckle. The two additional joints of the finger and the one in the thumb did not move at this time. This was added later in a 2005 patent. The patent cites six earlier patents, including two from industry (one filed by Otto Bock in 1989), two from US government researchers, and two from university researchers, one of which was authored by Dudley Childress. A second patent in 1998 consisted of a mechanical pivoting wrist, elbow and shoulder joint: The Edinburgh Modular Arm System [274].

There have been many multiaxis hands that could adapt to the shape of the held object and provide close to the full range of grips a natural hand. It has been the design goal for very many designers from the days of Götz von Berlichengen and before. Few designs have been robust enough to be used in the field for long. There have been a very few commercial hands that got close to this goal, the Becker hand was potentially the closest, it had the digits made from coil springs so the digits curled as they closed [275,276]. It was body powered and it could not form a lateral grip.

The electronics for the i-Limb hands drives all the fingers together. The drive for the fingers and thumb are arranged so that if the hand started with the fingers fully extended the tips of the fingers and thumb will come together. If any finger encounters resistance, the motor stalls and this is detected, and the finger stops. Meaning the hand can adapt to the shape of the target object. For the early i-Limb hands to change their grip, the operator had to hold back the fingers they did not want to close so that the grip was formed. This was tiresome. Over time the options for switching have become more sophisticated, but Touch Bionics have not used any switches built into the hand itself. There is a switch

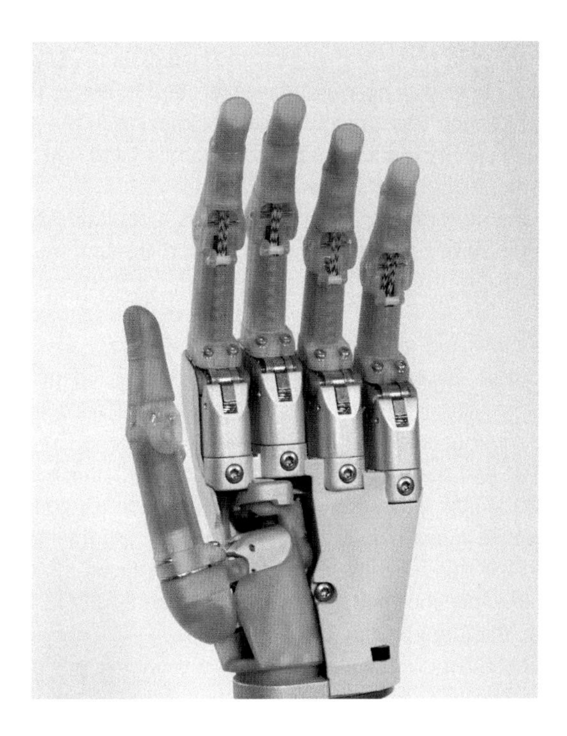

FIGURE 8.9

The i-Limb prosthetic hand. Using the ProDigits powered fingers to create a full hand it was the first multifunction powered hand prosthesis to be commercially sold after the ES Hand. (Used with permission of Össur.)

in the thumb of the bebionic hand, which means if the user pushes the thumb across to form a lateral grip, the controller knows and it moves the fingers to a preset shape to make the grip possible. For years there was no preset lateral grip for the i-Limb hand, so if the operator rotated the thumb across, the fingers and thumb would still open and close. With the thumb off to one side, all the fingers and thumb will move towards an invisible 'grip point' in space where all the digits contact. This means it is much harder for the operator to judge where the point of closure would be. For all multifunction hands, knowing where the location of the point it is essential because this is where the object to be grasped has to be placed. Miss that point, and the grip fails. Moreover you have to open the Touch Bionics hand up and start again. A later version of the controller allowed the user to select the grip so the fingers flexed to a fixed point and stopped. In this case flexion and extension was only for the thumb (if the user had moved the thumb across to oppose the side of the fingers). A more recent evolution is that this motion is powered too. This is a feature that had been crying out to be developed ever since David Gow conceived of the hand.

There are many reasons why the i-Limb was the first to be a successful product. The simple matter of the complexity of its construction may have prevented manufacturers successfully making the hand previously. There was certainly nothing totally new about the i-Limb. Most parts could have been built twenty five or more years previously. The hand benefited from better, smaller motors and microelectronics. While it did employ a new form of plastic for the shells, some of that was replaced by aluminium in later versions. There was no single breakthrough to create the hand. The breakthrough was the new motor geometry which was entirely novel and that made it a practical proposition. It was Gow and Touch Bionics that actually made a viable commercial hand with multiple fingers in the face of an industry that, at the time, maintained it was impossible.

Touch Bionics emerged as a company before there was a scramble for advanced hands. Indeed the firm may have created the desire in some quarters. Touch Bionics was a company before the US led wars in Iraq and Afghanistan provided the larger market for their devices. All these things came together to create this market for the i-Limb. While it was possible to make the hand a decade earlier (with the idea from David Gow and the money from the backers), it is not clear that it would have grown as quickly without all of these things (clinical environment, technology, finance, opportunity) finally coming into place.

The design of the EMAS Arms and subsequent i-Limb hand systems show their historic links to a clinical service in the PMR and the fact that their designer was a mechanical engineer. None of the systems built in the PMR workshops possessed internal feedback to the controller of the position of the digits, so the electronics cannot detect where the digits are and coordinate their positions. The fingers flex from fully extended to the tips meeting, only because the speed of the digits is programmed to achieve this from fully flexed. Once the hand has moved away from fully flexed, it is not certain whether the digits will get out of line, making grasping an object less easy. The previous generation of prosthetic hands from Otto Bock, Steeper, Motion Control or VASI did not use position feedback either, instead the gearing of the fingers and thumb guaranteed tips contact. For a complex hand with multiple joints, this is a shortcoming, yet the ability to detect digit position has never been incorporated. It is true that to add a sensor to reveal the amount of flexion of the fingers in an i-Limb hand would require at least one sensor per finger (two for the two axis thumb). This is likely to need an extra wire back to the controller per digit, and three wires running to every finger base. The additional parts mean more complexity and this means greater expense and often much poorer reliability. Once again it is the engineering trade off of complexity verses good enough. The Touch Bionic hands are already expensive devices and do have limitations in their reliability simply based on their number of components. However, the presence of a sensor would help the control of the fingers and their coordination making rapid and assured use more easily achieved.

Without internal feedback to ensure that the fingers and thumb tips meet in a tips grip, this has meant that the i-Limb hand must be fully extended before it is closed. The current sent to the motors and the gear ratios are set to time the movement of the hand to achieve this grip. If something impedes the fingers, or they do not quite get to the fully extended position, one or more fingers will miss the thumb or the object. This is an extremely annoying problem when the user is trying to efficiently and quickly perform a task. In contrast, the bebionic hand has internal feedback on each finger to ensure they are well controlled, and the Michelangelo fingers and thumb are mechanically linked so that they will always meet.

Gow had looked previously for ways of taking his ideas to market in order to make the products as widely available as possible. Other devices produced at the PMR had been bought cheaply by companies and not exploited. Gow and I had investigated the possibility that the English lower limb manufacturer, Chas A Blatchfords and Son (CAB) would make an arm which was the combination of the technologies from Edinburgh and Oxford. Established in 1890 [54] to manufacture all kinds of prosthetic parts, Blatchfords had chosen to make only prosthetic legs sometime in the 1940s. By the 1980s they made the most advanced prosthetics. They had introduced carbon fibre to prosthetics in 1981 [277]. CAB's patent for the use of carbon fibre in prosthetic legs came out a week before the helicopter company Westland's patent for helicopter rotor blades. They were the first commercial firm to use a microprocessor in a prosthetic limb. At one time in the UK, Blatchfords had a significant share of the market. Not through any particularly hostile commercial action on their part, but by simply supplying good devices that everyone wanted. Thus Blatchfords was the obvious choice to exploit this technology. However, moving into a new market (one admittedly much smaller than prosthetic legs) did not ultimately appeal to the company.

Following their involvement in the clinical trial of the children's hands in Nottingham, Steeper were given the opportunity to take on the technology, but this was not in that company's vision at the time. The only options left to Gow were to create a company to exploit the idea himself, or let it wither and die, like many a prosthetic hand project before it. So David sought external funding to take the idea forward.

To secure the investment meant forming a company and in-licensing the technology from the NHS. Real physical property is sold or leased. Intellectual property can only be assigned and licensed. Licenses, like leases, grant property rights and obligations for specific periods. In June 2002 Touch EMAS, was incorporated. External government grant funding was sought with support from the NHS commercial development group. The NHS legal office was the biggest impediment. The licensing of the technology to the company was, according to Gow, painful, bureaucratic, and slow. In their defence, it was a novel transaction for the NHS and the first company to be spun out of the Scottish NHS. All of the earlier work had been done under the aegis of the NHS and so they had some ownership of the intellectual property. SHIL became a significant shareholder because of its investment of patents, technology, and other intellectual property in the company.

Touch EMAS, which was a registered charity, became the first SHIL spin-out to receive significant funding. Funding came from a variety of sources. An initial Scottish Enterprise research and development grant of sixty thousand pounds was matched by an award from the NHS. This funding was used to develop the first version of the i-Limb. Further funding came from a leading business angel funding organisation, called 'Archangels' (£25,000 initially), and the Scottish Co-investment Fund (Scottish Enterprise based), who invested directly in Touch EMAS. Funding was used to develop shoulders, elbows, wrists and hands of the modular system, and to fund the two-year secondment (two days per week) of David Gow into Touch EMAS to serve as its first CEO. In this role, Gow oversaw the work by an external engineering design firm and subcontractors that took his clinical systems and developed prototype products for clinical trials. He also oversaw the development of a business plan, calling for recruitment of a CEO with commercial experience, allowing Gow to devote more time to research and development, user trials, international collaboration and the practical application of his ideas in the field [257].

8.2.2.1 Stuart Mead

In July 2005 Stuart Mead was recruited as CEO. Mead directed the early development work in consultation with Bill Dykes, a senior lecturer in prosthetics with many years of experience and a lecturer at the National Centre for Prosthetics and Orthotics at the University of Strathclyde, Glasgow (one of the bigger prosthetic centres in the United Kingdom and a leading institute for the teaching of prosthetics and orthotics). Touch EMAS used some of the national centre's expertise to get the product to a point where it could introduce it to some of its partners in the United States.

In contrast to the majority of the people who have contributed to the field, Stuart Mead does not have a degree. He was expecting to read Biomedical Electronics at Salford University but cuts imposed by Margaret Thatcher's government in the early 1980s meant the course was discontinued. Mead therefore needed a job *"to get enough money to buy a car"* [278]. As the company he joined *"looked after"* him, he stayed. Later he became a manager and CEO of different companies and while running one company in Scotland, the Archangels asked him to take on Touch EMAS. He took the company from a start up, with almost no full-time employees, to a leading producer of advanced prosthetic hands.

At the start, Mead said there was no company, just an idea, a patent, and a few prototypes. It grew to be what others called: *"The Apple of prosthetics"* although he was not sure about that claim himself. *"It was great to be in something that was so new and cutting edge, and so well received"* [278]. When they launched the product someone sent Stuart a copy of the Guatemalan Times, on the cover was a Touch Bionics hand. *"Quite a reach"* [278].

When he lost his chance to go to University, Mead was an early casualty of the 1980s recession in the UK. The Thatcherite rhetoric of the time was to reflect the attitudes of the 1930s and seek work. In the parlance of the day, he had 'got on his bike and looked for work'. As the 1980s drew on, the Tories had emphasised self reliance and the need to start one's own business to move up. They saw the need for serial entrepreneurs moving from small company to small company, as Mead has done. Accidentally, or deliberately, the cuts of the 1980s helped create exactly that sort of businessman in Mead.

To Stuart Mead, Touch Bionics was *"set up as a philanthropic exercise, originally"*. The investors in Archangels saw they could achieve some good with their investment, it was only later they appreciated it had significant sales potential. Mead felt that the simplicity and uniqueness of the patent meant that it got really good traction in the USA for people to buy into the program. From this promise the company grew quickly. Mead's first move was to create engaging images of the concept to take to potential backers and purchasers. He realised that people were aware of earlier work from Southampton and Edinburgh, but they did not believe that Touch EMAS was going to be a commercial operation [278].

His introduction to the market was the 2004 ISPO World Congress in Hong Kong. He was able to tour the trade show, meet the companies and see the state of the market. He remembered being surprised at how little effort had gone into sales, marketing and branding [278]. He would like to think that one contribution that Touch Bionics made to the field was to build a brand and image that people were proud of it rather than a device that was something to be embarrassed about. So the i-Limb hand moved away from *"horrible medical grade pink plastic"* [278]. He recalled a member of his sales team, who used a standard myoelectric hand. During meetings the user would hide his prosthesis underneath the table. Once he had an i-Limb fitted he used to leave it above the table.

The biggest challenge Mead said was making the hand robust enough. They took an early hand to the Walter Reed facility in the USA (one of the US's military centres), and fitted a veteran with the hand. The first thing he did was take the hand upstairs to the firing range and let off a Colt 45 pistol and an AK15 rifle. The hand *"just fell to pieces"*. They realised it was not robust enough for the market. This led to the introduction of Zytel, a new plastic from Du Pont which were looking for a high-profile application [278].

The early days were very much spent getting things to work. So if on the road something did not work, it was down to Mead to fix it. Prior to a demonstration of an early version of the hand: *"I spent...*

FIGURE 8.10

The launch of the i-Limb hand at the ISPO World Congress in 2007. It created considerable interest in the delegates. The display included many different nonanthropomorphically coloured gloves. (Permission: David Gow.)

the night before with a Dremel in my hotel room because it [the hand] broke when I took out of the bag. I remember being worried because the sparks from the Dremel were hitting the curtains".

October 2006 saw the European launch of the i-Limb, followed by the global launch in July 2007 (Fig. 8.10). The i-Limb could now be bought for about £8500 (approximately USD$11,100). By the start of the next year, annual sales were more than $2 million. They had an increasing network of clinics, and users in twenty-five countries. Sales in the USA represented 70% (driven by sales to soldiers wounded in the war in Iraq), with 5% in the UK and 25% spread across the rest of the world. Touch Bionics was also closing in on its two hundredth sale. It had the confidence of its existing investors, Archangel and TriCap business angel networks and the Scottish Co-investment fund, and projections of £5 million (approximately USD$8 million) in sales in calendar 2008. Gow had always taken the story of the Swedish ES Hand as his ideal (11.2.1). The ES is one of the only other high tech hands that reached the market. Gow's early ambitions were to achieve the same small sales numbers of this device. Once Touch Bionics exceeded this he was in unknown territory.

Knowing that a successful commercial launch would require introduction into the US market, Mead developed partnerships with several US based clinical companies. During this period the company also partnered with Livingskin from Middletown, New York and ARTech Laboratories in Midlothian, Texas, to take over the task to develop durable, lifelike cosmetic solutions. A market study group was formed from dozen users, a mix of civilian, military, and exmilitary wearers from the UK and the US.

In May of 2008, Touch Bionics bought Livingskin. The acquisition gave Touch Bionics intellectual property of silicone glove production and a greater presence in the US market. The gloves were necessary for the warranty of the hand. By April of 2009 sales exceeded 500 hands, and they had expanded into the Chinese market through a distributor. In 2009, Gow left the company after two years as director of technology. He considered that, by then, it was time for him to retire and pass on the baton [266].

Gordon McLearly replaced David Gow. He possessed years of experience in product development in large multinational firms. Although he joined the company after production had started, he saw the products needed to be redesigned following input from customers. A flaw in the design was it used Otto Bock components, leaving them dependant on a rival's cooperation. These were subsequently replaced by Touch Bionics' own systems, this allowed the company to expand its product offerings to other upper limb companies.

McLearly felt that the growth from 2007 to 2009 was almost too fast and exceeded their ability to plan a response [279]. In late 2007 he was roughly the twentieth employees. Two years later the staff numbered two-hundred. Initially all of the staff at the Livingston site (approximately 30 minutes from centre of Edinburgh) could fit into a single building. By 2010 there were separate buildings, for administration, research and development, and production [279].

As theirs was the first articulated hand in the market place Touch Bionics rediscovered the problem of introducing new ideas to the profession: The marketplace and users did not know how to use it, how to train for it, and how to get the best out of it. So the Touch Bionics team spent a number of years learning how to go about these tasks. They employed occupational therapists to train the wearers and the profession. It was a challenge to convince the market to fit their hands. The trend towards fitting based on empirical evidence has become the norm. This is a challenge for all manufacturers of multifunction hands as it is hard to obtain unequivocal evidence that they are more useful to a broad range of users than the older single axis hands [280].

Mead was proud of the way the custom glove for the hand (i-Limb skin) worked so well (Fig. 8.11). This is the nonanthropomorphic glove that fits on the hand. David's long years of working on silicones for the glove had demonstrated just how challenging it was and a solution still had to be found. They needed something they could produce quickly to protect the mechanism. Mead admits the inspiration for the glove was an iPhone case. The glove is generally unpigmented silicone, semi opaque white. This challenged the perception that if it the hand did not look natural people would not buy it. People liked the fact that it made the hand look 'techie'. It proved to be very popular. This would have been possible a few years before, but it is unlikely anybody would have tolerated it. Only with the current market did it look right. The drawback is that it might make it look dated as fashions change.

When asked about the introduction of the competition such as the bebionic and Michelangelo, Mead was positive. He thought that having a second hand was not a bad idea, as it adds validation to the idea. *"In many ways you want a competitor"* [278]. With four hands in the field, it becomes difficult for any other manufacturer to find a true gap in the market for another new product. Touch Bionics produced its own hands and promoted its products only. It was a one product company. The options facing it were to broaden their product range through acquisition or become someone else's expanded range.

David's strategy had always been to create the company in the expectation that one of the larger players would buy it out to add to their product lines. Mead's faith in the product helped to postpone this event for many years, but finally in 2016 Össur, bought Touch Bionics. Össur was the creation of Össur Kristinsson, an Icelander. It expanded by acquiring a range of products including: A microprocessor controlled knee from Hugh Hurr and a powered knee system from the Canadian company Victhom Human Bionics. Their earlier research also included investigating the ToMPAW arm around 2000. At the time the ideas and technology were not at all advanced. Waiting the decade or more meant they bought a developed product with distribution and users in place. As Össur lacked an upper limb product

(a) Tips Grip

(b) Adaptive grip

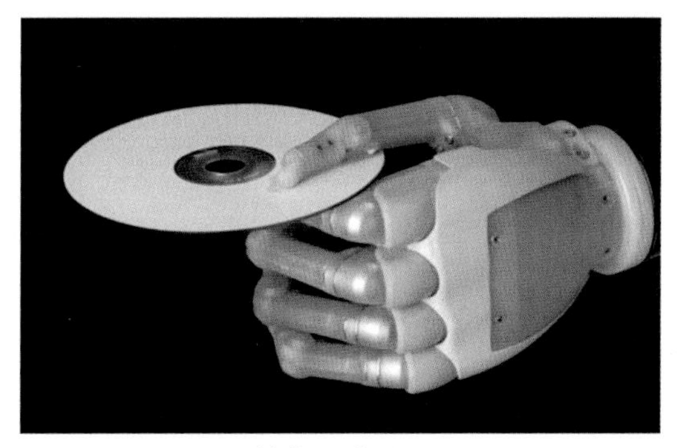

(c) Lateral grip

FIGURE 8.11

The Touch Bionics i-Limb hand showing its ability to create multiple grips. Hands in the upper images are using the noncosmetic i-Limb skin cover. (Used with permission of Össur.)

line of their own it was the ideal buyer to allow the i-Limb to continue to be developed. Initially Touch Bionics remained independent of the main company. Over the next four years the Touch Bionics brand slowly disappeared from the hand and its range. At the same time the product has continued to be developed.

8.3 The Scottish impact on the world

In his book *How the Scots Invented the Modern World*, US historian Arthur Herman charts the disproportionate influence that the sons and daughters of a small community of inhabitants of the northern part of the British Isles have had on the philosophy and technology of the world [259]. While his argument more reflects their influence on the United States and the US influence on the rest of the world, it is indeed true that Scottish thinkers and engineers independently have made an impact on the world, prosthetics being one such area. It is very rare that any one person really has a fundamental impact on a field, especially a commercial field such as upper limb prosthetics. Even the most important workers in this field tend to make change through their influence on others, rather than changing a single fact forever. Dudley Childress and Robert Scott have had a significant impact on the field, but this is mostly due to the many people they mentored over their long careers.

Many researchers (including the myself) had for years pushed the idea of multifunction hands and have even produced devices that have some limited practical use. Always the response from the clinical and commercial authorities that this was *"impossible"*. It was too complex to be built, or too hard to be controlled. No one would want one for long. While the eventual appearance of multifunction hands might have been possible without Gow, he made it more likely. Perhaps pressure from the research and engineering fields would have eventually allowed a multifunction hand to arise from the laboratories or workshops in the decade or two *after* the millennium. Both Steeper and Bock conducted research projects in multifunction hands while turning down the idea of complex hands offered from outside. They never showed any interest in marketing their own multifunction prototypes until after the Touch EMAS hands had begun to make a difference. There would likely have been very little pressure from the clinicians, as some have continued to say they could not understand the popularity of these hands with their wearers even as they fitted them. Harold Sears of Motion Control, who has been in part responsible for many innovations over his career, admitted he could not see the appeal of the multifunction hands. Hans Deitl appreciated that perceptions have changed since they started the Michaelangelo hand. Both Deitl and Ed Iversen of Motion Control admit they were wrong in not seeing that multifunction hands would be popular and they resisted their evolution for decades [252,281]. The truth is multifunction hands *are* popular. Both, the Vincent and Taska hand designs, are direct responses to shortcomings of the i-Limb design.

The introduction of the i-Limb changed the expectations of the users and the media, and made it hard for the companies not to respond. Before Touch Bionics, the market had moved forward slowly and cautiously for decades. Even if a company wanted to try something new (like Motion Control did) it had to be incremental. Campbell Aird's display of the EMAS arm at the 2001 World Congress produced a lot of press interest, but far less interest from the professional delegates. Six years later at the commercial launch of the i-Limb at the 2007 World Congress in Vancouver, there was a scrum of potential users, professionals, prosthetists and rival companies. Now the number of multiaxis hands from companies large and small is in double figures. The arrival of Touch Bionics was an explosion in the marketplace.

The word 'bionic' which has been loved by the media since the TV series *The Six Million Dollar Man*, and hated by many of us in the profession, including David Gow (8.2). The word was used loosely about any innovation in prosthetics, including the introduction to simple myoelectric hands. It oversold every incremental advance and raised expectations to a level no one could fulfil. Persons with recent amputations reported expecting more from their devices and being bitterly disappointed. This word has now has become unavoidable. After the creation of Touch Bionics it became almost impossible *not* to use the term somewhere. Steeper's hand was called 'bebionic' and some part of the Otto Bock line became branded as 'OrthoBionic'. The first powered knee was produced by 'Victhom *Bionics*'. Össur (now owner of Touch Bionics) sported the word in its literature. None of them would have dared do so before the i-Limb. Designer of the bebionic hand Mark Hunter is even clearer: *"You can level criticism of the TB hand, but he broke the mould"* [282].

The industry was not ready for multifunction hands and the people who had influence admit that they did not see the value in such hands. Only when they were in existence and in enough numbers, did they realise their mistake. With the arrival of 3D printing and the growth of citizen engineers, the balance has shifted. More people from outside the conservative world of the prosthetics companies are able to try multifunction hands for themselves.

It is therefore likely that, without David Gow, the market would only have responded to the users' desires for a more sophisticated hand in the medium term, if they ever did.

CHAPTER

9 Otto Bock

Otto Bock Limited is the largest producer/supplier of prosthetic and orthotic products. It is currently a privately held company with its headquarters in Duderstadt, Germany. Its research and development centre is close to the central station in Vienna, Austria. It invests about seven percent of its sales on research and development. Roughly one third of the company's employees work in upper limb prosthetics. Despite many changes in the market over its more than hundred years in business Otto Bock remains the dominant designer and manufacturer of electronic hands. Their single axis myoelectric hand is still the standard for externally powered anthropomorphic hand prostheses and the company has had a strong influence on the rest of the prosthetics market for many years.

9.1 History

The company was founded in Berlin 1919 by the prosthetist, Otto Bock [283], with the goal of supplying veterans of the First World War with prosthetic and orthopaedic products (Fig. 9.1). At the same time Hugh Steeper was responding to a similar demand in the UK (10.1). In the twenties Berlin was an unstable place to operate so Bock bought out the company and moved to his home town of Königsee in Thüringen. To address the greatly increased demand and the limited output that could be produced using traditional artisan methods, Bock applied modern methods of standardised manufacturing. This seems a straightforward idea today, but it was a revolution in the prosthetics manufacturing process at the time it was introduced. It transformed prosthetics as it had other industries. From then on, a person could order a replacement part, and fit it successfully needing far less technical skill. The production and assembly was also quicker, more efficient and cheaper. Bock brought this revolution to his company and to prosthetics, which at the time still relied on small batches and one person manufacturing all the parts. The large number of wounded after the First World War meant these more efficient methods were required all across Europe. Bock manufactured prosthetic components in series production and delivered these directly to the limb fitters for crafting the final prosthesis for the user, as is done today. Otto saw this as the cornerstone of the emerging orthopaedic industry [284]. He also experimented with the use of new materials in its products [284]. The company conducted tests on aluminium in the 1930s and incorporated the material into its products to offer lighter weight components. Likewise, plastic materials were incorporated in its components in the 1950s.

The second generation of company ownership and leadership rested with Dr. Max Näder, the son-in-law of the founder. Born in 1915, Näder was a doctor of engineering (Mechaniker Meister). He joined the company in 1935 training as an orthopaedic technician and sales representative. In 1943 Näder received permission from Otto to marry his daughter, Maria Bock. Näder soon became the

Making Hands. https://doi.org/10.1016/B978-0-12-820544-0.00018-4

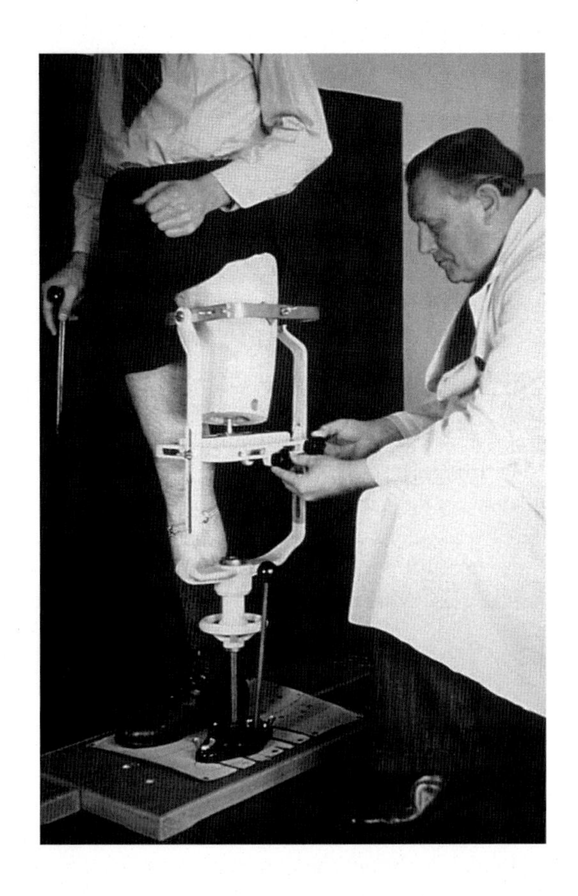

FIGURE 9.1

Otto Bock, founder of the prosthetics company. (Copyright: Ottobock SE & Co. KGaA, Duderstadt.)

driving force in the company and when they found themselves in the Soviet zone of postwar Germany, the family fled over the boarder and set up in Duderstadt, starting from scratch. When Näder's father-in-law died in 1953 at the age of 64, Max assumed official control of the firm. He was then 38 years old, and would direct the company until he was 75. In 1990, his son, the current president, Professor Hans Georg Näder, took over the reins.

During Näder senior's tenure the company grew through the adoption of new business methods and materials, the opening of new fabrication facilities, creation of branch operations, internal innovation, and the acquisition of technology development companies. The company's expansion included the central fabrication operations in the US and Canada from 1962 to 2002. They acquired other companies and facilities to expand their ability to make their products. Its major myoelectric acquisition was Variety Abilities Systems Incorporated (VASI) of Toronto in 2005 [285], which brought both enhanced research and development capabilities and new prosthetic and orthotic products for children. The VASI hand was the only competitor to Bock's own paediatric hand (Electrohand 2000). Dr. Hans Dietl, Chief

Technology Officer of Otto Bock and managing director in Vienna, said that the VASI hand was a good design and he wanted to make sure it remained an option for users [208].

Major technology initiatives included development of the MyoBock prosthetic arm system in the 1960s. The Greifer (a myoelectric gripper) was launched in 1979. In 2006 they announced a new myoelectric elbow, which could lift over five kilograms and swing naturally during walking. They acquired the bebionic hand from Steepers in 2016 (see 10.1). Otto Bock also produced the Michelangelo multifunction hand, launched in 2010.

Through these changes the company made the transition from start-up to the world's leading manufacturer of upper and low limb prosthetic devices. Contrary to the trend among many large multinational firms, Otto Bock did not divest itself of internal research and development resources. It instead retained a large upper limb group in Vienna, and continued its long-standing approach of adopting technologies and innovations developed outside the firm. It retains some of its small company feel, and a distinct corporate personality. This tactic, which includes open-innovation strategies, has served Otto Bock well, allowing it to avoid incurring the heavy expenses of fundamental research and pioneering of new concepts, while keeping it open to new ideas. It is able to exploit them with its development group. By 2020 twenty percent of the company was owned by investors and the aim was to prepare the company for public ownership within five years. This has led to a shift in the emphasis of the company, such that their popular exhibition at Science Centre museum for rehabilitation science and technology in Berlin was closed.

9.1.1 The evolution of the Z6 hand

The Otto Bock single axis hand has become the leading electric prosthesis in the market. The company was not the first to create an electric hand, and they worked with other groups to create a device that has been in the market for decades and was a sufficiently effective design that the engineers have been able to upgrade it numerous times while sticking to the basic form. Deploying and using a myoelectric hand required new ideas and approaches. To support the new devices meant changing the business model. Otto Bock moved from being just a producer of individual components, to a supplier of complete prosthetic systems, and thus a guarantor of the function and quality of an entire range of devices. To address this they created an international after-sales service. Otto Bock also had to adjust to increasingly complex product design projects, and an equally demanding internal and external (regulatory driven) testing of projects and clinical trials.

An insight into the way that Max Näder and the company saw the industry was gained when he was invited to give a keynote speech at the 1995 MyoElectric Controls (MEC) Symposium held at UNB (13.2.3). He divided the fifty-year history, from 1945 to 1995, into experimental and commercial stages, with myoelectric prostheses emerging clinically in the 1960s. Since then, Näder said, the fundamental ideas had not changed [286]. However, within a decade of his talk some things had changed considerably, both the dynamics of the market, and also the level of technology in the front line products. The presentation was a collaboration with his senior engineer, Hans Dietl, and reflected their current beliefs and interests [281].

Näder used the latest ideas to innovate design, while making sure the designer was aware of everything that had gone before, so as not to waste time in reinventing the wheel or creating products no one wanted [286]. According to Näder, what was particularly important in the development of myoelectric prostheses were the pneumatic powered prostheses developed in Germany immediately following

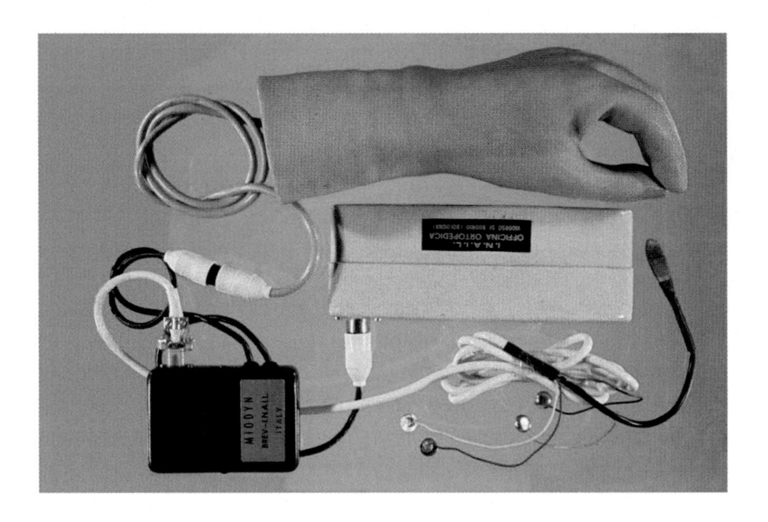

FIGURE 9.2

The S1 Hand developed between Otto Bock and the Italian prosthetics centre INAIL. (Copyright: Ottobock SE & Co. KGaA, Duderstadt.)

World War I and the proof-of-concept myoelectric systems following World War II. The pneumatic systems allowed researchers to study the difficulties of power supply and storage, as well as the problem of controlling an arm with a low number of control signals.

While earlier powered hands were steps on the way to a clinically useful powered prostheses, for Näder the critical development in the 1940s was Reinhold Reiter's use of myoelectric signals to control artificial arms (6.1.3). While he acknowledged the role of the Russian Hand and the teams in the UK and Canada that evolved the concepts, he also felt these resulted in *"exaggerated media reports"* and raised public expectations. This was not the last time that the media's response to the technology was instrumental in pushing the agenda forward (8.2 and 10.1.1).

This pressure was important for Otto Bock, because of a research and development project initiated by Mr. Hannes Schmidl of the Federal Vocational School for Orthopaedic Technology in Frankfurt, and a team from the Italian Workers' Compensation Authority (Istituto Nazionale Infortuni sul Lavoro or INAIL) in Budrio (near Bologna), Italy in 1964. Schmidl was a larger than life character, who went on to run the INAIL centre. His energy dominated any room he was in, while his assistants worked quietly in the background to ensure his requests were fulfilled. Näder said this project resulted in *"the first really efficient myoelectrically controlled prosthesis"* [286]. But it was not without flaws. Otto Bock had originally created hands by taking existing wooden prostheses (Hüfner hands), and hollowing them to take electronics. The device was too fragile to take the loads of an active hand, so in 1965 the company began development of a completely new prosthetic hand mechanisms specifically for myoelectric applications. The new hand consisted of the mechanism, an inner hand of soft plastic and the cosmetic glove, but not the electronic controls circuits, which were made by INAIL. This hand developed for INAIL was called the *Type S Hand* (Fig. 9.2). It used a DC motor connected to a self-locking spindle drive mechanism.

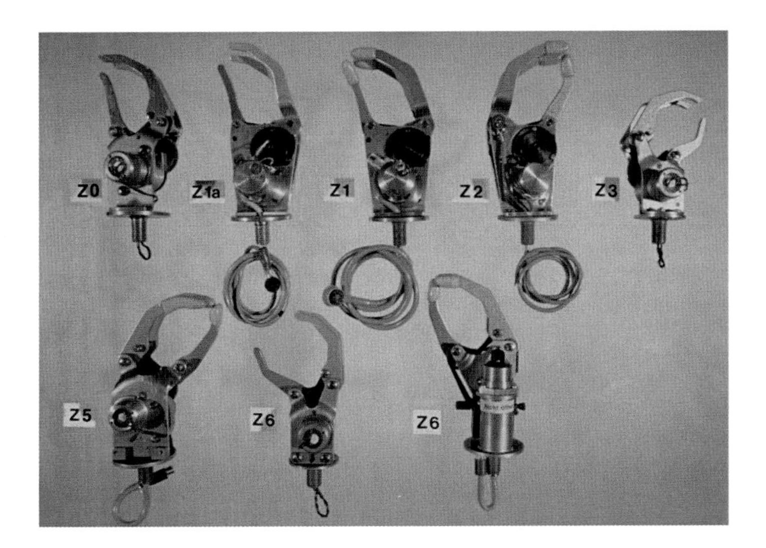

FIGURE 9.3

The evolution of the Otto Bock Z6 Hand. The Z2 is very similar to the Viennatone product. The Z6 is the basis of the most popular of the Otto Bock myoelectric hands. (Copyright: Ottobock SE & Co. KGaA, Duderstadt.)

Initial tests on the controls and user acceptance were conducted with two subjects one with a unilateral loss and the other with both hands missing. This trial was sufficiently encouraging for INAIL to produce fifty hands for testing to identify the weaknesses in the design. After a year, when they were convinced that it was a useful device and would give users an advantage over other prostheses, they began mass production [287].

While the initial fittings were for below elbow wearers it was always Schmidl's intention that it should be part of an arm with other powered joints, so INAIL also investigated multichannel EMG controllers. For the next three years they worked on the controller and by 1970 they were confident enough to go for application of the multichannel controller and prostheses with up to eight movements [287]. By 1973, INAIL were using a different design of electric hand, mass produced to specifications from Otto Bock (8E8 Model Z6 hand) [287].

At the same time, another Austrian, a Dr. Zemann, was working on a myoelectric control system as a result of publicity surrounding the Soviet myoelectric hand [286]. Zemann contacted Otto Bock for a suitable electric hand. Näder said that Otto Bock developed a new hand for Zemann, this one called the Otto Bock Systemhand Z1. According to Hans Dietl, the modern Kesheng Hands resemble the Z1 in many ways [208]. As with the prosthetic hand for INAIL, it used the newest nonferrous magnets. According to Näder, this hand was subsequently built by the Viennatone Company, an Austrian manufacturer of hearing aids, who also continued developing myoelectric control for Dr. Zemann. Otto Bock continued with its development of the Type Z hands. The Viennatone Hand was launched around 1967 [288] and was the first readily available commercial system [289], with the Otto Bock myoelectric hand system appearing soon after (Fig. 9.3). The Viennatone hand also possessed a one way nonbackdrive mechanism. This meant that the hand could be closed by the motor, or an external force, but would not open again without the motor driving it, allowing the hand to close on an object and stay closed even with the motor turned off. A Viennatone hand modified to take multiple sensors was the basis for the first clinically fitted microprocessor hand [269] created in Southampton in 1989 (10.2.1).

The first hands had significant disadvantages. They ran off a twelve-volt battery which had to be worn on the user's waist or in a chest pouch [290]. The usual problems of wire breakage and electrical interference were common with this solution. While the hand was made of aluminium, making it light in weight, the fingers and thumb had a relatively flat design formed through stamping the profiles from sheet metal. This meant the digits were easily bent or broken. The hand was also slow, and the grip force limited. Finally, there was no way to manually open the hand in case of an emergency. As a result in 1967, Otto Bock began work on a new design. The result was the most widespread electric prosthetic hand, the Z6 [286]. The redesign effort was worth it. The basic appearance and design principles of the hand mechanisms (as opposed to the electronic control systems) have remained essentially unchanged to the present day. Small modifications to the structure, and the Quick Disconnect Wrist (QDW) have been added, but even the shorter, transcarpel hand shows its lineage back to the Z6. The mechanism has been able to take repeated upgrades through newer, faster more powerful motors. During the same time, the electronics for the hands have been changed numerous times to keep pace with innovation. Sensors have been added, but the Z6 is still clearly visible in the latest single degree of freedom hands.

Another feature of the Z6 derived hands is the two-speed gearbox. This allowed the hand to open and close quite quickly. Once the fingers were resisted by the presence of an object the mechanism was able to change to a higher gear ratio creating up to 120 N pinch force. This ratio change was seen by Näder and Dietl as the same as Childress's 'Synergetic Prehension': The ability to control grasping speed and grip force separately [291]. It is the same as the 'HOLD' mode in a Southampton Hand [6] and the 'FLAG' feature that appeared in the Motion Control hand (see 7.2.2). Otto Bock achieved this by the mechanical action of the gearbox, as opposed to Childress's solution of two separate drives for fingers and thumb with different ratios.

The threshold for the point where the gearbox changes ratio is controllable to some extent by how much the motor is screwed into the gearhead. This changes the pressure on the mechanism. It is another highly precise component manufactured by Otto Bock which if made poorly would fail often. Hence the quality of manufacture at Otto Bock must be held to high standards.

As with other myoelectric control systems and electronic hands, development of improvements was an ongoing activity. In 1973, a six-volt battery was introduced, which meant the battery could be built into the forearm of the prosthesis. At the same time Otto Bock was able to produce a smaller hand size (6 3/4). In 1975, an electric wrist rotator was introduced.

9.1.2 Industry standards set by Otto Bock

Over the decades there has been a tendency for the market to move towards default standards, often shaped by what the largest manufacturer is producing. These moves are generally for pragmatic reasons. Myoelectric electrode/amplifiers from different manufacturers have tended to use the same footprint and connectors to allow them to fit in an existing socket and plug into a different manufacturer's controller. Similarly, as part of a competitive market, Steeper was already making plastic shells and high definition silicone gloves for Otto Bock hands, so when they decided to replace their existing adult hand and make new powered hands themselves, the result bore a strong similarity to the Bock product. Indeed it has to in order to fit the internals of the glove (see 10.1). The Steeper hand has one or

FIGURE 9.4

The Quick Disconnect Wrist (QDW) on the end of a single axis hand showing how it fits into the distal end of the forearm. Twisting the hand a little over one rotation will cause the wrist to release the hand allowing for simple replacement of the hand with another terminal device. Compatible versions of this device is being manufactured by many of the major manufacturers. (Copyright: Ottobock SE & Co. KGaA, Duderstadt.)

two differences, including a finger release for when the hand is unpowered (this feature is in common with their previous model hand, although an entirely different mechanism). Unlike the Z6 it does not possess a two-speed gearbox.

Another default standard is the Quick Disconnect Wrist. This was released by Otto Bock in the early 1980s (Fig. 9.4). The wrist allows a full 360 degree rotation of the hand along with a coaxial plug running down the centre of the wrist. This removes any problems of wires twisting and breaking. The design is elegant, but complex. The hand part contains twelve teeth around the circumference of a ring that must engage with twelve teeth on the distal plate in the arm. This pushes a ring containing ball bearings out to lock into a groove in the wrist. The load is thus spread all the way around a ring four centimetres in diameter. It is a larger area than pegs or pins could achieve, reducing the forces at any one point and increasing the robustness of the device.

The resistance to rotation can be adjusted through use of washers. Rotation is also impeded by a series of bumps within the structure. This means that when it is rotated, it will give a series of audible and tactile clicks. This is not welcome to all users and can be replaced by a smooth ring. For many years the only way to use a QDW on a different manufacturer's product was to extract a wrist from an Otto Bock component and modify the target hand to take the wrist. This was very successful in allowing interchange between hands. The mechanism itself is a high precision device and it took many years before the other manufacturers began making their own versions of the wrist, it was never patented. Motion Control, a long-term user of Otto Bock hands, was the first to take this step. They also extended the number of conductors in the wrist from four (two power and two EMG) to six (adding two electronic bus communication lines). This allowed microprocessors to be used and reprogrammed in the hand and arm, without the need to disassemble the arm. Steeper and Touch Bionics later developed their own QDWs. One problem is that the design requires high precision manufacture for it to work reliably. Occasionally an Otto Bock produced wrist may not work, especially when worn. The failure to release in rival versions seemed to be more common in the early days of production. There have been times when the wrist of one manufacturer did not work with the proximal ring of another, although this became less of a problem as the manufacturers learned how to create this device [20]. More recently Hy5 has created a variant that is simpler to produce and easier to adjust (see 14.1.5).

To release a hand from the QDW, the user must grasp the terminal device in their other hand, and rotate it a little more than one revolution. It will then come off. Placing a new device on the end is a matter of pushing it straight into the ring, guided by the central coaxial plug. Once powered wrists were added it became possible for the QDW to inadvertently release the hand. This would happen if the hand is holding on to a fixed object, to drive the QDW round, rotating it until the hand is released. It is not clear how common a problem this is. One response to this could be to restrict the rotation of the powered wrist unit. The need to have a wrist that can rotate more than 270 degrees is debatable as the natural wrist rotates about 180 degrees. Instead Otto Bock has moved towards a much larger wrist unit, which is oval in cross section and more anthropomorphic and releases via a button on the side instead of rotation.

Pressure to create a paediatric solution came from the parents of children in countries across Europe (10.1), and in 1985 Otto Bock developed their first children's hand prosthesis. The design offered new compact electrodes. The electrodes subsequently became their standard for adult prostheses too. Their first device was replaced by an innovative design for their second children's hand, the Electrohand 2000 (launched in the late 1980s), with reduced weight and optimised gripping and handling capabilities.

9.1.3 Hans Dietl

The electric hand was one of the projects overseen by Hans Dietl, who initiated and ran projects for Otto Bock for many years. Dietl came into the industry by accident. He was working on his PhD in biofluid dynamics, which included studying the respiratory system and also creating objective measures of smell. Circumstances meant he did not pursue an academic career, instead taking a position in Vienna. When Hans started with Otto Bock his boss *"really challenged him"*, getting him to work on different projects [281]. Once at the company he was in some way prepared for his work because the 'objective smelling' work of his PhD needed to deploy a range of biological measures. His equipment combined dosing systems of the chemical to be smelt with EEG signals and analysis for objective detection of the smell, *"it was quite a strange machine"* he commented. During his career in Otto Bock he worked on many of their innovations, particularly the electric elbow, the child's hand system, and he also headed their advanced prosthetic leg project (the C-leg).

The Electrohand 2000 was the outcome of research and development to create a prosthesis that would fit children from ages three to six. This had followed work in centres such as the clinic in Örebro, Sweden and the Ontario Crippled Children's Centre that showed that improved acceptance and functional ability by users was the result of early adoption of the hands [292]. The new hand was quite different to the format generally adopted for powered hands. Conventionally, the drives for the fingers and thumb lie on a single axis that aligned at right angles to the long axis of the arm (across the hand). Instead, a compact motor and gearbox were placed at roughly forty five degrees to this axis, giving the hand a more anthropomorphic grip line. It was designed this way to reduce the compensations required at the elbow. Using the diagonal of the hand also gave the designers greater length than simply across the hand. They also created a configuration rarely seen, a pair of motor and gearboxes in a single driveshaft. This increased the power output without making the motors wider in cross-section, saving space. The supply voltage needed was also reduced to 4.8 V, which allowed batteries to be smaller and so could be embedded in the compact forearm of an infant prosthesis. Subsequent advancements

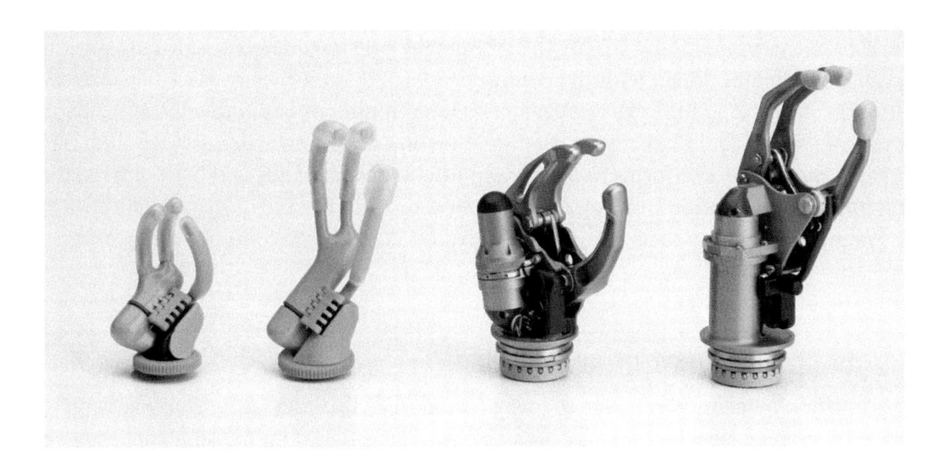

FIGURE 9.5

Range of single axis hands offered by Otto Bock. Left to right: The two smaller hands are Electrohand 2000, for paediatric fittings and are very different to the transcarpel hand or the Z6 derived adult hand. (Copyright: Ottobock SE & Co. KGaA, Duderstadt.)

in materials, engineering and manufacturing techniques resulted in further reduction in the size of the Electrohand 2000, making it available for children as young as eighteen months. Fig. 9.5 shows the range of sizes of electric single axis hands that Otto Bock offers.

Otto Bock changed its manufacturing process to incorporate a unique grade of aluminium for the finger and thumb segments. Production involved heating, moulding, and then machining and milling the aluminium components to the hand's design. The electronics were changed to use a microprocessor, resulting in reduced power consumption and wear of the gears. Battery charging was also modified and the addition of a LED to indicate when charging was completed, this allowed better control of the cycling of the battery and so extended the battery life.

At the time of his MEC presentation, Max Näder did not believe that much emerged from university research that was applicable to the industry. Since this time, Bock has worked closely with a number of universities, including Johns Hopkins University, Northwestern University and Vienna Medical University. Moreover, a number of features of the product line come from university research. Most recently they have deliberately fostered academic groups who are doing work that is seen as important to Bock's vision of the future of prosthetics. For example, the establishment of the laboratories of Dario Farina in Göttingen and London was supported by Otto Bock. It was Dietl's idea to establish the Max Nader Lab at the Rehabilitation Institute of Chicago. It performs research fittings with objective measurement, to inform prosthetic prescription. Similarly, Oscar Aszmann's lab in the Medical University of Vienna, Austria, which is undertakes research on surgical interventions in upper extremities enjoyed support from the company. Once he was partially retired Hans continued to collaborate with this group [281].

It is interesting to note that, particularly during the earlier period where Otto Bock did not appear to invest as much in external contacts. It appeared from the outside that the innovation in Otto Bock products was also less apparent. They initially turned down the offer of microprocessor controlled knees and later had to catch up with their rivals Chas Blatchfords . The first microprocessor based myoelectric hand controllers were programmed simply to combine the previous range of self-contained electronics controllers into a single package, with the controller selected via a coding plug. This saved on inventory and the cost of making different electronics sets, but it did not begin to exploit the capabilities of microprocessors, which could have been used to allow the prosthetists to customise the device to the user more fully. Similarly, new forms of shorter anthropomorphic hands and built in wrist flexion units were first introduced by Motion Control, later followed by Otto Bock. Ultimately Otto Bock did begin to include features in their products that really do exploit the potential of a microprocessor controller.

9.2 Twenty first century innovations

Otto Bock's position in the market is due to many factors. The fact they provide a 'one-stop' service for the full range of supplies needed by a practising clinic means that both the small and large clinics can see advantage in using one principal supplier. In addition, the quality of manufacture of the prosthetic components and the large distribution network mean that it is seen as less of a risk to fit an Otto Bock device than something from a smaller supplier. For many years there seemed to be only incremental innovation from Otto Bock. From 2010 onwards Otto Bock's products began to reach beyond this range of devices. Two devices that typify this change are the C-leg and the and Michelangelo Hand. The former is outside the scope of this book, except to say that while they were not the first to market with a microprocessor controlled leg that adapts to its environment, but a more recent generation leg from Bock, (The *Genium*) was the first commercial prosthesis to be waterproof.

9.2.1 The DynamicArm

Before Otto Bock launched the Michelangelo Hand, which it describes as the *"heart of its new prosthetic system"*, they launched the DynamicArm. From one perspective, the DynamicArm is an incremental progression of their ErgoArm product. From another, it showed their vision of where they would go with the use of advanced microprocessor technology.

The ErgoArm uses a 'force balancing' system, where a spring assists an unpowered elbow rise against gravity (see Fig. 2.6 for both the ErgoArm and its powered variant the DynamicArm). This means that the force required to raise the arm can be very small. Through movement of the shoulder and upper arm, the operator can 'throw' the arm upwards and 'catch' it at the top of its swing by applying a brake (through a switch in the harness). Thus, the motions used to operate the arm by a skilled user can be slight and fluid. It is also possible for the operator to dial in the amount of gravity compensation with a wheel on the underside of the arm, so that the spring can compensate for a heavier load, or a stiffer sleeve. One consideration that was apparent was that if it requires a small force to raise the arm, it follows that it will require a smaller motor to raise a powered version. This was an obvious modification to the arm, but the manner in which Bock carried this through showed the care and thought that typifies Otto Bock products.

Dietl admits that he had not expected that there would be a need for a powered elbow [281]. He thought that the inconvenience and weight would make it unpopular. However, the prosthetists who were influencers in the field liked the challenge and so they wanted to fit powered elbows. For a long time that meant they fitted someone else's elbow and an Otto Bock hand (see 7.2). With the introduction of the DynamicArm this no longer needed to be the case.

From the outset it was clear to Dietl that the arm had to be fast and capable of lifting loads without stalling. The compliance of a person with a high-level loss is low, if you *"have to wait all the time for your arm"* the compliance would be even worse [281]. A quick and powerful arm was required.

The result of powering an ErgoArm is the *DynamicArm*. It keeps the force balance system and has the motor drive the flexion in such a way it can still be employed like an ErgoArm too. A myoelectric input, or cable pull, can then be used to turn the brake on when they have thrown the arm up, or turn it off, allowing then arm to descend. The myoelectric control of a hand and wrist can then be added. A user can be as fluid in use as they would be with the ErgoArm. The user can also cycle to the elbow control and use the myoelectric signal to command the motor to drive the arm up or down.

The hand, wrist and elbow communicate with each other via an electronic bus (the *Axon Bus*). The DynamicArm was the second prosthesis to be used in the field that employed a bus system after the ToMPAW Arm (10.2.1). Using a bus means the supervisory electronic controller of the arm can be informed of the status of all the different systems, including the status of the batteries, and adapt the arm's response to circumstance. One use that the bus can be put to is to allow the arm to degrade performance gracefully. If the batteries fail, then a prosthesis stops moving. At this point, a conventionally designed arm could not be lowered, or the hand would not let go of an object. This might be inconvenient or at times dangerous. The design of the DynamicArm has allowed for the graceful degradation of the system as the batteries become exhausted. The main power drains of the prosthetic arm are removed from the control of the user in descending order. The power hungry elbow is disabled first, then the hand and finally the brake motor is turned off. Towards the end of the charge, the hand is allowed to open and so release an object, but not close and irretrievably grasp something. As the qualities of the original ErgoArm are maintained in the DynamicArm, it means that even when there is little power left in the battery (and no motor to assist the wearer), they can still use the elbow. Even when the battery is too depleted to operate the electric lock, the lock can be operated mechanically, so the arm will not remain in a flexed position and can be used as a prop or to steady objects. The arm (if not the hand) remains functional and useful to the operator even when the batteries are fully discharged. Given that the primary purpose of a prosthesis for someone with a single sided loss is support, it means a fully depleted DynamicArm remains useful to its wearer.

To achieve this requires *"a very, very complex gearing system"* [281], which has cost implications. Not only does the gearing cost money to build, but it adds mass and potentially noise to the arm. It can also reduce the reliability of the arm. Its reduction is a worthwhile thing to explore. Brushless motors can operate with far less gearing than standard DC motors, if they are run properly. As a result the engineers at Otto Bock have investigated ways to reduce the amount of gearing used in the arm.

One aspect of the construction of the DynamicArm is an example of the Otto Bock approach. The brushless motors they adopted for the arm are designed and built for them. Brushless motors are more efficient than conventional DC motors, but they need a computer to control them which has previously restricted their use. Otto Bock specified their own design, which is more expensive than just using an off the shelf motor. By controlling the design and manufacture of the motor, they can achieve the performance they desire. Similarly, the way Otto Bock maintains the quality of their other motors is by extensively testing each motor when it is received from the manufacturer. The motors used for all other products are from the high quality suppliers in Switzerland or Germany. However, Otto Bock runs every motor in an environmental chamber for eight hours in a range of climate conditions (temperature and humidity), through a standardised testing protocol. If they run without any glitches they are accepted. Otherwise they cannot be used in an Otto Bock product.

9.2.1.1 A communications bus for prosthetic limbs

The introduction of a bus to control a number of systems is an important step forward. It uses only a small number of wires (often two) to talk to any number of distant systems (such as hands and elbows). The most familiar form of bus is the internet, where many computers (nodes) can talk to each other over the same pair of wires. Without a common bus, the alternative would be pairs of wires running to individual computers (or prosthetic parts). By using just two wires, the complexity is cut down and the chances of failure reduced. With a bus system in place, it is possible to assemble an arm from components, much like Lego, and then the prosthetist can customise the arm's behaviour to match the user.

A bus is not simply a pair of wires, but a way of different computers 'talking' to each other. When a bus is specified it includes the 'language' spoken, the length of any message, and many other factors, especially electrical conditions, such as the voltages used. If the standards are followed, one computer can understand another. This is the underlying strength of the internet: It is a party anyone can join. Engineers do not need to know how someone else builds their computers or writes their programs. These standards also make it easy to spot mistakes or reduce the effect of electrical interference, and so improve the reliability of the system. They also mean that one node can talk directly to another without any of the other nodes needing to respond. The internet protocols are such standards. However, if the standard is kept secret it can also be used to cut out other people from using the bus. In this case prevent the prosthetist mixing and matching the parts to the user's needs. If someone employs a proprietary bus structure, such as Otto Bock's Axon Bus, it means that no one else can hook their device into a DynamicArm elbow, unless they know how to talk 'Axon bus' or have devices that will translate Axon into some other format they can understand. So if no one outside Otto Bock knows Axon bus, then no one but Otto Bock can link their hands to the elbow.

This is the opposite to the Internet. The Internet works because there is a simple language which allows every computer to talk to each other. Without these freely open standards modern systems would not work together to create the Web. Without the World Wide Web, it would be like Microsoft and Apple having totally different internet systems and these two would also differ from the internets of each the manufacturers of mobile phones. It would be a world where one could only use Nokia phones and computers with a Nokia mail system, and only correspond with people who also owned Nokias. The Axon bus is like a world without the Internet Protocol (IP). This is a world where it would be necessary to have different networks for every company, different wires to carry your 'MS network' from your 'AppleNet'. It would mean different suppliers of software and separated groups of users and venders. It should be obvious that if that had been the case in the 1990s, the wired world we know today could never have happened.

As part of the Otto Bock one-stop service they have not only created their own communications system that no one else can use, but they have combined their training with the configuration software and the ordering software into a single package. A prosthetist can now check out the potential user, try them with a myo tester/trainer, determine all the components the user will need *and* the resins and casting materials the prosthetist will require to make the device and order them all from their local Otto Bock supplier. It is simple and easy. It cuts out any concerns about the compatibility by ordering different parts from different manufacturers. It reduces paperwork, and streamlines work flow. It ensures all parts meet the standards of the industry. For a busy clinic it is a boon. However, might reduce the choice for the users.

For a company keeping one's own standards, this approach has advantages. It reduces chances of a problem by maintaining complete control of the system. For a prosthetist assembling a prosthesis for a wearer, if they use more than one manufacturer's devices and one part fails, they have a problem of knowing who is responsible, and who will fix it. Hybrid systems made from components of different manufacturers may have unexpected interactions between products. For example, a prosthesis built using a hand and a wrist from different manufacturers will use a single battery as a power supply. It is possible that the hand may cause the output voltage of the battery to drop below a level that the wrist can cope with. Yet a hand from a third manufacturer may work flawlessly with the wrist. No manufacturer can guarantee all combinations, and few will assure any combinations but their own. Whatever the motives, the result of this approach is competitors are kept out.

Prosthetists who do not want to take the responsibility for failures may well adopt the safer approach. This restricts the choice to the prosthetist and user. One company that took on the task and the risk was Liberating Technologies. They often found themselves making hybrid systems. LTI's Bill Hanson commented that *"I know why they [Otto Bock] do it, it makes sense from a business marketing perspective and an engineering perspective. It is easy to qualify the product when it's all one manufacturer"* [20]. Dietl is more definitive about the position of a manufacturer: *"No company is in control of the product life cycle development of other devices on the market. So how to do a proper system validation?"* Without full knowledge of its development he stresses it is impossible to be certain how the device will behave under every circumstance and this risks potentially dangerous failures, something Otto Bock will not tolerate [283]. The descendant of the Boston Elbow from College Park was built to be agnostic about which other components it uses. In the USA, LTI assembled and tested a particular combination if you choose to be supplied by them irrespective of the original source of the components [232].

9.2.2 Michelangelo Hand

Rumours of an Otto Bock multifunction hand predated the official launch of the Michelangelo Hand. That Otto Bock was considering such a device evoked a lot of interest in the profession. News and images of the Michelangelo Hand came out with updates of the progress of the DARPA projects from 2005 onwards. While the participants of the project got to see the hand in detail, it was only with the official launch of the hand at the ISPO World Congress in Leipzig, Germany, in 2010 that it was widely seen (Fig. 9.6).

The Michelangelo Hand is an advanced function hand, but it differs from Touch Bionics, Steeper, Vincent and other hands, in not having separate drives for each finger. The fingers are also without the two interphalangial joints that curl the hand when the fist is closed. The Michelangelo Hand instead has a single motor that drives the fingers together, making the closed hand look a little awkward. By contrast when the fingers are extended they are arranged in a highly anthropomorphic form. The proximal finger joints are not aligned for mechanical simplicity but follow a natural looking curve, and the thumb axes are displaced, so that the open hand resembles the natural hand, far more like a cosmetic prosthesis, much more hand-like than any other powered terminal device.

The fingers are a flexibly coupled single driven unit, so that they can partially adapt to the shape of the grasped object. This means that the Michelangelo Hand is not really in the same category as the other multifunction hands: It is more an evolution of the single motion hand, with added compliance and a single second grasp form, the lateral grip. These compromises that have been made are typical of

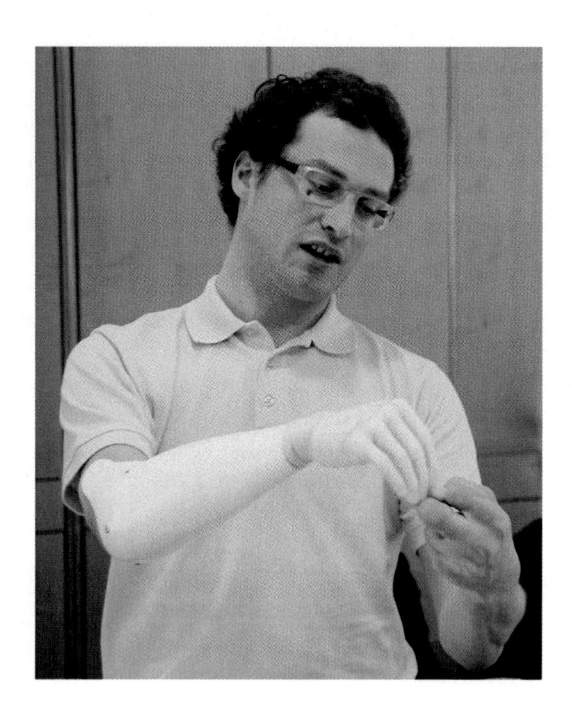

FIGURE 9.6

The Michelangelo Hand being used at a press conference in 2010. The user shows off the precision grip and the anthropomorphic lines of the hand. (Copyright: Ottobock SE & Co. KGaA, Duderstadt.)

Otto Bock. The design decisions allow them to make a high quality product which is robust enough to satisfy their exacting standards. Hans Dietl was surprised by the popularity of the other multifunction hands. When asked about curling fingers he said: *"the customer requirements changed a lot"* [208]. Despite the fact that the fingers on this hand are more easily broken they are still acceptable to a higher percentage of the user population than expected. Dietl admits he did not predict the trade-off between appearance and function that led to greater demand for the true multifunction hands [281]. For their design Otto Bock chose to use stiff rugged digits, rather than adopt less robust, curling fingers.

The anthropomorphic line at the base of the fingers means that as the fingers close, they also adduct together, the tips contacting. The power of the motor is sufficient that it can continue to close the hand despite this resistance. By doing this, the fingers become united in a single group. This group is stronger than any one finger, which makes for a more robust hand and grip. This geometry also means the fingers are also pressed together when the lateral grip is chosen. This makes for a much more rigid platform to oppose the thumb, thus a stronger and more secure lateral grip than any of the true multifunction hands. All others simply oppose the thumb tip with the side of the index finger. This design requires the single finger to be able to resist large side loads, a demanding requirement. The Michelangelo Hand's combined fingers are far stronger.

The motor in the Michelangelo Hand is a flat disk in the palm of the hand. It is similar, but not identical, to the motor in the DynamicArm. The movement of the fingers and thumb comes from the main motor. If the motor rotates in one direction, it draws the fingers and thumb together in a tips opposition grip. If it rotates in the other direction, it forms the lateral grasp. To switch to a lateral grasp, the hand opens and then the drive of the thumb goes round more than 180 degrees, so that the motor still draws all the digits in together. However, it pulls the thumb from a slightly different angle. Combined with the small angle shift from a secondary motor, this means the thumb now approaches the fingers from the side, forming the lateral grip.

The fingers and thumb tips will *always* arrive at the same point. The advantage of this is that the operator knows where the tips will be and can be confident of where to place the hand or the object, ahead of the fingers getting there. It results in a hand that can be easier and faster to use. Other multifunction hands have independent fingers. As they close they might be delayed by obstructions and so miss each other. When a user of a Michelangelo performs a grip, the operator can be certain where to place the hand to perform the action easily. This is an advantage that cannot be overemphasised.

The Michelangelo Hand is supplied with an entirely new wrist format, which cannot mate with the existing QDW. It was Otto Bock's long term aim to supply a powered wrist to mate with the hand, so getting round the potential problem with the QDW of 'spinning off' the hand. The new wrist does not have this problem. A second difference of the wrist is that it is not circular. A circular wrist means the join between the hand and the forearm is always the same size, and it looks slightly unnatural all the time. An oval wrist is only lined up in one (or two) orientations. This means there are orientations where there may be a step between the end of the socket and the start of the hand. Combined with the more natural thumb orientation, when in a neutral position, the entire hand looks far more anthropomorphic, enhancing the passive cosmetic appearance.

Prior to the launch of the Michelangelo Hand the unofficial response from representatives of the company was always that this form of sophisticated hand was unnecessary, unwanted or indeed impossible. Following the controlled market launch of the new hand they stated that *"Detailed monitoring, especially at the start, is important even for a technically mature system in order to ensure first-class fittings and to create a transparent process for everyone involved". "...We offer the technology which the market demands"* [293].

Otto Bock's research efforts have rarely been made public until a fully finished product is ready to be launched. This is distinct from the Motion Control approach that tends to use field testing as a way to promote their product and get feedback from their users. This policy for Otto Bock changed to accommodate the opportunity presented by the projects funded under the USA's Defence Research Agency. As a publicly funded project in the United States, it is likely that the need for disclosure of a more advanced hand meant that Otto Bock had to reveal the Michelangelo Hand's existence some time before it would normally do so.

9.2.3 Other developments

One of the other results of the DARPA collaboration was control of the elbow and hand using targeted muscle reinnervation (TMR) [294,295]. The first Otto Bock TMR upper limb prototype was demonstrated in Vienna in 2007. It allowed the user to move seven joints in real time. The version designed for everyday use, offering control of three joints, was introduced in 2010. This has continued to be developed and refined.

The next step on Otto Bock's research and development roadmap was to develop commercial products that can use sensors on the nerves of the user to not only take motor control signals (e.g. signals from brain to limb via implanted chest muscles), but also signals which transmit sensory perceptions in the opposite direction. According to Dr. Hubert Egger, head of Otto Bock's *"Mind-controlled arm project"*, he described Otto Bock's development activities as:

> Microscopic sensors which record the temperature, the grip strength and the surface properties of an object being touched are integrated into the fingertip of the prosthetic hand. The measured data is 'translated' into suitable signals using a microchip and transmitted to actuators on the skin. These actuators generate lifelike stimuli for the sensory nerve endings, leading to restored perception in the brain: Prosthesis wearers can feel things as they did with the index finger of their natural hand prior to amputation.

In the past it would have been unusual for a representative of Otto Bock to make such a broad and long term prediction about the future of prosthetics. In a 2010 news release Hans Dietl, at the time CTO of Otto Bock HealthCare Products, said Otto Bock *"expects to bring the sensory arm prosthesis to market in approximately four years"* [296].

Writing in 1995 Näder said: *"The history of myoelectrics points to the designers eagerly awaiting the invention of the transistor. Only the transistor allowed the realisation of their ideas"* [286]. The multichannel system with proportional control and pulse width modulation (both innovations from University research) began with Schmidl, according to Näder, in response to the user needs and to the unique capabilities of the INAIL. It was followed by advancements in proportional control from Childress and others in 1970, and then application of pattern recognition, and subsequently implementation of a multiple-sensor concept by Nightingale and Todd [297].

For Näder, these advancements stand in contrast to the many research and development projects during the period that: *"did nothing but fill up valuable laboratory space"* and *"most never got beyond the stage of a laboratory sample"* [286]. The one exceptional university project he cited was the introduction of integrated circuits and development of the Utah Arm at the University of Utah. He called it a *"revolutionary component concept"*, and said it was way ahead of its time and *"...the only product where the complexities of a university research project were successfully transferred into everyday prosthetics"* [286]. However, by the time that Näder was presenting this, Chas A Blatchfords already had their second generation microprocessor controlled knee in the market. This concept had emerged from research at a Japanese university in the early 1990s.

At the end of a thirty-three year career with Otto Bock, Dietl is able to reflect on the decisions made. He recognises that he misjudged the market's reaction to the multifunction hands. *"This was one of my mistakes"*. He had taken a pragmatic attitude to the provision of hands. He always focused on the functional benefits to the users. All the time the company did (unpublished) evaluations of the advanced hands, he could not detect any functional advantage to the user. He considered that perhaps the outstretched index finger for keyboarding or use with a smart phone might have a little benefit. In general he expected the advantage to be low. *"I was fully convinced that the multiarticulated hands don't bring any rehabilitation benefit"*. Other research has supported his assertions that the more complex hands do not have greater functional capabilities [280]. Hence the Michelangelo hand had the rigid fingers and limited motions. He admits he: *"underestimate[d] the emotional effect of the multiarticulated hands"*. Moreover there was no cosmetic cover that worked with the multijointed hands and he still cannot see any innovations changing this. However, he did not expect there would be the broad acceptance of hand types where you can see the mechanism. When it happened, it became clear that Otto Bock needed to either develop their own hand or acquire one. Hans saw that the market was not

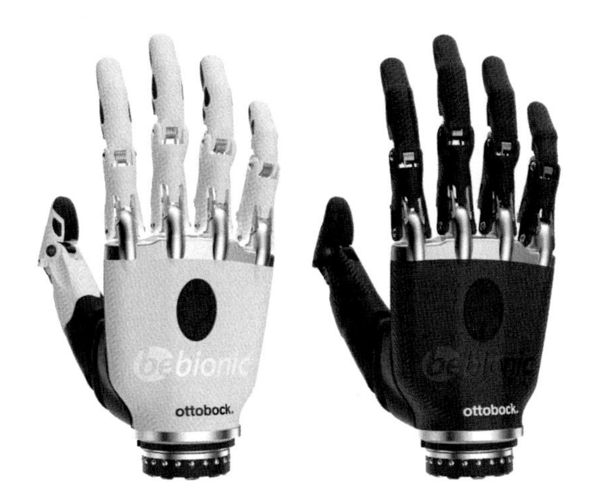

FIGURE 9.7

The bebionic Hand as produced by Otto Bock, was the second modern fully multifunction hand to the market. Originally developed by Steeper it was acquired by Otto Bock in 2016. (Copyright: Ottobock SE & Co. KGaA, Duderstadt.)

big enough to support a large number of models so he decided it would be better to buy an existing one rather than take time developing one in-house. It was for this reason that Otto Bock acquired the bebionic hand [281] (Fig. 9.7).

Over the past twenty years Otto Bock has worked with a number of universities, and ideas originated in university research have become embodied in their prosthetic limbs. However, Dietl cannot see many of the possible innovations making it to the market. The development, testing and regulatory restrictions will make many of the current ideas (such as implanted electrodes) uneconomic to turn into a product.

"Upper extremity is a really cool research field..." "It was a playground for researchers, and in the future it will be a playground for researchers" [281]. But the move from laboratory to clinic requires effort. For every new feature you add to a clinical device you need extensive testing, validation and documentation to satisfy legislation. *"A huge huge effort"* according to Dietl [281]. To introduce the single feature of water resistance, Harold Sears said it cost least $10,000 per test for Motion Control to get this approved. If the device had failed it would have to be retested at a similar cost. All of this needs to be recovered in the sale of new devices. Adding implanted electrodes and other high-tech features to limbs will result in many tests. A manufacturer needs to be sure of a large market before they invest in this. A complete clinical study would cost many millions of euros, there is little chance of getting back their investment [283].

While he can see the possibility of implants making significant improvements in rehabilitation, Hans cannot see a way to make it commercially viable. In the past Dietl worked on a 'drop foot stimulator'. This is a device that allows people with strokes who cannot pick up their toes to stimulate the muscles to do this every stride. Otto Bock got as far as making the approved production facilities. He said that from the tests he had hardly ever seen better clinical results. This experience includes subject compliance, something that often limits testing. Otto Bock had 500 patients with implants who did not want to give up their stimulators. It was the lack of a way to get reimbursement for the devices that killed the project. Moreover, this is a population of people with strokes. It is much more difficult to create viable solutions for this patient group. They are people that have a great deal more to gain from the intervention than someone a missing a hand. They are a far larger group than the prosthetics user population. Taken all into account, they look like a much more profitable proposition. The conclusion is that if a device such as this could make a major contribution to the quality of life of a large number of people, yet they cannot find a way to pay for it. How can prosthetics benefit from technology? Dietl has a lot of experience with implants *"but my enthusiasm that this is [the] next big thing for the companies, is almost completely gone, because of [the question]: Who pays for it?"* [281]. It is likely that quality of life improvements alone will not be able justify the expense of development of implants and it will be only in applications where lives are saved/extended will anyone see any commercial application in the foreseeable future [283].

In contrast, Hans believes that the current market of a number of start ups is not unusual in the history of prosthetics. There are many occasions when there were small companies all vying for a share of the market. As before, some will fail to make money and fade from existence. Others will produce useful products and get absorbed by the larger players. The smallness of the upper limb market means that he thinks that there is only room in the long-term for two major suppliers and he firmly believes that one of them will be Otto Bock.

CHAPTER

Prosthetics in the United Kingdom

10

Founded on global trade, supported by significant levels of slavery and exploitation, the Industrial Revolution in the UK changed many aspects of production from the one off, craft based, to the mass produced with all items being the same. These innovations were slow to find application in prosthetics. In response to overwhelming increases in limb loss during the First World War, Otto Bock in Germany (see Chapter 9) and Hugh Steeper in the UK, scaled up production of limbs for injured soldiers.

10.1 Steeper

During the First World War two Belgian émigrés, the Vindevogal brothers, were living in a flat above Hugh Steeper's family owned water heater business in central London [298]. The brothers were creative and when a friend suffered an amputation of his hand, they carved a new hand for him [298]. The Vindevogals set up a clinic in the lounge of Steeper's flat and began fitting hands (Fig. 10.1). This was the era when many small companies were started by individual limb makers this led to a cottage industry in making hands from wood and then metal. Meanwhile Hugh Steeper was finding this much more interesting than the family business. When the Vindevogals wanted to return home to Belgium in 1919 and the company was liquidated, Hugh bought them out and created the company that bore his name in 1921. A century later the company was still led by the same family.

Steepers originally made both arms and legs. However, following the Second World War the UK Ministry of Pensions played a role in managing the industry, regularly reviewing component prices, profit, design, and issues within the industry as well as monitoring repair and fitting rates calling the firms to account if needed. This oversight also extended to guiding the industry and removing what they saw as ineffective players by mergers. Thus Hugh Steeper Limited acquired Masters and Mackay and were persuaded to specialise in upper limbs whilst Blatchford, Vessa, Hanger and Vokes specialised in lower limbs. Privately all the companies made both [55].

When Hugh Steeper retired, the company continued to be run by the family. In the early 1960s company facilities were centralised in a new location in Leeds [55]. Over the years the Steeper company changed and diversified and in 2006 the Dunedin private equity group acquired the company and started looking for new opportunities. Steeper had looked at adaptive hand designs over the years, but none had proved successful, so when Mark Hunter approached the company with the idea for a multifunction hand he found a receptive audience:

> "Mark approached us when we were looking for new opportunities to expand our prosthetics business, private equity owners are typically looking for areas to invest in to take the company forward.

Making Hands. https://doi.org/10.1016/B978-0-12-820544-0.00019-6

FIGURE 10.1

The Vindevogal brothers and Hugh Steeper performing a fitting (Copyright: Steeper.)

Mark gave us a call, I had a meeting with him, and the rest is as they say, history. It was right time right place" [298].

This led to the bebionic Hand.

In December 2015 the city of Leeds where Steepers is based suffered considerable flooding. At the time, Dunedin were aiming to sell the company and this development jeopardised the negotiations. It was clearly going to take time to rebuild the company, however the investors were still keen to sell, so a compromise was sought. At the time they were in negotiations with Otto Bock, but as recovery from the flood complicated any purchase, the full sale was restructured so that Otto Bock acquired the bebionic Hand. A subsequent offer was made to Dunedin by the CEO, Paul Steeper and the COO, John Midgley for the remaining business of prosthetic and orthotic patient care products and services supplied to UK and international markets. The result was the business came back under family control. This transaction also allowed the company to move into a brand new state-of-the-art head office and manufacturing facility, re-establishing the identity of the company [298].

On the loss of the bebionic Paul Steeper is philosophical: *"Whilst sad to see it go, it created an opportunity that would not be there otherwise"* [298].

10.1.1 Robin Cooper

One of the forces in British upper limb prosthetics in the latter half of the twentieth century was Robin Cooper, who worked for Steeper for most of his career. He was born in 1943 in Warwickshire UK. His father served in the Royal Air Force as a technician. Cooper senior found himself at the cutting edge when he installed the first UHF sets in seventeen and nineteen squadrons for use during the Battle of

Britain. Later in the war he went on to radar systems and at the end of the war he was part of a team led by R.V. Jones which went to Germany, to assess the level that German technology had reached. Following the war Cooper senior worked for the engineering firm Pye, where he met Dennis Collins, which led to his son Robin's connection to Steeper.

The Cooper family had moved to Thriplow near the Duxford airbase by the end of the war. It was a village with a population of around two hundred. Cooper recalls it had a *"Little village school, all ages, crazy teacher"* [182]. After this he went to a Secondary Modern school in Cambridge. Robin took A levels at the Cambridge Technical College (now Anglia Ruskin university). Cooper likes to recount that it was at the college that he made friends with a group of students who had formed a band. They played at his sister's eighteenth birthday party and charged twenty-five pounds for the gig. The band offered Robin a job roadying for them, which he declined. When members of the band moved to London they formed *Pink Floyd*.

Following college Robin worked for the radio astronomy group at Cambridge University. This included working on an early satellite, EDSAC 1, his job included data entry from the satellite to a computer. Later offers of jobs from the aerospace industry (De Havilland and Martin Baker) did not appeal. At De Havilland the job was to be in the Stress Office working out the stresses in airframes. This was a bad time for the industry. The pioneering jet airliner, the De Havilland Comet, had crashed due to metal fatigue. The role of stressing airframes was now known to be vital. Cooper decided this career was not for him, so he took a job with Cambridge Instruments. The work was varied; he installed gas analysis equipment in the Hyde Park underpass in central London and he worked on the first portable cardiograph. This was not an area that he felt confident in as it required a level of knowledge of X-ray physics. Having finished his A levels he did not want to commit to additional evening study [55].

Through his wide reading, Robin became interested in many of the projects that the company Steeper were considering, including the design of adaptive hands. The designer was Dennis (John) Collins who worked for Pye designing handling equipment for nuclear power stations. He was using his knowledge to look at prosthetic hand design in his own time. One evening in a Cambridge pub Robin Cooper was introduced to Collins, by his father. Over a pint or two they discussed the developments in prosthetics. Robin had been reading about the Belgrade Hand that had a glove with contact sensors built in. It was an opportunity to work in an area that interested him, and earn money in the evenings developing his skills, and so Robin joined John for six months. Initially he machined parts for the small multifunction hands.

The UK government's Ministry of Pensions supported the work, but it was not fully part of Steeper's portfolio at the time. Collins produced the hand in his garage and Cooper made parts for it out of Perspex. The parts were then sent up to London where a group of *"guys in a Nissen hut"* assembled them [182]. The hand mechanism was made to use three different power sources: Gas, electric motor and body power. This was an interesting idea and had the potential to make it a very good commercial proposition; three products, one set of parts. In later years Jan Cool from Delft (see 5.4) and David Gow (see 8.2) both proposed making a modular hand that could use the same parts but different power sources. No one has yet turned the idea into a commercial proposition. The result of the work was a hand that was compact and worked well for demonstrations. A film from the UK Ministry of Information was made and shown widely. As a result they got *"loads of begging letters from around the world"*. The first twenty built were distributed within the UK, Sweden and Canada [182].

In early 1966 Robin got a full time position at Hugh Steeper. Although hired as an engineer he was asked train as a prosthetist. There was no qualifying course of study for prosthetists at the time and so he was signed up for a *Surgical Instrument Makers* course. This was three years of studying a part time qualification. The program included human dissection and anatomy and surgical observation. At the end he was qualified as a 'Limb Fitter' which started his *"ever upward progress towards retirement"* [182].

During his training as a limb fitter he travelled to Germany and Sweden to support small scale trials of the hand with the Thalidomide groups at the University of Munster, University of Heidelberg (Prof Ernst Marquardt) and the Swedish Handicap Institute (Henry Lymark). Later as a prosthetist he worked in different centres in the UK including time in Scotland where he worked with David Simpson and David Gow.

In early clinical trials of the Collins Hand, they were all returned broken within two to three weeks. *"You give something to people that is more capable and what do they do? They go and break it"* [182], a lesson repeatedly relearned by modern prostheses builders. The gloves were made out of latex, a rubber that breaks down quickly, so they rapidly yellowed and began to smell. However, the work served its purpose; to get a different government department interested, the *Department of Health*.

The market in the UK changed following the McColl report in 1986 [299]. According to Cooper, Professor Ian McColl was a prominent surgeon, a Tory supporter, and had the ear of government [182]. He was appointed the head of a committee to look into prosthetic provision. The rest of the panel included two prominent businessmen who had limb absences. They both later went on to run major nationalised organisations: Bob Reid (Shell Oil) and Marmaduke Hussey (BBC). At the time prosthetics services were not fully part of the National Health Service (NHS), but they were contracted in. The McColl committee determined the service should be brought within the NHS, and a tendering arrangement was created. The results were stark. The UK company J E Hanger of Roehampton went from having eighty percent of the market to eighteen percent overnight. One of the results of this change was that Steeper received many more contracts. This made Steeper the largest supplier of upper limbs in the United Kingdom [182].

Eventually Steeper was acquired by private equity. The investors streamlined the business to provide more focus and deliver growth on prosthetics and orthotics patient care services. The manufacturing, research and development were reduced. Robin cut his staff from thirty to ten overnight including his own job. He worked at the Crystal Palace prosthetics centre in south London, until in 1997, Paul Steeper (grandson of Hugh) invited him back to manage upper limb clinical sales. His work focused on upper limb prosthetics including launching the bebionic Hand. He retired at seventy.

10.1.1.1 The origins of Steeper myoelectric hands

Over the years the company had investigated different aspects of prosthetic design and control which included an anthropomorphic robot and a prosthetic arm with powered pro/supination at the wrist and elbow flexion [55]. Their most significant contributions to the field are in the development of two powered prosthetic hands with origins outside the company: The Systemteknik Hand and the bebionic Hand. In the late 1970s the Systemteknik Hand was being fitted in Sweden. Enormous pressure in the UK was created after Rolf Sörbye appeared on the BBC's weekly technology program: *Tomorrow's World*. The show had a high profile and would have been watched by millions. A team from the UK's Department of Health and Social Security (DHSS) and Steeper, (including Cooper) was rapidly put together and sent to Sweden. The visit was partly a fact-finding mission and partly made to offer cooperation to the Swedes. On their return a program was put together to fit the Swedish hands in the UK. Initially based at Roehampton and Manchester it was later extended to Edinburgh. Hands were purchased and protocols for user selection were created, and manufacturing and training schedules were devised.

In reaction to the publicity from the BBC program, parents with children affected with upper limb differences created the charity *Reach*. It aimed to put pressure on the government and companies to improve the service to their children. While the original organisers were all the parents, now many of the officers of the charity have come from the ranks of the children themselves as they have grown up. Reach has funded research, their aim was to exercise more in control of the direction of research compared with existing short-term funding from governments and other charities. Some early work on David Gow's ProDigit design was partly financed by Reach (see 8.2).

At the first official meeting of Reach, Cooper was the only representative from the industry. Years of frustration in the parents and the perceived lack of action from the industry made them a formidable opposition. For Cooper *"It was very much a 'to the barricades' affair"* [182]. As he had already fitted a couple of Systemteknik hands he was able to show some movies of the hands being used. The parents could see that at least something was happening, Steeper and the DHSS were delivering. Cooper was able to say that there would be a program to fit hands, regulated so that the children with the appropriate type of loss, level, age and access to centres would be given priority [182]. From 1978 to 1981, seventy-six children were fitted within the scope of the trial (and others to levels not originally covered by the protocols). The hands were used for a long time, though many of the children rejected them when they became teenagers [182]. Steeper eventually made Systemteknik an offer for the design, and manufacturing was transferred to the Steeper factory in Leeds.

The team identified children who matched the criteria to be fitted and Steeper made the hands. They began with one size, which was scaled up to second size as the children grew. The trial began after they had ten hands and three centres (Manchester, Edinburgh and Roehampton). Steepers organised training courses for the prosthetists and therapists. The OTs got together and worked out what they needed to do in a formalised way. The protocol was to fit as many as they could and then work out if, when, and why people dropped out.

Techniques of socket design had to be changed from the body powered devices to allow good contact between the skin and the electrodes. The load of the prosthesis was spread evenly (and so lightly) on the skin. The design of the socket made sure that the socket did not impede the motion of the natural elbow joint and give the children the greatest range of motion of their elbow. The wires to external batteries were the source of greatest number of failures. The solution found was to create special cable, which was mode from a spiral of wire wound around a central core. The power-carrying wires were wound outside that and it was encased in silicone. The cable could respond to changes in the distance between the ends of the wires without being damaged [182].

To investigate the age limitations, Steeper followed the Canadian experience, fitting all ages down to eighteen months, rather than the Swedish practice, with first fitting at age four to five years. This move also responded to the parental demand. They gained much good publicity with politicians and popular television programs. It was *"always a good news story"* [182]. By 1968 it was clear that the Systemteknik design would only scale up so far. Steeper needed to design their own hands. By 1980 they had five sizes of hand, a year later the trial finished. By 1984, the design of adult hands was being manufactured.

10.1.2 Mark Hunter and the bebionic Hand

Historians of ideas (especially science and technology) like to relate their analysis to the ideas that come from outside the range of existing ideas (the 'current paradigm'). The usual source for new ideas for prosthetic limbs has tended to be from people within the industry or users themselves. David Gow changed the market with ProDigits and the company Touch Bionics, but he was clearly from within the professional arena. Bob Radocy was a user first. However, Mark Hunter is very much outside this realm. He is in the category of 'concerned observer' trying to make a difference. There are many stories of a parent making something for their disabled child or a neighbour helping the child in a wheelchair. Engineering friends of persons with an impairment usually solve the problems of one individual and cannot (or do not) extend their interest beyond this personal horizon. Hunter was different in his approach. Even his background does not fit the usual pattern: Mark has a bachelors degree in 3D (interior) design from Newcastle Polytechnic, UK. He had two years making custom furniture, which taught him how to make things really accurately, before joining Jim Henson's Creature Shop.

Hunter's prosthetics story starts when a friend's daughter was born with a congenital absence of part of her hand and the parents asked Hunter if there was anything he could do for her. Working for the Creature Shop meant he had a great deal of experience in making things that looked natural and moved well, including hands, claws and flippers. He looked at the market, and like many, he was surprised to see nothing effective available. He felt he had the skills to correct this absence and that he was a compassionate person, so he started to work at the problem [282].

Jim Hensen (the creator of the Muppets) built the Creature Shop to make his puppets for the film industry. Its work is typified by the practicality of the task. It needs to make creatures that will look a certain way and perform in a certain manner, on a certain date. The creatures were puppets controlled by a puppeteer, which explains the organic performances they have [282]. The machinists and technicians created devices that looked right on the screen and worked well for the cameras. They specialised in making things that work, but are perhaps not to continue to work for a long time. This may have mitigated against Hunter in his early dealings with a sceptical prosthetics industry, who may have not seen a possible connection.

Hunter felt this job in the Creature Shop was an *"Engineers' dream"*. None of the engineers were doing formal engineering, they were making it up as they went along. The biggest lesson Hunter had to learn as he moved into prosthetics, was that the film business needs something to work just long enough to get the movie made. In contrast, a prosthesis has to work in the field, unserviced, for many months. As he was a free lancer he had time between jobs to work on his own ideas. Some of his routine jobs included making hands for the puppets, even those assignments added experience. As part of the learning process, Mark also identified those who were working in the field and visited them, first in the UK and then beyond. In the late 1990s he approached to the only two engineers doing prosthetics research in the UK. He spoke with me and did some manufacturing for David Gow, helping to make parts for the arm Cambell Aird wowed the press with in Glasgow in 2001 (see 8.2.2).

Eventually Mark decided the project needed all his efforts, so in 2005 he sold up, left the film business and moved to Los Angeles. There he learned to surf, met his wife in January 2006, and on their honeymoon they toured prosthetic arm centres, including Chicago and Utah. In October 2006 he attended a meeting of the DARPA arm projects. He considered pursuing the DARPA money, but *"it didn't feel right"* [282].

It was Steeper who expressed a commercial interest in the hand design. Cooper said that Steeper had been working on their own multifunction hand for many years. When Hunter presented his work they

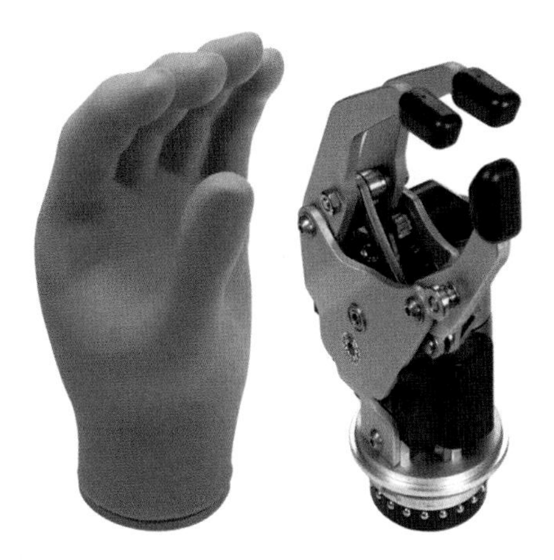

FIGURE 10.2

The Steeper single axis hand launched in 2007. Made to match the stock gloves they have means it is very similar to the basic Otto Bock Z6 mechanism (Copyright: Steeper.)

were receptive. They moved rapidly to a working prototype. A video from July 2009 shows a device with an electronics package taped to the back of the hand, while the user demonstrates its capabilities using the SHAP test (see 10.2.1.1). The device looked a great deal like the eventual product, yet this was less than a year since the start of the collaboration.

Regarding the launch of the bebionic Hand, Cooper said *"It was, I think, a surprise all round to people"*. This is not an understatement. To the outside, Steeper had not seemed to innovate for many years. Their devices lacked cutting edge technology at that time. In 2007 when Touch Bionics was launching its hand at the Vancouver ISPO World Congress, Steeper showed off its new single axis hand. While it was silver, not gold, and had a single slow single speed, not a powerful fast two ratio gearbox, it was unmistakably similar to an Otto Bock Z6 type hands (Fig. 10.2). The explanation is that Steeper sold spare parts for the market, including ones to match the Otto Bock range. In the same way as many companies have thrived selling parts to replace the alternator or battery in a Ford car. Steeper already made a wide range of gloves for this hand shape. If they were going to make a new hand it made sense to use the parts already in existence, and as new moulds for new rubber parts involves a considerable investment in money, new mechanism must be this shape too. However, this takes the justification too far. To fit the glove the new hand might have superficially been physically like the Z6, much as the Motion Control single axis hand is. The new Steeper hand looked only a little shy of a straight copy. Thus a microprocessor controlled multifunction hand, as functional as and easier to use than the Touch Bionics hand, but made by Steeper, was the last thing on anyone's radar.

Cooper maintains that the bebionic Hand was not a direct response to the threat of the Touch Bionics hand. Steeper had a long history of working on multifunction hands starting with the Collins hand [182,298]. Hunter has a different perspective: *"Although they were competitors I will be forever grateful to David Gow for changing the industry"* [282]. He clearly thinks the existence of Touch Bionics made his work to persuade Steeper easier.

Hunter came up with the bebionic name plus its logo. He wanted something simple and it took a while to arrive at this. The name is based on the hypothetical question and answer:

> "If I was an amputee what would I want (to be)?
> I would want to be bionic" [282].

The resulting hand has separate drives for each finger. Each finger has linked proximal phalanxes and a fixed distal one. Unlike the Touch Bionics hand, the drives are in the palm so that it cannot be used as separate individual fingers. The compliance is on the joint closest to the palm (not further up as in the rival) so a knock on the back of the hand will cause the hand to close and will not break the fingers, as is more possible with the i-Limb. This also means that the user can easily fold the hand up and rest on their knuckles on the table to rise from a chair. This feature could easily add to the hand's endurance. This is something that Paul Steeper saw as very important in the design of the Hand:

> "Key in compliant hands... is: reliability, durability, functionality and simplicity of use. Clearly existing compliant hands are more fragile than myoelectric hands. To some degree they are a victim of their own success. If they are more functional people will use them for a lot more things, not necessarily within the design specification, and therefore you get a lot more breakages. With the best will in the world where you are building more complex products with more joints in the fingers you are building in weak points into the design" [298].

While the thumb rotation is not powered, the digit clicks into a specific place to oppose the tips of the fingers. When the user wants to reach for an object they know *exactly* where the fingers and thumb will end up. This is something the original i-Limb hands could not do, resulting in user frustration and poor performance. Additionally, when the bebionic thumb is rotated to oppose the side of the fingers, a switch in the thumb triggers the hand to go into the lateral grip automatically. There is no need for any other actions to signal the change, as there would be in an equivalent i-Limb hand. There is an additional switch in the middle of the back of the bebionic Hand. This can be programmed to trigger a specific grip, making switching into one of the many possible grips faster and simpler than the Touch Bionics hand of the same time. Finally, the controller can detect the degree of flexion of each individual finger. Although they will stop closing when they touch an object, when they open all fingers return to a line together, so the digit tips always oppose the thumb tip, something only guaranteed with the i-Limb if the fingers are fully extended first.

The original model was a little fragile and the components broke under comparatively light use. Over the years the team at Steeper learned from what broke and strengthened it in the critical areas.

> "It took a long time to take it from bare bones to something effective prosthetically. About two and a half years to get it in the market and then second and third iterations followed fairly quickly over the next couple of years. To get to a product they were happy with" [298].

Hunter aimed for simplicity and he admits there was a great deal to learn as he built the hands, such as how to make it simple, and how to design for manufacturing at a lower cost. Some of the learning was the additional aspects of being an independent businessman: Finding partners you can work with, learning to negotiate terms for exploitation, licensing, etcetera. Also he learned how to patent an item: This is a skill Mark is immensely proud of, he can read a patent and judge its strength as well as draft his own so reduce the work a patent lawyer needs do to file it [282].

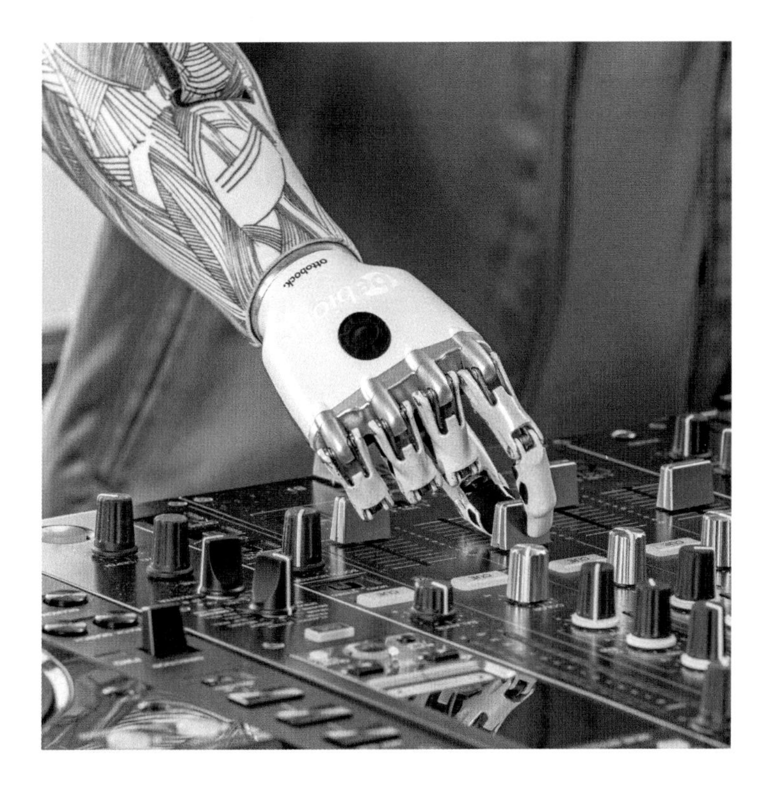

FIGURE 10.3

The bebionic Hand is able to achieve a range of grips as each finger is individually actuated. (Copyright: Ottobock SE & Co. KGaA, Duderstadt.)

In 2016 Steeper sold the bebionic Hand to Otto Bock so that two of the four multifunction hand designs then on the market were produced by the same company. The two were clearly aimed at different places in the market. Otto Bock also retained the name for that model (Fig. 10.3). Paul Steeper acknowledges that the complex hands are more expensive and less robust. Such things do not fit with the Steeper company emphasis on the practical:

> "There will always be a place for compliant hands but factors such as simplicity of function for the patient, durability and cost will determine whether there is real potential in the future for multi-articulating hands to fully replace myos. A myo is a very simple device, but it is very functional, very reliable, it's very durable and it does what it says on the tin" [298].

Steeper had been a close collaborator with LTI for many years selling each others products in their local markets. In 2020 they acquired the Espire powered elbow from College Park.

10.2 Adding a computer to prosthetics

We live in a world where computers are ubiquitous. It is perhaps hard to remember a time when the addition of a computer to a system was novel, even ground breaking. Many of these innovations occurred in the 1980s. Originally it was very hard to sell to prosthetics industry the idea of a computer controller for a prosthetic limb. This might have been because prosthetics originated from a craft base, with the engineers, directors and managers originally being qualified as clinicians, used to making practical devices out of conventional materials. This made it harder for them to be confident in applying new ideas from other arenas to their businesses. However, from the early 1960s a group of engineers from Southampton, led by Professor Jim Nightingale, attempted to show the way to the future.

10.2.1 The Southampton Hand

This project was started at roughly the same time as the Sven hand (see 11.2) and developed mostly in the city of Southampton, but it has an extended research history far longer than the Swedish experience.

Southampton is an ancient port from which King Henry the Fifth left England to fight for lands in France. The city sits at the junction of two rivers and at the head of Southampton Water. The Solent area, which includes Portsmouth and the Isle of Wight, is an exceptional series of sheltered bays and inlets that benefits from a strange double high tide. Southampton is still a major port for the movement of goods. In the city is Southampton University, which has close links with the sea. It offers degrees in maritime themes, from boat design to marine biology. It was also the birthplace of the Southampton Hand.

Jim Nightingale was the originator of the Southampton Hand. He was an unusually bright student and was supervising PhD students before he finished his own doctorate. Nightingale's reputation as a supervisor and lecturer was daunting. He had a razor sharp mind, always asking the hard questions. It is a mystery is that while much of Nightingale's work was heavily mathematical and theoretical and he had many control engineering projects with funding from defence. Yet the Southampton prosthesis projects, were always very much more practical. This project aimed to make a control system for a prosthetic hand/arm that was easier to use than the conventional controllers. It led to devices that could be tested with users, and became small enough and light enough to be used clinically.

The principle idea behind the Southampton Hand/Arm project was to reflect the structures used in biology. The human central nervous system (CNS) is arranged in a hierarchical manner (top down), so that the conscious part of the brain is not aware of what the other parts of the brain are doing to control the arm and hand (Fig. 10.4). Instead, it gives fairly simple commands and the rest of the CNS sorts out the way the arm moves and what the hand does to hold the object. The effect of this is that when there is an amputation, prosthesis users have to take over conscious control of the reflexes, with an unwelcome increase in concentration. Transferring the control of the reflexes away from the user reduces the mental effort they need to use to control their prostheses.

The ideas are based around work Jim did with Hugh Glanville (first professor of Rehabilitation in Southampton University) and Mike Sedgewick, (who later became professor of Neurology in Southampton) to create an engineering model of the human central nervous system [300].

The *Southampton Hand* contains an electronic controller for the prosthesis and uses sensors in and on the device to detect what is happening. The controller automatically selects the correct grasp for the target object and if the held object slips it automatically corrects the grip. The result is that the user of a Southampton prosthesis need only to give simple instructions and leave the rest of the operation

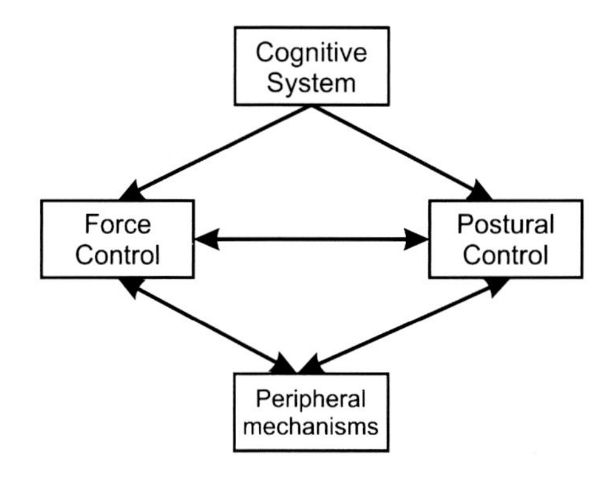

FIGURE 10.4

The principle of the Southampton Hand controller. The control is hierarchical in a similar manner to the central nervous system. (Adapted from [6].)

to the machine intelligence. In this way, the user does not have to concentrate on the way the fingers are coordinated, or how much pressure the hand needs to hold an object. In the natural arm all this is done through the large amount of information being fed up the arm to the lower levels of the brain, which rarely needs to interrupt the thoughts of the higher brain. The importance of this control can be observed in the amount of the CNS given over to control of the hands compared with any other part of the body. Remembering Pearson's homunculus, (Section 3.2), is representation of this phenomenon: The areas to control the hands are very much bigger than any other part of the body except the lips and tongue.

Jim Nightingale began work on the control of prosthetic arms in the early 1960s, when Thalidomide was big news. The idea Nightingale's team worked on evolved into the concept that; *if control of a complex hand is difficult and may well take up too much attention, it is likely that it will be rejected by the average user*. The control must be made as simple as possible, and the choice of grip the hand needs to make is decided based on the shape of the object. A Southampton Hand is fitted out with sensors on the palmar surface of the prosthesis and at the joints of the fingers (Fig. 10.6). If a cylindrical object is best held in a power grip, then sensors in the palm would trigger the controller to tell the hand to go into a power grip (Fig. 10.5 (e)). Alternatively, touching the finger tips leads to a precision grip (Fig. 10.5 (a)), and contact on the side of the index finger results in a lateral grip (Fig. 10.5 (f)). This is context based selection and needs more computer processing power to decide than was available, at that time. Contemporary prosthetic hands were either simply switched to open and close at a fixed speed, or the most advanced research hands moved the hand at a speed proportional to the tension in the commanding muscle. This idea of using the context of the limb to effect decisions on what the prosthesis will do is now being used in other prosthetic systems. This includes the advanced prosthetic legs. Here the pattern of strides leading up to, or on, steps tells the leg's controller that it is approaching a set of stairs and the knee's stiffness changes.

For hands, one extension of this idea is to use a camera built into the hand, which would allow the computer to recognise the shape of the target object and decide automatically which grip is needed to hold it. When Nightingale started, the idea of a camera small enough to fit within the hand, or a

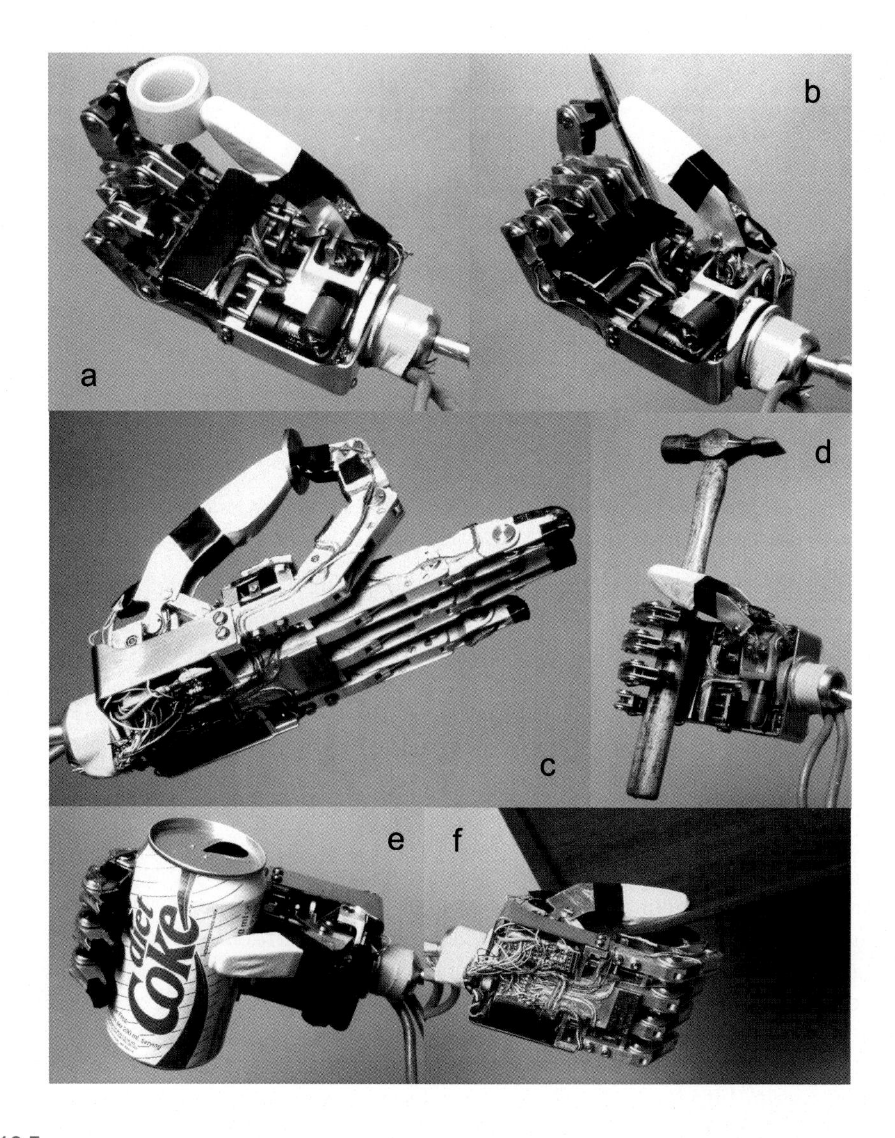

FIGURE 10.5

The grips available to the four axis Southampton Hand. (a) Precision, (b) Two jaw chuck, figures flexed, (c) Two jaw chuck fingers extended, (d) Small Fist, (e) Power grip, (f) Lateral grip.

computer fast enough to recognise the shape and control the hand, would have been laughable. In the twenty-first century it is almost trivial to achieve this and other groups are now investigating it. Such a hand would still be using a simple extension of the Southampton philosophy.

A second aspect of the Southampton Hand controller is that the hand detects if the held object slips within its grasp. If it does, the hand automatically increases the grip force without the operator's

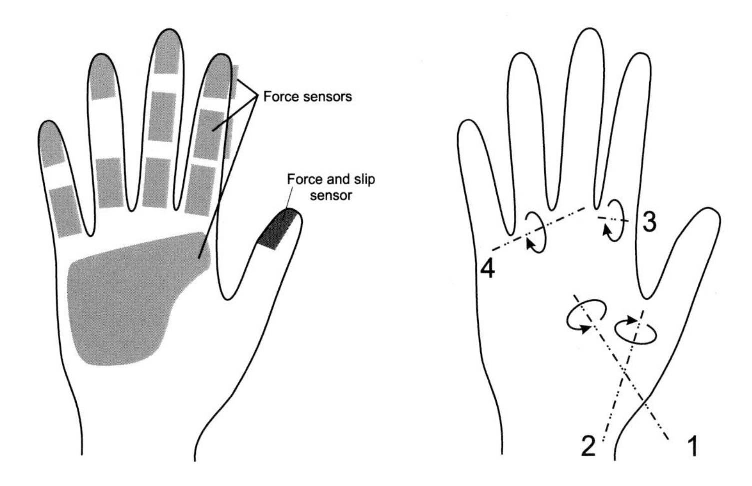

FIGURE 10.6

Sensors on the Southampton Hand controller. Sensors on the digit tips and the palm detect the points of contact between the hand and the object and dictate the grip force and shape. Sensors in the joints allow the controller to maintain the right grip shape. The four sensors on the right are for a conventional four degree of freedom Southampton Hand. (Adapted from [6].)

intervention. This had been considered by Salisbury and Coleman [301], but there was a need to devise a sensor that could reliably measure slip from within the cosmetic glove.

These ideas were developed over a number of years by Nightingale's students. His first was at Leicester University (Harold DeSousa) and second at Manchester University (John Baits), before Nightingale took up a chair in Control engineering at Southampton University.

The first hand that embodied all of these ideas was designed by Bob Todd in 1967 [302,303]. Typical of Nightingale's relaxed supervision Todd was given the vaguest brief and left to get on with it [303]. Todd was a graduate of Southampton University's Electronic Engineering department in the mid 1960s. His was the first year *not* to be taught to design circuits using valves (vacuum tubes), although Bob had already tinkered with them as a hobbyist. Bob had been offered a job as a Medical Engineer at the Roehampton Hospital, but Nightingale made a counteroffer and Bob studied prosthetics in Southampton.

Nightingale gave Bob him a few contacts (such as Alistair Bottomley who was working on powered prosthetic hands), and left him to determine his own course. Todd does not feel that the earlier work by Baits contributed much to his thinking and he came up with the hierarchical controller and grip decisions pretty much on his own [303].

Todd used motors and gearboxes mounted in the forearm to control the motion of the hand. An early idea for the speed reduction came from Nightingale. It consisted of winding cords round capstans (drums) to get enough reduction and drive flexibility to control the hand. The capstans reduced the speed of the motors without using gearboxes, which are heavy and noisy. It was not successful. For his next mechanism Bob only used the cords routed through and around the finger, making a flexible drive that allowed the fingers to curl as they closed.

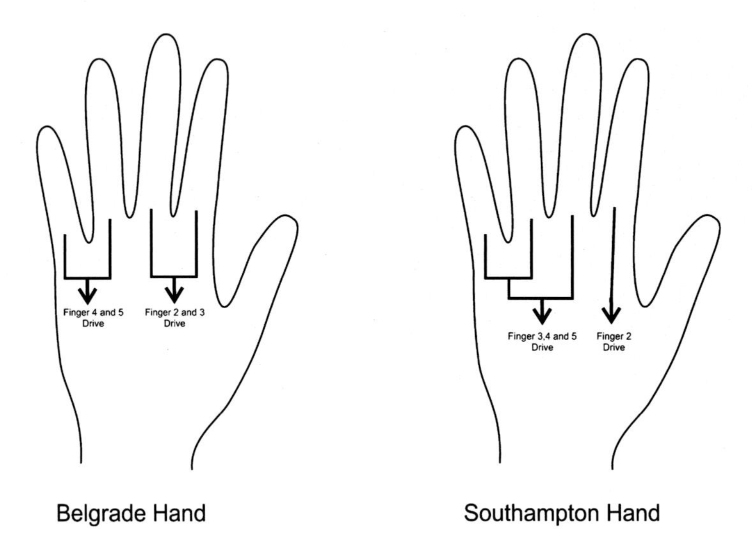

FIGURE 10.7

To create an adaptive grip while using fewer motors than number of fingers a whiffle tree is used. The Belgrade Hand used a split where each pair of fingers was on its own tree. The Southampton Hand split the drive into a drive for the index finger and the others on a whiffle tree. This allowed an adaptive grip and independent control of the index finger.

The drives for the hand placed in the forearm in a similar way to how our own arms are driven. This is, not a very practical design for a real prosthesis, but in the mid 1960s it was a pragmatic way to investigate the idea for a tiny budget. It was clear at the time and the longer-term goal was to make it more compact. Which is why it is all the more surprising that this solution continues to be suggested by others.

Bob built two hands, the second mechanism being able to achieve all the targets set for it. The mechanism had four fingers and a thumb. The fingers curled into the palm. The index finger was driven separately to the other fingers, which were linked through a three branch whiffle tree mechanism (see 3.3.1). When the same structure was used to pull fingers closed, the fingers adapted to an uneven shaped object by wrapping around the perimeter of the target object. With a simple hand the fingers are rigidly linked together and because of this if it is used to hold a ball, the hand will stop closing when one finger has contacted with the ball. Fingers closed via an equalising mechanism means that every finger closes on the object, the force is more evenly spread and the grip looks more natural. Natural hands rarely use large forces because they are able to spread the load more effectively in a similar manner to this [88].

Todd chose to split the drive of the index finger from the action of the three others. By choosing this 'one and three' split, a user could choose to perform a tip pinch (two jaw chuck) (Fig. 10.5 (b)) or a three jaw grip (Fig. 10.5 (a)) depending on which fingers were used. This was unlike the prostheses that came out of Yugoslavia, the Belgrade Hands, which had a 'two and two' split [304], where each pair of fingers are linked (see Fig. 10.7). When a Belgrade Hand closes on an object, there is little force from either finger until both fingers are pressing on a surface, or have reached the end of their ranges. By using the one and three split, the index finger is free to grip quickly and the other three digits can be extended (Fig. 10.5 (c)) or flexed (Fig. 10.5 (b)) out of the way.

In this hand the thumb could both flex and extend and rotate at the base so that it could oppose the tip or side of the index finger, giving the hand an extra grip shape. Both of these motions were powered. Only when the Touch Bionics hand had been on the market for a decade (in 2013) did it have this useful feature, giving some idea of the difficulty in making something like this work well enough to be in a product.

One aspect that is almost unimaginable in the twenty-first century, is that as there were no compact computers available to control the hand so the electronics had to be built specifically for this purpose. This is a form of 'fixed logic'. This is a simple program, a sequence of mathematical operations. When the circuit is turned on, a fixed logic device performs exactly the same sequence of additions, subtractions and comparisons every time. To change the sequence the designer must change the wiring. A computer has a similar arrangement, although now in silicon. Its hardwired controller performs all the basic functions, including loading the operating system program when we want to run it (this is part of the 'booting up' when the computer is first switched on). This arrangement is again a hierarchy. It is a little like the staff of a kitchen who know how to cook almost anything, but it is the menu that defines the recipe of what they will do on any particular night and the clientèle that actually request the dishes. The wiring is the cooking skills, the application programs are the items on the menu, and the customers' orders determine which application runs.

At the very base of the computer, and other logic devices, are switches. How these switches are arranged determines how the computer works. Modern computers use switches made in silicon. These switches are transistors, and because it has become possible to make them smaller and smaller, computers have become smaller and faster in a way that is staggering to anyone who knows the history. Todd's controller did not use transistors. At the time he could only have used individual transistors, not large numbers of them as in a computer. Bob found transistors too expensive and not that reliable. They tended to fail after a short time. Instead he chose relays, which are individual *mechanical* switches. This makes them larger, slower and noisy. No sound recording exists for what the controller sounded like, but Bob said it was *"Quite a lot of clackety clacking"* [303]. The controller was a series of modules mounted on a wooden base. The relays closed when a decision was made, and much of the rest of the electronics was analogue.

Cine recordings were made of the hand in action (Fig. 10.8). A professionally shot film shows the hand moving and reacting to the inputs. An amateur 'Super 8' cine film shows a tantalising glimpse of the controller, lights blinking on and off across a panel in the way that the 'computer displays' of the original *Star Trek* television series (about the same era) blinked and flashed. The difference for this controller was that this display was not random, but the lights and the sequence mean something.

The movies show the hand working. It is driven by a proportional EMG signal, so that the hand opens in proportion to the extensor muscle tension. This is the POSITION state. Throughout the history of the Southampton Hand the different states of the controller are referred to in full capital letters. It is probable that this was a necessary way to emphasise the state names in an era when the thesis (and any publication submissions) would have been typed on a manual typewriter (bold text was difficult to create, and different font unavailable). When the operator relaxed the hand closed and if it had closed on object sensors in the finger tips told the controller to stop closing the hand and apply the lightest possible touch (Fig. 10.9).

FIGURE 10.8

The first Southampton Hand, behind are the electronics required to control it. (Taken from the Super 8 movie of the demonstration of the hand.)

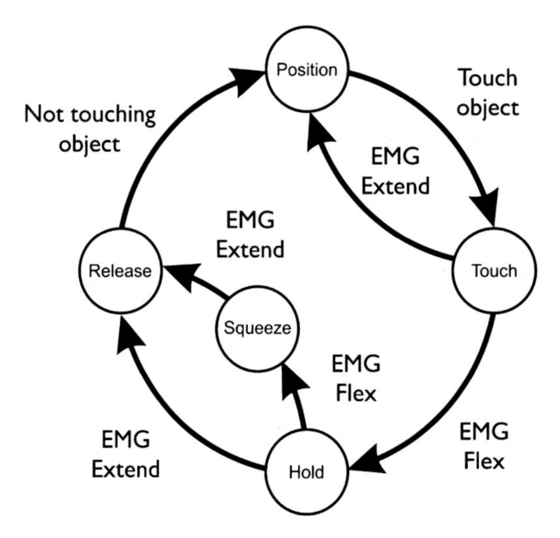

FIGURE 10.9

The control of the Southampton Hand. The hand moves between states, switching based on input from the sensors or the EMG input. At each stage there are a limited number of choices which the user needs to make. These are shown by the side of the arrow. Control starts at the top with hand empty. (Adapted from [6].)

At the start of any grasp the hand is in a full tips type grip with the finger tips meeting the thumb tip. The position of each the digits is measured by an ingenious sensor that was based round the barrel of a disposable ball point pen [302]. The barrel supported a coil of wire, and the finger's motion was linked to a piece of metal that moved inside the barrel as the finger closed and opened. The presence of this metal within the coil changes the electric field in the coil and this was detected by an electronic circuit. As the finger flexed and extended more or less of the metal was left in the tube, so the electric field change and the flexion angle was detected.

A further sensor was built around the stylus of a record player [302]. It detected vibrations as the object slipped. The more the object slipped the harder the hand gripped the object, stopping the slide. There were also eight force sensors in the tips and inner surface of the hand [297], they recorded where the hand was touching the object, which determined the appropriate grip. A contact on the tips resulted in a tips grip, touching a sensor on the palm resulted in a power or fist grip and contact on the side of the index finger triggered the lateral or side grip.

Once the hand had closed on an object and the controller 'knew' it was TOUCHing an object, application of the flexor muscle (not so far used) switched the hand into the HOLD mode. It was here that the slip/grip reflex was employed. If the person did not want this reflex to be on, or wished to override it, a second application of the flexor muscle turned off the reflex and made the grip force proportional to the muscle command. This was the SQUEEZE state. This is different to conventional hands where the closing velocity is made proportional to the flexor muscle and there is no change in the relationship between effort and force once the hand has closed. Forty years later, in 2008, Engeberg added sensors to a Utah hand and also found this change in the relationship between EMG command and motor response was useful [256].

In either HOLD or SQUEEZE, if the operator wished to let go of the object the extensor tension RELEASEd the object before returning to the POSITION state. In the POSITION state, using cocontraction made the hand go into an open palm form and switch off. This would be thought of now as a 'stand by' mode. It also made putting the hand down sleeves easier. The user could also select in and out the control of the other three fingers; middle, ring and small fingers. The fingers could be held close to the palm or outstretched. This made performing detailed precision tasks easier as the other fingers could otherwise get in the way and obstruct the user's view of the target object (Fig. 10.5 (b and c)).

If the object was held in a power grip and the object was small, the digits would close around the object, but the contact area between the object and the hand then became small too. To increase the contact and make the grip more secure the hand detected the grip size and automatically rotated the thumb around at its base towards the lateral type grip, so it would press against the handle or other object (Fig. 10.5 (d)).

These are the essential features of the Southampton Hand, what became known as the Southampton Adaptive Manipulation Scheme (SAMS). What is important is that it encapsulated the ideas of how the hand was controlled *not the particular device*. The idea could be applied to any prosthesis, and subsequent PhDs took the idea and used it when exploring different aspects of the idea or its application.

Two additional features were not carried forward beyond Bob's work: Firstly the hand had a powered wrist with pro/supination and flexion/extension. The wrist was ultimately seen as a feature of the arm control and so moved to another project. The second was a MANOEUVRE state. This was used when the hand was TOUCHing the object. The controller caused the fingers to vibrate a little with the aim of allowing the user to adjust the grip orientation of the hand without letting go. The modern version of the hand is programmed to only apply a light touch, maintaining contact of the fingers with the object by detecting the contact force with the sensors in the finger tips. The object can be moved or rotated within the grip without the slip reflex being triggered.

After Todd completed his PhD, he supervised a second project undertaken by Roger Codd [305, 306]. Codd's project explored the practicality of controlling a hand this way. As a result he worked on a hand that was compact enough to be worn (but had an external controller). The hand was contemporary with the Sven hand. He was able to build the controller with transistors. It was at this point that Dr Tom Sensky was recruited to try out the hand. Sensky usually used a body powered hook but he was fitted with the hand and tried it with a number of different tasks that simulated everyday activities to see if was possible to complete them (Fig. 10.10). Sensky could control the hand easily and on the movie he was able to show some fluidity in operation. He performs a series of tests. Some address his ability to move an object from one level on a set of shelves to another without inadvertently triggering the hand to open. Others that simulate domestic activities, from cutting objects to getting cigarettes from a packet to picking up the receiver of a telephone.

This last operation is the one from which I learned my first big lesson about prosthetic usability. I had seen a copy of the movie and was impressed by it. I was told that Sensky reported that under normal circumstances he would not have carefully manipulated the receiver into his prosthesis from the cradle, as he did there, but simply picked it up with his good hand and jammed it into his hook. This was a lesson that users will do what is practical, not necessarily pretty nor what the engineer wants them.

Sensky was a medical student at the time and following his experience with the Southampton Hand he wrote a short article for the Lancet entitled 'A Consumer's Guide to the "Bionic Hands"' [307]. A knowingly provocative title, it still makes compelling reading in the era of multifunction hands. Sensky sets out his resistance to the advanced hands. Reflecting his assumptions about the *"dishonesty"* of anthropomorphic hands. He feared that if he was wearing an anthropomorphic hand when he met someone for the first time the only way they might realise that he had a false hand is when he encouraged the new person to shake it, when the person would be startled by the dead feeling and the glove. While this is true, many people find ways to get around the moment. A person with a loss of their right hand, can reach across with their left inverting it as they go.

After a few years teaching at Southampton, Bob Todd became interested in alternative energy, an idea that was starting to take off. Gerard Morgan-Grenville had the idea of working on new ways to look at current problems of energy generation. He eventually founded the Centre for Alternative Technology (CAT) in a valley in the Welsh hills. It attracted Todd's attention and after he visited the centre in its very early days Bob found himself being hounded by Morgan-Grenville to join the centre and work on the electronics. Eventually he agreed and joined, although the head of department in Southampton thought he was crazy. Initially the CAT existed in a few leaky buildings in a rainy part of the UK. Over the years it has established an international reputation and the site includes a state-of-the-art building with a lower energy requirement than any commercial building of its kind in the UK. Bob has never regretted the decision and even after retirement he worked for one of the spin-off companies designing electronics at the cutting edge of commercial viability [303]. When asked in 2015, Todd still regarded the prosthetic hand project as a hard one. He could see it was difficult to create a device that can work for people for extended periods of time.

The next hand at Southampton, was built by David Moore [308]. He aimed to develop a more compact hand. David managed to fit all the drives into the overall profile of a large male hand. He found a new plastic (Delrin) that he could make the hand from. He also built much of it himself. He

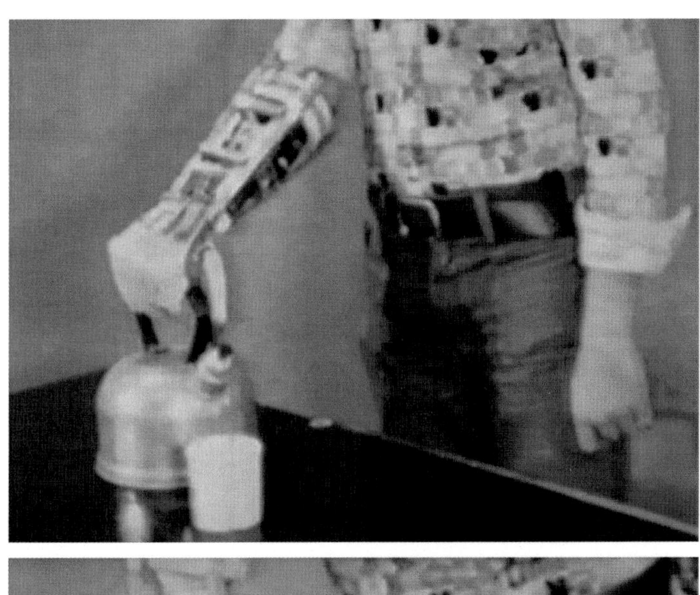

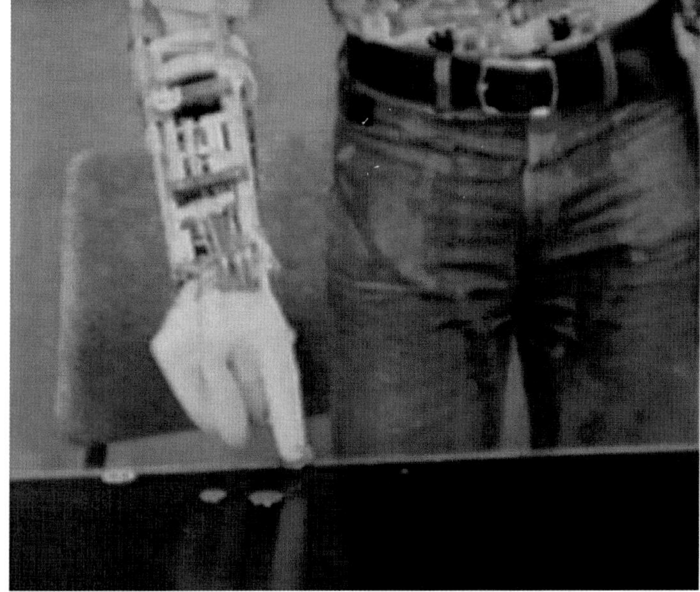

FIGURE 10.10

The Southampton Hand being used to perform a range of different tasks using a power grip and a precision grip with two fingers. (Taken from the movie of the trial.)

had to take a heavy lathe up several flights of stairs to his flat in order to make it at home. Silicon technology had moved on, so while the controller still employed fixed logic (still needing to be rewired to change the program) it now used integrated circuit technology, requiring a smaller electronics board. Sensky again performed the same tests with the new hand mounted on his forearm. The controller, however, was still in a box at his feet.

Around 1980 Mohammad Barkhordar began his PhD. He used Moore's mechanism but replaced it with a microprocessor controller. Finally, the hand was programmable (even if it was a more involved process to change the program, compared with today). This controller was quite compact, although not quite portable. However, another part of Barkhordar's project was to reduce the controller for a single motion commercial hand (Viennatone 9.1.1) with sensors built into the tip of one finger, so that the slip/grip controller could be incorporated in a simpler, portable, computer based controller. This was tried in the lab in 1984. It was probably the first prosthetic hand to be controlled using microprocessor and used by the wearer of a prosthesis.

At this time the rapid advance of silicon technology was becoming apparent even to the nontechnical. To support the need for increased public awareness the BBC commissioned programs to teach the public to program a computer. They showed examples of what this technology could do. One example they showed was Moore's hand with Barkhordar's controller and EMG amplifier/processor. The control was sufficiently simple to learn that the hand was controlled by the anchorman for the show.

In 1985 I started my PhD in Southampton. Aside from working on a new design of the hand (Fig. 10.5) and a controller with a more modern microprocessor, I also took the design of Barkhordar's prototype hand controller further. I removed bugs from the electronics and software and created a printed circuit board version (Fig. 10.11). This made it sufficiently robust that it could be used in the field (Fig. 10.12). One user took it home with him, which was the first time such a microprocessor controlled prosthesis was used in the field (upper or lower limbs). At the time it was not appreciated that this was such a breakthrough and very little was made of the fact. The main part of my PhD was to create the controller for a more modern design of four axis hand using a 'microcontroller'. This is a computer chip with much of the interfacing built into it [234]. This technology was a significant advance at the time, and in the subsequent decades pretty much every microprocessor used for applications from toys to food mixers is to some extent a microcontroller.

My own route into the field was circuitous. I arrived on the planet in the middle of the time the Thalidomiders were being born [179].

To me, the Thalidomiders seemed to be a constant presence in the media when I was growing up, and they were visible examples people needing solutions to assist in their daily activities. In the popular media, persons without arms or legs are otherwise limited to Captain Hook or others similarly unrealistic portrayal, often with negative connotations. The Thalidomiders were different, they were real people with real problems. A rare positive image for an amputee was the *Six Million Dollar Man*, but then he was 'cured' by use of advanced technology and became a prosthetic superman. In the late 1970s the BBC's flagship documentary program *Horizon* made a program that took the idea of the *Bionic Man* seriously. It asked how close modern technology was to the fictional account. It was this that made a deep impression on me. I could see that there was an interesting problem to be solved and we were a long way from achieving it (the bionic man drama made it look like it was all completely solved). The program included an interview with Steve Jacobsen and shots of an early, but recognisable, Utah Arm with a young Harold Sears in charge.

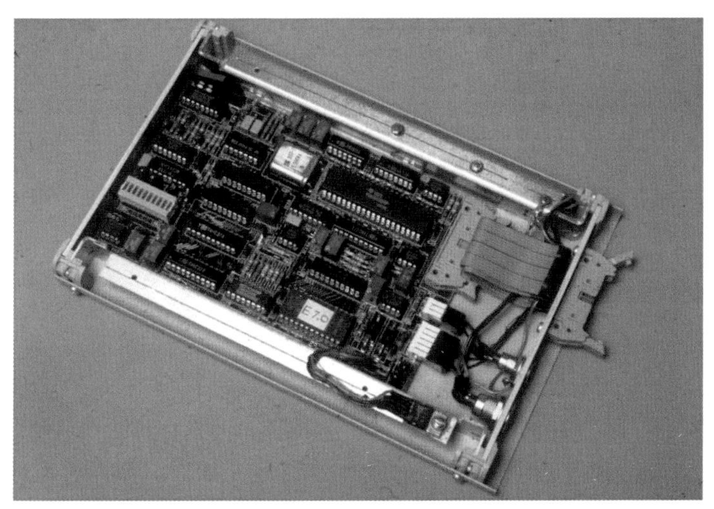

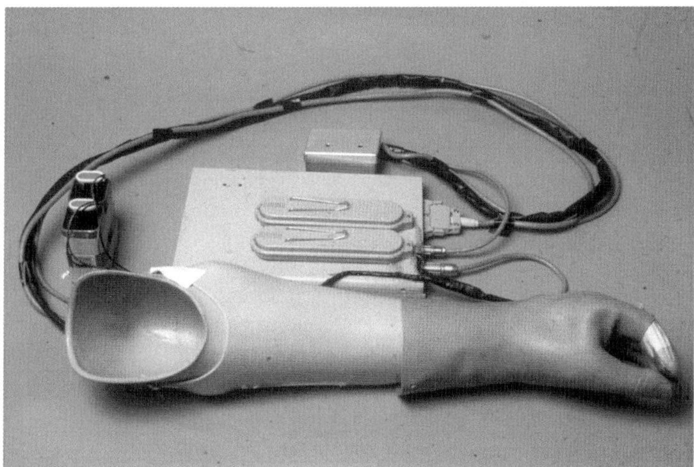

FIGURE 10.11

The first portable microprocessor based hand and controller. Based on a modified Viennatone hand, it was able to realise the Southampton Hand control format.

After a science degree at Durham University, I was finishing a master's in Electronics when Jim Nightingale was recruiting for the next student to work on the Southampton Hand. To me it seemed to be the dream choice for a PhD that simply dropped out of the sky into my lap. Before, I had never actively considered it as likely for future work. Now, nothing else has ever seemed possible. In 1990 after my PhD, I moved to Oxford, to the Nuffield Orthopaedic Centre hospital. Over the next decade he worked on other aspects of Orthopaedics, but came back to prosthetic hands when funds became available. In 1989 I met David Gow at a conference and our common interest in prosthetic hands made us natural allies and collaborators over the years. Later, we attempted to get commercial support from

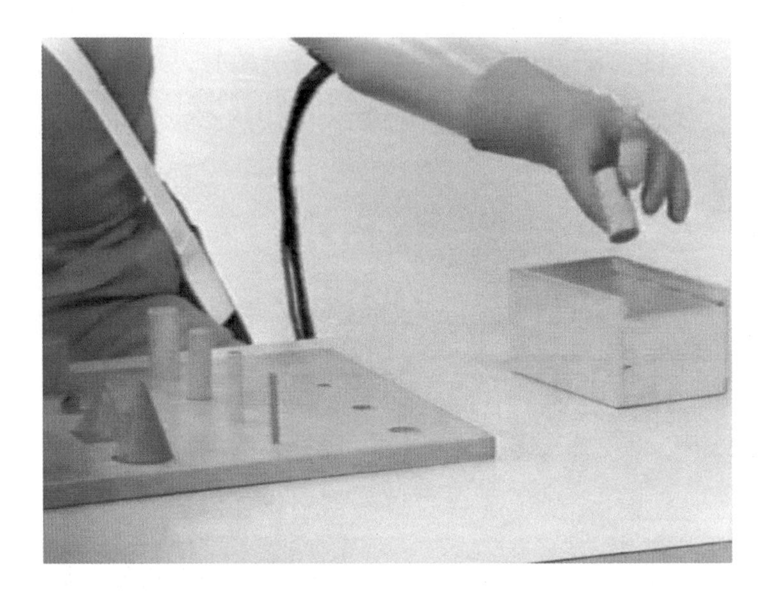

FIGURE 10.12

The portable Southampton Hand being tested. The user was able to perform a range of tasks using the hand.

Blatchfords before gaining money from the European Union to explore prosthetic arm design with a team from Sweden that worked with Peter Herberts from the Sven hand project.

One outcome from this later project was the first prosthetic arm to use a local area network to communicate between the different joints in the arm, and to be used in the field (known as ToMPAW). The arm was the natural extension of Gow's idea of a fully modular arm. The network allowed the different parts of the arm to only use two wires to communicate between the sections, cutting down on wires, complexity and potential failures. Each joint had its own controller and an arm was built up from different parts, being put together like Lego. While networks seem obvious today it was a step forward for prosthetics [272]. At the end of the project David went off to form his own company (TouchEMAS), while I took up a chair at the Institute of Biomedical Engineering at UNB in Canada.

Subsequent to arriving in New Brunswick I became part of a group that has aimed at bringing standards and validity to hand function measurements. The team was composed of members in North America and Europe and created a process to bring agreement within the profession (see [94,96]).

10.2.1.1 SHAP

During my PhD, Jim Nightingale retired and his role as supervisor of the project was taken by Dr Paul Chappell. A few years after, I was succeeded by Colin Light. Colin studied the problem from the perspective of the design of more compact fingers for a multifunction hand. However, his most significant contribution was to look at the assessment of prosthetic function. An engineering perspective on hand function might focus more on the things a user can do with the hand, rather than actual use the prosthesis. Certainly while the hand remained in the lab there were limitations on what could be measured. Light made a study of the tools used by clinics at the time. He found that if the therapists used an existing tool they tended to only use *part* of it to save time [93]. Regrettably this invalidates the tool as we have seen this validation can take many years to complete (see 3.4).

Knowing that the process to develop a new tool was a long one, Colin chose to use tools already known to be more reliable. This meant he already had some of the testing done by others. He then constrained the tasks to a small number set within a small volume with fixed starting and ending points on a marked board (known as a 'form board'). Since self-timing was known to be more reliable than tester timing, he used that too. If someone else times a task, they have to judge when the person has started, which harder to judge. However, the subject *knows* when they have started. Colin then developed a score out of one hundred, where a hundred was the top mark and slower/poorer performance had lower scores. The score was based on how much slower the subject was than someone in the physically typical group, but it was also scaled based on how frequently a particular grip was used in daily activity. A poorly executed task employing a less used grip had less of an impact on the score than a slow task with a popular grip. The tool was intended to be used for a wide range of impairments and aimed at testing the hand and not the user of the hand. It came as a series of objects and a board in a single flight case to make setting up and carrying around simpler than its rivals. It could take less than twenty minutes to execute fully, and gave a single number at the end, which appealed particularly to the engineers. It became known as the *Southampton Hand Assessment Procedure* or SHAP [309].

SHAP's simplicity and immediacy has meant that it has gained popularity with engineers and therapists, although the prosthetics OTs always remind one, that in daily activities a user with a single side loss will employ their prosthesis in a support role and SHAP makes them use it in a dominant role. This is because SHAP aims to look at the activities, not what the user actually does with their prosthesis. If the intention to understand the user better, they should use the ACMC (11.1.1). SHAP works because its design controls the sort of movements a person makes, so that differences between movements made by two people can be made clear. If the activity is different for every subject it is hard to compare two people and how they achieve a task. As SHAP constrains the hand and arm it has been used for measuring the motion of the arm during activities. SHAP has also been used in experiments to track eye gaze [104,310,311].

10.2.2 The Southampton Arm

When Neil Storey joined the team in October 1972, he thought he would be working on a hand controller but Nightingale had different ideas. He wanted to investigate controlling the rest of the arm. The hand could respond to the shape of the object, but someone with a higher loss still needs to get the hand into the right place to grasp the object. They need to control the hand in six dimensions (three straight line motions; up/down, left/right and in/out and three orientations; yaw, pitch and roll). At that time, the only way to control that many axes was with one channel at a time, switching between them. This would be slow and hard to do. A study of the motions of the arm showed that we are really only interested in the location of the hand in space, not what the joints are doing to get it there [312]. A controller built to look after those aspects of arm control would make using a complex arm that much simpler.

Had Neil started the project at the time he finished it, three years later, he thinks the controller would have almost certainly have been a microprocessor based device, but at the time they were unknown. Even so, the use of microprocessors would not have changed the core of Storey's PhD. His project was about the *control philosophy*, which was demonstrated fully [312]; the implementation was tricky, but only a secondary challenge.

The application hardware was completely different to how one would go about the problem now. Storey built a mathematical (digital) simulation of the system on one of the department's 'minicomputers'. To control the arm in real time with computers was way beyond the capabilities of any digital computer available to him. Having worked out what mathematics was needed to control the arm, he then built the controller entirely in analogue circuits. The arm had seven motions so the three dimensional position of the wrist (X, Y and Z) was calculated using *analogue multipliers*.

Analogue is the older and more traditional method of representing items. A quantity of something can be represented by an amount of something else. It could be the amount of water in a bowl representing the amount of money in a system, or it could be a voltage level representing the sound made by an instrument. If the thing one wants to control is represented by a voltage, the bigger the voltage, the larger the value. If the voltage represents the arm's position in the up/down (Z) direction, the bigger the voltage, the higher the wrist. If you wanted to take an input and make the Z value double that in size you could build an amplifier that doubles the input voltage (a *gain* of two). You can also make circuits to add to, remove from, multiply or divide the signal, as appropriate. For example: Twice the size minus a little bit would be an amplifier with a gain of two. After that a circuit will take the measured amount off the voltage (a subtractor). This was the way that both analogue electronic and mechanical computers worked in the past. Modern computers work differently, they take in an analogue signal, convert it to a number (digitise it) and then perform the mathematics on numbers before outputting the result to a screen, or to a driver that would move a joint. For the programmer, digitisation makes the operations easier. It is far simpler to change the operations, or add an extra step, when you don't need to build a new circuit each time. You just write a little extra code.

The first Southampton Arm took simple instructions and converted them into motions of the arm. Instead of needing each joint to be controlled separately and then the user switching between joints, the arm was given simple instructions about where the hand needed to be in space and the computer worked out how to move the joints to enact the wish. The arm used what was in effect a signal from a joystick at the shoulder (angle) and turned it into voltages at the motor. If the operator raised the shoulder the wrist would be driven to move up. If they pushed their shoulder forward, the wrist would move forward. Finally the side to side motion of the hand was related to the amount the trunk swayed left and right. Three motions could control the wrist in space.

With shoulder flexion and abduction, humeral rotation and elbow flexion, this is not a simple *add one and double it* relationship. This is complex mathematics of the angles of the elbow and the shoulder. If you want to reach forward, your shoulder goes *up*, but your elbow goes *down* a different amount. This has to be calculated. The mathematics is straightforward, building an analogue version is less so. Storey's task was to work out the mathematics needed to control the arm and then build twenty-eight analogue multipliers with six predetermined fixed numbers (*presets*) per amplifier to perform the calculations. The presets, defined the proportions of the amount of movement and where the joints started their motions. This means that if the output was double the input one of the presets would have a value of two. The number that is added or subtracted from the signal determines where the wrist starts, e.g. in the middle of the range and not at one end. All of this would have taken time to set up, but Storey said it was quite stable thereafter [312]. To provide the three coordinates for the wrist he needed two shoulder motions and lateral sway, *"so essentially the arm moved in an exaggerated motion of the shoulder"* [313].

As an analogue system it moved smoothly, rather than the more discrete jumps that a digital system is prone to suffer from. The arm would not return to a precise starting position, it would go back to roughly the same region. This is in fact acceptable for a prosthesis, as long as it's predictable. A greater potential problem with any system that aims to get to a target point is a possibility that it will oscillate. This is because as it might overshoot a target and then system tries to correct, overshooting again, and so on. One way to make the system more stable is to slow it down so it does not overshoot the target too far but can stop in time (5.2). Storey's arm moved slowly to make sure it was stable. This was a success and to Storey *"When you were controlling it, it was surreal that it was doing what you were thinking. It was easy to control"* [312]. There was quite a lot of play in the mechanics which meant it did oscillate a little on settling to a target. To test the performance of the arm Neil fed the controller inputs to move from one position to another and then back to the start. The precision to which it could perform the task and return to the start was a measure of the controllability of the arm.

The mechanism Storey built to test the ideas was a very basic mechanical device. The only approximation to a prosthesis was that the motors were in the arm, and not external to it. He did not consider the mass of it (it was very heavy), although he did build it to about the right size. What surprises Storey now is that the mechanism went on to be used for a second PhD and also at the start of a robot arm project years later [312]. For the second PhD a wrist with flexion/extension and pro/supination were added and the shoulder was rebuilt to remove some of the backlash.

The tests of the arm were recorded on cine film. This was an insurance in case of failure on the day of his PhD exam. It also would serve as a record to show his sponsors (the government body, the Department of Health and Social Security). However, there was no way that Storey could get a copy of the film and take it with him. It was some thirty years later before he saw a copy. It shows a young man with hair, beard and tie appropriate for the era. He is working a series of loops of cord that run around his shoulder to sensors on a stand he sits beside (Fig. 10.13). These loops capture the motions of the shoulder joint and behave like a joystick not easily obtainable at the time. The arm is shown moving in response to the inputs. He shrugs up and down, the wrist moves up and down. Then he moves his shoulder forward and the wrist goes forward and back. Then he moves his shoulder in a circle, and the arm follows. Finally, the camera pans across to the controller. It shows a series of complex electronic circuit boards taking up more than a square metre of bench space (Fig. 10.14). It was twenty years later when a final year student at Reading University's Cybernetics Department, Ben Jones, had access to a ToMPAW arm and the controllers, and was able to reproduce this controller in software, with all of the electronics mounted on the arm. After his degree Ben Jones went on to work for Touch Bionics.

At the same time as Storey was conducting his PhD Nightingale's group started the first PhD on Functional Electrical Stimulation (FES) of an arm. This struck Storey as a neat idea. FES is, in effect, the reverse of myoelectric signal detection. If a muscle is stimulated by an electric current it will contract: If the right muscles are stimulated a paralysed arm or leg would move. Nightingale's idea was to use the same sort of sophisticated controller as used in the prosthesis to drive a paralysed arm [312].

A later student, Ian Swain, progressed both these ideas further. Swain had studied as an undergraduate in Electronic Engineering at Southampton and Nightingale had been his supervisor. During lectures he also used examples of the control work in prosthetics so when he offered Ian a PhD place it was not difficult to say yes, *"it was better than designing control boxes for fighter aircraft"* [313].

Swain explains the work he did for his PhD:

> "To control an arm you need seven degrees of freedom and you don't have more than three to actually control it with. If you want to avoid sequential control you need to reduce the problem to controlling the position and orientation of the hand in space. In a digital computer, Storey's work needed more

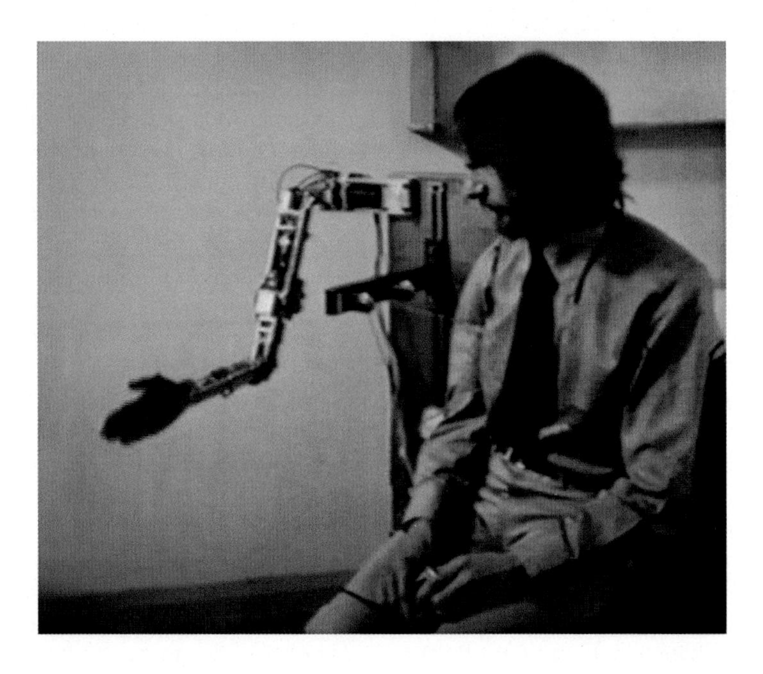

FIGURE 10.13

The first Southampton Arm. Neil Storey, the designer controls the end position of the arm by movements of his shoulder and trunk. (Taken from the movie of the demonstration of the arm, Permission: Neil Storey.)

FIGURE 10.14

The first Southampton Arm controller. It performs analogue mathematical operations. The electronics to the right are a series of potentiometers that set the parameters for the controller. The simple electronics to drive the motors is to the left. (Taken from the movie of the demonstration of the arm. Permission: Neil Storey.)

> than two minutes to calculate the angles. By simplifying the equations he could get to calculate the joints down to the wrist in real time".

Ian looked for information not needed when positioning the hand at an end point. If he removed the need to supply this information, or to calculate it, then the operation of the arm could be simpler. He also limited the movement of the elbow to a range that matched human use. This reduced the problem to a small enough number of calculations to allow the computer to calculate the necessary joint angles in real time to control the hand in space as it was used. For example, to move the hand to a particular position there are many positions in which one can place ones elbow (above, to the side, below). However, if the solution is limited to natural orientations (the elbow is pointing downwards), there are many fewer solutions to the problem. Hence, limiting the elbow simplifies the equations.

Swain also needed to calculate the orientation of the hand relative to the wrist. He thought of a way, which occurred to him on the bus from Southampton railway station up to the University, he regarded it as the one real novelty in the thesis. *"To use the curvature of the path you move to in order to control the two degrees of freedom at the wrist. And so as you could define the path by how you moved to an object you could get the hand roughly where you wanted it, in terms of orientation"* [313]. Simply put: The orientation of the hand can be interpreted from the approach the hand makes to the object. You are likely to move the palm of an open hand towards the object. Thus the arm follows the direction of the palm. Once the hand was close to the object the controller used the touch sensors on the hand to fine tune the orientation to maximise the contact area between the hand and the object. If the side of the hand closest to the little finger touched the object first, then the wrist would rotate round to bring the index finger towards the object too (pronate). If the index side touched first the wrist would rotate the other way (supinate).

Swain found it *"quite entertaining"* to have Jim Nightingale as a PhD supervisor. Jim's philosophy was; *"it was your PhD not his"*. He already had one and did not need another.

> "If you had a fair idea where you were going then you didn't have any great problems... If you had any problems, you went and talked to him about it and he was exceedingly good at the mathematical parts. He didn't give you many ideas of where you were going... The more practical engineering based students were usually fine. Those who needed more guidance probably struggled a fair amount" [313].

The maths Nightingale came up with in tutorials was always hard, and the group had to work it out between them after they got back from supervisions. The group was substantial, the year before Swain arrived six new students had started. This was the *Control Engineering Group* and the work of the group was very broad and included control of combustion in car engines, image processing for nuclear medicine and rehabilitation. As there were few of the tools available that are today considered essential, one of the students, Martin Trotter, wrote the programming language which Swain used to control the arm. In its form it resembled the language that came out later called Pascal. However, Trotter was less motivated by marketing and history and his name for the tool was *"Soddit"* [313].

While today there are computer-based tools to help solve complex mathematical formula, back in the 1980s Swain and Storey preformed the calculations and then checked each other's work. One of the ways that Swain used to speed up the calculation of the angles needed for the arm was to use the same basic formulae that a computer used to calculate sines and cosines. Ian went back to the original papers written by researchers in IBM to see the methods used and found short cuts. This is remains necessary, unless you use dedicated microprocessors designed and built to perform mathematical calculations quickly and efficiently (these are the Digital Signal Processors used later by Williams of LTI and Englehart of UNB).

Following his PhD, Storey worked for a year in Trinidad, before returning to the UK and getting a job at the University of Warwick, where he spent his career working on many aspects of computer control. Ian Swain stayed in the field. After his PhD he remianed in the University department before, in 1982, getting a position to support the activities of the new spinal injuries unit at the Odstock Hospital in nearby Salisbury. There he spent thirty-two years as an engineer and ultimately the Director of Clinical Science and Engineering, looking at aspects of control with FES for persons with spinal injuries. Early on, it had become clear to him that there was a bigger need and more to do than in prosthetics. One aspect of the work in Salisbury is that they produced a spin off company, Odstock Medical, which was the second to be come out of the UK National Health Service, (NHS) . It was the first in England. The only previous company was David Gow's Touch Bionics, a small irony that both companies came from people who had worked on arm prosthetics.

The work on the Southampton Arm was reduced following Ian Swain. Only robotic applications were pursued in Southampton. David Brown, a graduate of Regent Street Polytechnic, did not want to simply modify the Storey arm. He chose to build a smaller device which used a harmonic drive for the elbow and a swash plate to cause the wrist to rotate. It used less gearing and a motor that aligned with the forearm, and not across it, taking up less space. He also used two motors for the elbow, as one powerful enough for the role would have been too large to fit within the forearm space. He then took the ideas for the arm and sold them to the Atomic Weapons Research Establishment (AWRE) [314]. The idea was to produce an anthropomorphic arm that would fit inside the glove boxes used in the nuclear and defence industries. The underlying problem is that the radiation load for any worker is limited, which determines the amount of time the worker can be exposed to the radioactive sources in any month or year. When the dose was likely to be high, a worker has a short time to work and then a long time away from additional exposure. At times like this, a robot could be used in the same place as the human, until the chance of exposure was reduced. To be a direct replacement a robot would have to be able to stand-in for a human and work through the same apertures as the people. These are glove boxes, where people put their hands into thick long gloves that poke into the dangerous workspace and are able to manipulate the objects naturally. The constraint of using the same aperture as the human operators meant that the robot had to be the same size and capability as a human. The robot was a spectacular creation with harmonic drives in every joint and the latest in off-the-shelf controllers for its movement. However, none of the lessons of prosthetics were employed in the control of the arm. This was not the first foray into anthropomorphic hands for AWRE. In the early 1960s they had supported a collaboration with Hugh Steepers to create a prosthetic hand for experimental use [55].

Following his time in Southampton, Peter Kyberd worked at the Oxford Orthopaedic Engineering Centre, on a number of different aspects of orthopaedics and engineering. When an EU initiative was launched with the aim to encourage research in engineering for the disabled, the obvious choice for a project was to look at aspects of design and control of prosthetic arms. Any collaboration in Europe involves a number of different centres across Europe. In this case Peter worked with his second PhD supervisor, Paul Chappell in Southampton, a small robotics company in Northumberland (TAG), and two partners in Italy, the University of Pisa, and the INAIL rehabilitation centre. This project was followed a few years later with a second EU collaboration, the ToMPAW project (see 10.2.1). The arm was used in Sweden and the UK. One user was so keen on his arm that he retained it and used it until 2012, when Peter produced a new version of it for him. He uses that arm still.

Between the two projects Peter also gained money from the Leverhulme foundation to make his own hand with separate two motions. This 'Snave' hand had curling fingers and a separately moving thumb, plus touch and slip sensors in the fingers and thumb tips. This allows the hand to select precision and power grips and perform the automatic force control of the Southampton Hand. It was this hand, driven by new electronics and united with wrist and elbow from Edinburgh that became the hand for the ToMPAW arm.

One of the predictions that Ian Swain had made for the use of complex technology in computer-controlled arms, was that it would be accompanied by the addition of other sensors, such as accelerometers [165]. With accelerometers, the arm could know which way was up and could create a self levelling platform. When he made these statements such devices were expensive, large, heavy and power hungry, all things the enemy of the prosthetics engineer. By 2010 most mobile devices (phones, laptops, tablets) had multiple accelerometers built in, so that when he created the second generation ToMPAW arm in 2012, Peter was able to simply add these features to the hardware.

Computers are built in to every device possible now. The idea that they can all be linked through the internet (the Internet of Things) is a popular projection of where the technology is going. The manufacturers are still learning what the technology can do and the users/consumers are discovering what they want. The same is true with prosthetics. New designs and ways to control prostheses are being developed routinely, but we still have to determine which aspects of advanced computer control are genuinely useful to the wearing public.

CHAPTER

11 Research in Sweden

The pioneers of paediatric prosthetics and osseointegration

Sweden is a small country with a public based healthcare system. They pioneered research into pattern recognition control of prosthetic hands, producing the first commercial prosthesis to be controlled this way. Engineers and clinicians also lead in the development of paediatric myoelectric hands, and in the assessment of user control of their hands.

11.1 Örebro

Örebro has been an important source of ideas and techniques in upper limb prosthetics. Through the work of Rolf Sörbye and Liselotte (Lottie) Hermansson, the clinic in Örebro became one of the most influential clinics in Europe for paediatric fitting and, more recently, assessment of the outcomes of fitting prosthetic hands.

Rolf Sörbye came to Örebro to practice neurophysiology from Göteborg where he had trained. In 1971, a mother of a two-year old girl with a congenital absence who used a body powered hook, heard of the powered prostheses being used by adults and wanted something more for her daughter. At the time there were no paediatric hands available at all. Sörbye tried to talk her out of her idea. No one had ever tried to put a myoelectric hand on a child before. The mother was insistent, so Sörbye agreed to give it a try. The device they put together was based on the smallest adult hand available (a Viennatone size: $6\frac{3}{4}$ teenage hand), and used external batteries. The child used the same movement patterns she had to operate the hook to move the hand and this worked. *"The mother was right"* [22].

Initially, the work expanded with word of mouth referrals, but when it gained some press publicity, the fitting began to attract more interest. Systemteknik was a company that made load cells for cranes and worked with the Swedish air force. Robin Cooper (from Steepers) speculated that they were making *"some sort of handling device"* at the time and so were persuaded to make a prosthetic hand for children [182]. Support came, in part, from the Swedish government. Torbjoön Holmqvist led the design the hand was ultimately made in two sizes [55]. It was in 1976, five years after the original fitting, that a child's hand made by Systemteknik was ready. Following success in Sweden hands were tried in Japan, Russia, Canada and the UK [55]. Over the years the range of hand sizes grew and eventually the English company, Hugh Steeper Limited bought the hand and continued to make and sell it. When tracing its history it is sometimes hard to separate these eras, as people continued to refer to it as the *Swedish Hand* even when it was being made by Steepers.

Until 1978 Sörbye was conducting the training and assessment of the children himself, but in that year he got his first OT and in January 1980 Lottie Hermansson came to Örebro. At school Lottie liked creative subjects and had indicated she wanted to work with people, so she was directed towards occupational therapy by a careers advisor. Occupational therapy was not a widely known profession in

Making Hands. https://doi.org/10.1016/B978-0-12-820544-0.00020-2

Sweden at that time, and before working in prosthetics Lottie had worked in other places, conventional occupational therapy, neurology, hand surgery, rheumatology, psychiatry, and rehabilitation. When a position in the limb clinic became available, she applied.

Hermansson first visited UNB in Canada when she was invited to give a talk at the 1983 MEC symposium. She worked with the therapist at UNB to help create the test aimed to assess the function of children with their prostheses. The test was based round the performance daily activities. The activities were chosen to be age appropriate, giving four different age categories. It is this practical functional approach that typifies Hermansson's methods. Unfortunately, the test is not generally applicable across the world as it is aimed at the children and uses activities that are appropriate for a Canadian user in the 1980s. It cannot transfer to adults or children outside North America. Indeed more recently the UNB test has been updated because the tasks are not appropriate for the modern Canadian child either.

Lottie's interest in assessment was because she wanted to be able to objectively measure progress of a user. She wanted to answer the question: *How does someone use their prosthesis?* Conventional tests, such as the Jebsen test of hand function [315] (well known in hand therapy), were irrelevant to her. They only told her how the hand works, not what someone could do with it.

Over the intervening decades Lottie has learned a great deal about assessing someone's function. Her first attempt at a more sophisticated assessment was the Skills Index Ranking Scale - SIRS. She admits this is not the name she would use today as it is confusing [22]. SIRS attempted to capture the steps made by a wearer in increasing their skill level when using a prosthesis. Hermansson learned about instrument development and Rasch analysis [316]. Rasch Analysis is aimed to allow the designer of a test to make sure the test produces meaningful results. When scoring is not arbitrary, a specific change in score means knowing more than simply that something has changed. For example, if a test involves a series of tasks and one task is twice as difficult as another, it should get twice (or half) the score the first task does. If it was not scored this way, then a five point increase at the bottom of the scale would mean something completely different to a five point change further up. It would be nearly impossible to interpret the scores of one person against another, or even the same person if later they had improved their ability. A simple analogy is if a running race was held over a fixed time, not a fixed distance. If the race was on a flat course then running twice as fast would allow a runner to cover twice the distance, this is an easy comparison. However, if the course had a steep hill climb in the second half of the race, the distance of the slow runner would no longer be half that of the faster runner. In manipulation tasks it is harder to see if there is a harder part of the test. Rasch analysis allows the designer to identify the 'hills'.

Lottie applied the Rasch analysis techniques on the data she already had from the SIRS. From this she was able to develop her new tool, the ACMC *(Assessment for the Capacity of Myoelectric Control).* It was the result of twenty years of work. Her epiphany came when she realised what she was looking for when she studied a really skilled user: *"How competent they were at operating their hand?"* [22]. It was this that she felt she needed to measure. This problem was made more complex as she wanted to be able to capture a meaningful score from the very best user to the very weakest operator without getting ceiling or floor effects. If the test is designed well, even those at the edges get a change in score with improvement in performance. An example of a person who would hit the top end of almost any scale of functional control of a prosthetic hand is one of Hermansson's colleagues. She is an occupational therapist who uses a myoelectric hand. Without feedback the colleague can tie a bow behind her back. She cannot see it, she can only feel her arm's position and the vibration of the motor through the socket. This is certainly something that requires considerable skill to do and is hard to capture objectively.

11.1.1 Assessment for the Capacity of Myoelectric Control

The name of Lottie's test, the *Assessment for the Capacity of Myoelectric Control* (ACMC) is most specifically chosen. It is the person's ability (or capacity) to control the hand that is being measured. Nothing else is measured. It is also why, despite requests, she has not approved its use to measure body powered prosthesis control. It cannot be simply and easily extended without much more work. By 2019 there were hints that she had taken on the big task to extend the test to body power.

The ACMC is an observational test where the tester is trained on what they should look for in a systematic way, to see and note instances of control (or lack of it). To perform a test a person is asked to undertake a task they know well, there is no need for them to learn a new task. The tester watches the subject and notes when they spontaneously use the prosthesis to pick up an object, use it to carry an object or simply use it to hold an object passed to it by the other (intact) hand. The ACMC attempts to capture the ability of a person to perform an action with myoelectric control. It tries to measure what the person really does with their prosthesis, rather than using arbitrary tasks set for them by the tester. The tasks are chosen by the subject and the rater observes the person performing the task. As such, the ACMC is a good guide to help the therapist record how good an individual is and compare it to some other time, before or later. It is not designed to compare one user relative to another, as this is not particularly relevant information to the therapist.

The score is a complex measure and is not easily interpreted. Other professions would prefer a simple metric, such as a number. This is especially true of engineers. One easily recorded number is time to completion of a test, but this may not measure what one thinks it does. Ed Biden often referred to Bob Scott's resistance to timed tests, as Bob believed the time simply measured how competitive the subject was and did not reflect any other aspect of their ability to use the hand [172]. As Lottie has refined her test, she has gained a perspective that timed tests are not as poor as she might have once considered. From the outside, timed tests look like a competition rather than a reflection of the person's ability with their hands. Data shows a correlation between functional ability and speed of execution [317]. Basically, if a user has good control they will naturally be faster (or perhaps more significantly, a subject with poorer control will not be as fast). Thus timed tests embody aspects of the persons' control within the test, and produce a simple result; the time. *"The information I got from watching children doing these timed test taught me more about how they can develop their skills in operating their hands"*. Skills such as when the subjects open their hands and how much they do so, just enough to perform the task. *"If they are very skilled it will get quicker"* [22].

To perform the test, an activity that the user has some experience of is selected and they perform it as the OT observes. The observer then scores based on the various ways the subject performs the activities. For example; the subject is asked to make a cake, by using prepared cake mix in the kitchen in the clinic. The subject must gather the tools required, plus the ingredients. They then remove the ingredients from their packaging, mix them, put them in a cake tin and place it in the oven. It is observing this that gives the therapist their insight. The subject will have to search the cupboards, open the packages and pour and mix the contents. All of these tasks can be performed single, or dual handed. The observer sees if all the actions are done with the sound hand. They try to answer questions such as:

- Is the prosthesis used to stabilise the object?
- Do subjects open two cupboard doors with a door in either hand?
- Do they open with the sound hand only?

- Do they then pick up an object with the prosthesis, or the sound hand and pass it across?
- Do they walk across the room with the object in the prosthesis, or in the sound hand?
- Do they drop objects?

The aim is to determine how able the person is and how spontaneous they are with their myoelectric control. The different methods get different scores, which leads to the ACMC rating.

In essence the ACMC is a framework to guide the way a person looks at the user. It attempts to encapsulate years of experience so that one can, at least in part, see the activity of the user with the perspective of someone with Lottie's experience.

The ACMC can only be measured reliably by a trained therapist. It was observed that the most consistent scorers are those with the most experience. ACMC is designed to allow a trained observer to make valid observations. It has been made and tested to ensure that the measures remain valid. Initially only experienced OTs could take the course and be certified. I was the first engineer to observe the training. With this I gained a great insight into the way that OTs view users and how users work.

11.2 The Sven Hand

The influence of the work in the Soviet Union in the 1950s spread throughout the prosthetics community and led to many different projects. In Sweden, Carl Hirsch had learned of the Russian Hand through contacts in the Russian orthopaedic profession and was interested to progress the ideas it presented. When a young doctor, Peter Herberts, arrived in Göteborg to start a PhD, he was informed it would be working on improving myoelectric control [318].

Göteborg is a port city on the south west corner of Sweden. It is a pretty city with wide streets and canals that run through the town. It was built by Dutch engineers, but despite this the locals refer to Göteborg as *Little London* due a resemblance that is centuries old and was probably lost when both cities were redeveloped.

Herberts trained as an Orthopaedic surgeon, and he studied for his PhD between 1965 and 1969. In 1971 he was appointed associate professor and was responsible for the multifunction hand project in the 1970s. Later, he was put in charge of joint replacement at the Sahlgrenska Hospital. This was the time Herberts' focus shifted to clinical work. In 1976 he started a project to record the details and outcome of every prosthetic hip implanted in Sweden. This work became known as the *Swedish Total Hip Registry*, and he ran it for 30 years. The registry is such a significant resource of information on different designs of hip arthroplasty that researchers from other countries come to sift through the data and learn about which designs and techniques are successful. It is also the model for other countries and other areas of medicine, even though it took (for example) twenty years for the United Kingdom to follow suit. Since this time, one hundred other medical registries have been set up in Sweden to extend this knowledge to all parts of their healthcare. As a result the Swedish healthcare system has been reorganised and is described as *value based care*. Clinical decisions are made on the masses of data which give doctors the possibility of being able to calculate what benefit any intervention may have for a particular patient [318]. Internationally the number and type of registries has grown. Two of the more recent registers that have been set up are one on osseointegration (see 11.3), and a global registry on limb prosthesis fitting.

At the start of Herberts' PhD in 1965, it was clear that the initial goal of his project had to be to create a system with proportional control from the EMG signal. Longer term they considered a multifunction hand a possibility. At the time, the Soviet prosthesis used the myoelectric signal simply as an on and off switch for the motor to rotate at a constant speed, and so the hand opened and closed at one speed. To make the hand controllable it had to move slowly, if it were run too fast it would repeatedly overshoot the target position. The first part of Peter's project was to look at the myoelectric signal to see if there was more information in the signal that could be used as a proportional control input. His first study used a dedicated piece of equipment to study the power in the different frequencies in the signal [319]. This so-called 'Power Spectrum Analysis' is a simple function in any suite of signal processing software today, but in the 1960s it required a custom built machine for the purpose [318].

As we know, the myoelectric signal is very variable and changes over time, so Herberts' first tests looked at the signal as the person fatigues. What he found was that as the person grew tired their signal changed with the high frequencies decreasing and the lower frequencies increasing.

From this he created electrodes that could filter the higher frequencies that gave a more constant signal for control. Electrode design and construction was undertaken at the Chalmers University of Technology in Göteborg. The work on the design included looking at ways to make the electrode implantable with wireless transmission of the signal through the skin. These electrodes were tested in people and the report in the prestigious *Journal of Bone and Joint Surgery* in 1968 was widely read [320]. This work showed that such electrodes were possible, but they were *"a little too big"* to be practical [318]. Indeed it is only in the last decade or so that this sort of transcutaneous design has become even close to being usable.

In the early 1970s the group in Göteborg began discussions with the Swedish defence agency. The latter was interested in the design of a multifunction hand and so the *Sven Hand* team was formed. The aim was to create a complex hand system which could be controlled easily. The work was funded from major grants from the medical research council, and Chalmers School of Technology enjoyed grants from the Swedish technical research council.

It was realised that the hand would need a rubber glove to cover it. This glove would need to be especially flexible. As a result, the condom manufacturer Centri was approached. Ultimately Centri began making other prosthetic systems, which included one of the few children's hand designs on the market. They also produced an adult hand which was different in design to most of the other single axis hands. Most of the others in the market then or since superficially resemble the Russian Hand, with one motor driving three fingers in a precision grip. In the Centri hand the wrist and fingers were moved together but they were linked to the wrist as well, so the wrist flexed as the fingers closed, putting the hand in a good position to pick objects off a horizontal surface.

The team quickly realised that without the buried electrodes specifically and reliably placed within the major muscles of the forearm, the possibility of controlling the hand by simply picking the major muscle that performs that action (wrist rotation, hand opening etc) was low. They developed the idea of pattern recognition as a means of control. The team from Chalmers: Roland Kardefors, Christian Almström and Peter Lawrence (visiting from Canada's Institute of Biomedical Engineering), used signals from six muscles to detect six motions from the subject's phantom hand. This technique required a novel socket design to hold the electrodes precisely. They also needed, (what was for the time), considerable computing power to detect the person's intention. It was able to work but according to Herberts: *"An experiment like that took half a day"* [318]. The results were prone to electrical and mechanical disturbances and interference from the mains electricity, but it did work.

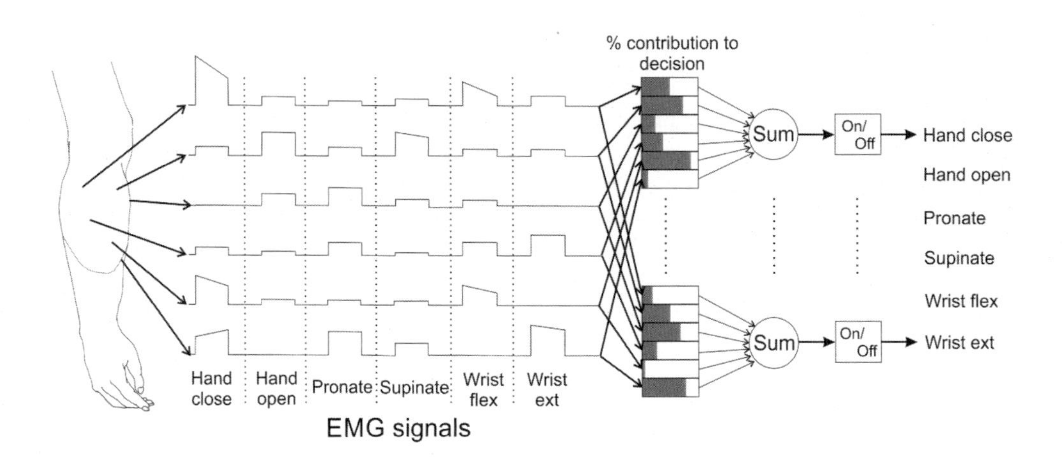

FIGURE 11.1

Sven pattern recognition system. The control of any single channel is simple on or off and is the sum of the contribution of six electrodes on the forearm of the user. (Adapted from [7,8].)

After Lawrence left it was Christian Almström [321] who continued the work and developed the software. Once the idea was demonstrated with a desktop hand and the user attached to the computer, Christian evolved electronics small enough to be carried by the user so that functional tasks could be performed in the lab.

The next stage was to use it to work with a hand prosthesis. The hand had three axes, and so six motions: Fingers opening and closing, wrist flexing and extending and rotating left and right (pro-supination). As such it needed six commands. In the history of powered prosthetic hands there have been very few wrists that flex. Even today it is not one of the motions that is widely available in a powered form and is rarely conceived of in research devices (Fig. 11.2). Herberts was concerned that the hand was too ambitious and it could have been simplified in the first instance, but the engineers wanted to make the perfect device, so they persisted with the complex hand.

11.2.1 The Sven pattern recognition system

To control the hand, the team created a controller that used the results of the computer analysis. Each of the six outputs was the sum of the activity of all six muscle signals. The computer analysis revealed how much emphasis would be placed on each signal for each output. For example, the output to drive the first axis (output one) could mostly be input three, with a little of two and five, so the sum would be: 'five times one, plus two times two, plus two times five'. The computer analysis of that action determined that this sum.

Output two could then be mostly four, with some of one and six ('five times four, plus one times one, and two times six') and so on. This was effectively a *weighted sum*, all added together, but each channel given a different weight (or emphasis). The system adds all channels together with some in greater proportion than others (Fig. 11.1). This means that it was possible to build a simple set of six amplifiers with settings for the level of amplification for each of the inputs and outputs. Instead of a computer that took up half a room, it was eventually a box the size of a hardback book. By the middle of the 1970s the Sven group had shown that pattern recognition control of a multifunction hand was possible [7,8,224].

The occupational therapist for the project was Kerstin Caine-Winterberger who worked on the project for 18 months from 1975 to 1977. Conventional prosthetic use training began with intensive work in the clinic. Following this, the wearer was able use the hand at home and improve with practice. Any subsequent visits to the clinic were to help refine the user's technique or adjust the settings. As the Sven hand could not be taken home, the four subjects came into the centre twice a week for two months for training. Once the first session was over and the base settings known for the arm it took only fifteen minutes or so to set up the arm at the next visit [322].

The control of the hand was sequential. Each axis had to be moved alone and then the motion switched to a different axis. This is still a problem that many in the pattern recognition field are still working on. Kerstin said that some of the subjects found it difficult to coordinate the separation of the movements, wanting to grasp and pronate or supinate at the same time, but they all improved with practice [322].

A video of the hand was made with one of the users showing how he could move abstract objects about (including very small ball bearings). The user made a sandwich and, as he was an interior decorator, hung a piece of wall paper. The video is spectacular, with the size of the controller made clear and the ability of the hand, and most especially the ability of the user revealed. Herberts said that the subject was very laid back and did not have much training in using the hand [318]. According to Caine-Winterberger, the gentleman did not use any prosthesis normally, but he can be seen to employ it very effectively [322]. In the video, the early tests are with the abstract objects or in the kitchen. The hand is the primary manipulator, or at least giving strong support to the intact hand. When the task moves to the user's own profession, he becomes more one handed, and shows himself to be a very good decorator. If one was scoring his capacity for myoelectric control according to Hermansson's *ACMC*, then this lapse into working with his natural hand and simply passing the prosthesis the brush etcetera would give him a lower score. The kitchen preparation would be rated far higher.

When I saw the video for the first time in the late 1990s he was struck forcibly by how useful wrist flexion and extension was to remove the compensations needed from the shoulder (Fig. 11.2). It underscores the idea that flexion is an important motion of the wrist. Later the group at the Scuola Superiore Santa Anna in Pisa has shown that with a simple (nonmultifunction) hand and a two-axis wrist, there is far less need to compensate than with a fixed wrist and a multifunction hand [323].

The resulting Sven hand garnered a lot of interest locally and internationally. The hand and user were presented to the Swedish king. The day before the meeting, Christian Almström had been labouring hard to get the hand to work at all, and with little time to spare they placed the hand on the user. The user realised that some part of the controller had been reversed, so the hand opened when he instructed it to close and vice versa. The user was quick enough to learn how to overcome this flaw. When the time came the hand opened and closed on cue and the subject shook hands with his monarch [321].

The British Broadcasting Corporation (BBC) asked Peter Herberts to travel to the UK to demonstrate the hand, but Herberts was unhappy with the idea that such an appearance might raise expectations in the user community [318], a question that those of us in the field have to struggle with, even today. He knew that it was a long way from a commercial product and so was it not right that he should appear on television where the presentation of the work might lead potential users to think that they could have one very soon. In fact the problems associated with the delivery of the system were still considerable. It took a long time to set up the system for the user. It was hard to make a socket that

FIGURE 11.2

The Sven Hand being demonstrated by one of the experimental users. The use of a wrist flexion unit allows the hand to be placed in much more natural positions to manipulate objects. (Taken from video made of the demonstration.)

would hold the electrodes in place and create a stable control signal. The amplifiers still needed to be improved to get more information from the muscles. Finally it was clear that battery technology could not provide enough power for long enough to make the hand useful [318]. Over the last thirty-five years these problems have been reduced or overcome, but it is clear he was right to be cautious about how fast such progress would be made.

After a pause, in 1980 Ulf Bengtner joined the team and produced a microprocessor controller which was tried by users before 1984 [324]. The result was an improved controller, making the adjustment to the wearer's signals much quicker. They used a newer mechanism that still had the original six movements (hand open/close, wrist flex/extend, wrist rotate), plus an ability to ab/adduct the wrist (Fig. 11.3). The work allowed the recording and analysis to be done more quickly than before on a computer. The transfer of the settings to the controller was done by writing to reprogrammable memory. Finally the Motorola microprocessor (68HC11) and electronics was fitted within the forearm space of the prosthesis in a cylindrical space about five centimetres round and up to ten centimetres long.

At this time at Chalmers University of Technology, also began to look at using EMGs to study work and occupational based problems, with multiple fine wire electrodes placed in the body. *"It was all right up to four electrodes, after five everyone fainted... (it was) was kind of a side effect"* [324]. The testing of the hand was on 'a few' amputee users and colleagues in the laboratory, but it never progressed beyond the lab, due to the limited number of hands available for testing and the difficult nature of getting the settings correct and stable (a problem only solved by Blair Lock in the early years of the twenty-first century).

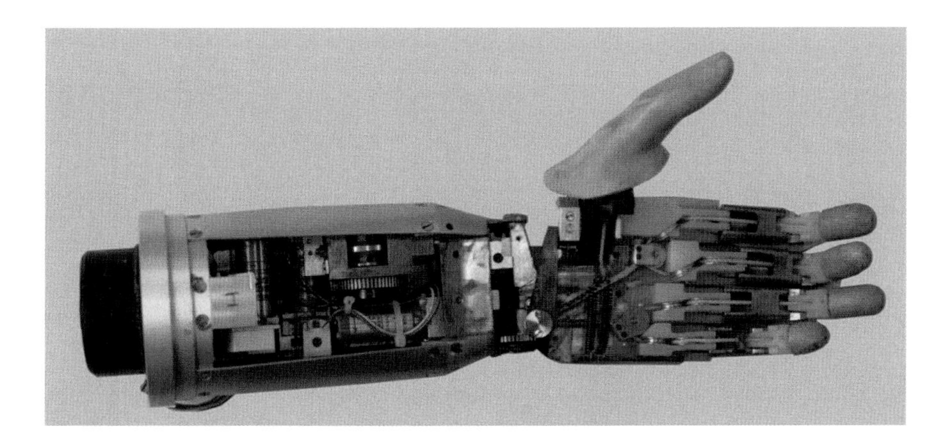

FIGURE 11.3

The Sven Hand. The mechanism has continuously curling fingers driven through two whiffle trees, powered thumb abduction, wrist flexion and pro/supination.

Unfortunately for prosthetics, a boom in the Swedish industry drew engineers away from clinical research to more lucrative jobs (including Bengtner before he finished his PhD). At the same time a downturn in funding from the Swedish research councils made projects scarcer. It was at this point that Herberts and Kardefors decided that it was not practical to continue and so they moved on to other areas of research.

At the end of his career Herberts still believed that the basic need was for a light, easy to use prosthetic hand. The development of the modern multifunction hands is welcome, but he cautioned against the high level of complexity if it comes at the price of too much weight. Later work produced the second design by Henry Lymark, for which Ulf Bengtner made the controller. The last version of the hand was produced by the two companies: Een and Holmgren Ortopediska AB and the company producing the children's hands for Sörbye in Örebro, Systemteknik AB. The resulting hand became known by the initials of the two companies: *ES Hand* [318,321].

This was the only powered commercial multifunction hand to reach the market before the Touch Bionics iLimb. Fig. 11.4 shows the fingers curled as they flexed and the thumb passively abducted. The fingers were pulled open by springs on the dorsal surface of the hand and the finger drives were arranged in a whiffle tree arrangement to allow the fingers to adapt to the shape of the target object.

Kerstin's career has spanned all the way to the modern multifunction hands, which she regards as different from the Sven. When asked about her impressions of the Sven hand, Kerstin felt that the hand looked more natural when it was being used well, but her emphasis remains on finding what is practical for the user, which might not be an advanced hand at all [322].

11.3 Osseointegration

The idea to attach a prosthetic device directly to the bone is one that has occurred to many people over the years. A problem is the same as for the other external surgical techniques such as transplants; the

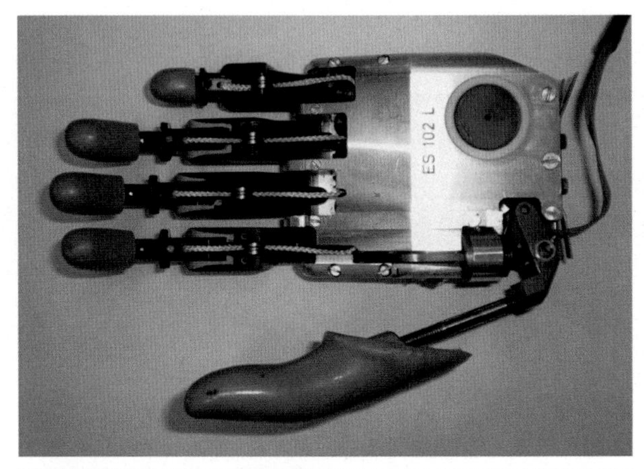

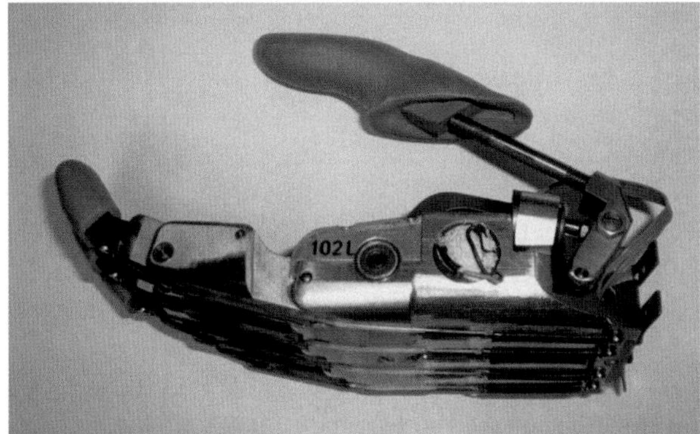

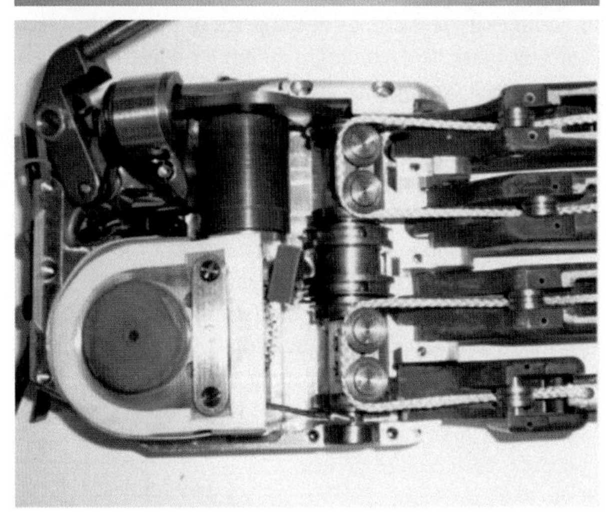

FIGURE 11.4

The ES Hand, the only multifunction hand to be marketed before the Touch Bionics hand.

body will recognise the object as foreign and reject it. The most significant problem is that a device passes through the skin, creates ideal route for infections to enter the body. Infection in the bone is very serious. Poor access to the bone from the outside makes it hard to control any infection with antibiotics, so that one of the few options open to medicine when there is an infection deep in bone, is amputation. This has meant that the idea of introducing a foreign object into the shortened bone that might then cause *even more* of the limb to be amputated is not to be considered lightly. The Swedish team that succeeded were neither the first, nor the only, team to try this. In Utah, Jacobsen's early work was in this area, but without sufficient success he moved on to other things. The difference was that the Swedish anatomist Per-Ingvar Brånemark, had two elements essential for a breakthrough; a little luck and a prepared mind. He came by the first clue to success accidentally.

For his PhD, Per-Ingvar was working in Lund in the 1950s. His area of research was looking at micro circulation in bone [325]. In order to distribute the building blocks for life and take away the waste, the body needs fluids to flow inside the bones. To observe this flow, the tissue must be alive and part of a living animal. To observe fluid flow in vivo in another part of the body, a window is made by creating a thin area in the local tissue (such as skin) and shining light through, the tissue. A light microscope can magnify the image and the tiny vessels and cell movement can be observed. Bone is of course opaque, so this method cannot work in bone. To observe the flow inside bone, the experimenter has to grind down the bone walls to a wafer thin region that is mostly transparent and then they can use the microscope. With this, much progress in studying the flow was made, but the method had its limitations.

Brånemark then moved to Göteborg in 1960 to work at the University of Göteborg's Department of Anatomy. Here, Brånemark developed an implant with a window in it that screwed into the bone. The device was illuminated from below and viewed from on top creating a way to observe the flow in detail. The same window was left in the subject and it was then used over time to study healing and growth. When designing his implant P-I asked an orthopaedic colleague what material should he use. The friend suggested the new metal; titanium. It is light and quite inert. The oxide that covers the outside of the metal is less reactive than gold oxide. P-I produced his first implants, which successfully allowed him to study the circulation. Later when he was trying to retrieve the implants at the end of the studies, he found they were stuck firmly into the bone. They did not just fall out as he expected. At the time this aggravated him, and it was later he realised that titanium's ability to fuse with bone might be a useful trait [325].

He started to collaborate with colleagues in different areas and exploit titanium as an implant material. The first application was in plastic surgery, repairing cleft palates in adults. This disorder is where the roof of the mouth does not form properly in the embryo and the person is left with anything from a small hole in their palate to a large gap in the midline of their face that includes the palate, the upper lip and parts of the nose. At the time any reconstruction was only performed when the patient was an adult, so there was a need to attach new bone or find replacement material for the gap. Many of these people had problems with their teeth, so being able to firmly implant false teeth to their jaws was useful and far more practical than conventional false teeth which attach to neighbouring teeth. Early osseointegration experiments were conducted in Beagles with a follow up of up to five years to see how firmly the teeth had attached [326]. This established that the implants were stable long term and did not need special care to ensure they survived. With this Brånemark could move forward to human subjects [325].

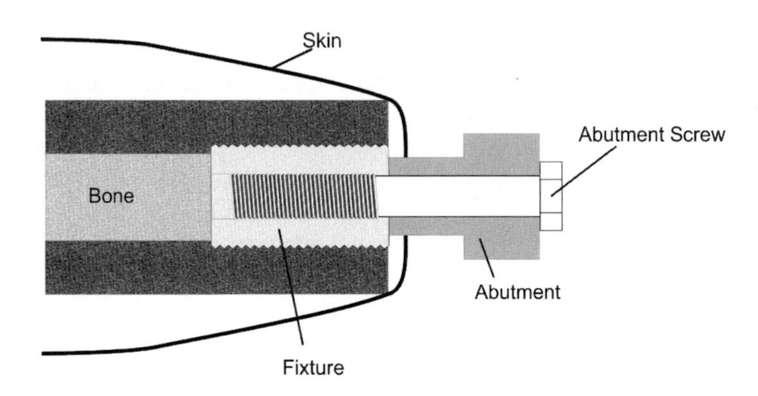

FIGURE 11.5

Brånemark osseointegration. First a titanium fixture is screwed into the bone and allowed to integrate before an abutment is added to the implant and very thin skin draped over the end of the bone. (Adapted from [9].)

In 1965 the first person to have a dental implant, was an adult male with a cleft palate [325]. It had not been simple to make a new implant as this was a completely new application of dentures and no one had ever made such devices previously. To get to a satisfactory result the team had gone through fifty designs in the animal studies and then the best went on to be tried in humans, improving the design further.

Over the years the design of the implant and its insertion has been refined. The application of an implant uses a two stage process. The first stage is when a threaded plug made of titanium is very carefully screwed into the bone after a hole has been slowly hand drilled into the jaw and a thread tapped into the bone. If living cells are heated there is a risk they will die, so power tools are not used and the bone may even be cooled to make sure that as few of the cells are damaged as possible. Brånemark was very careful with the technique making sure that it is performed consistently [327].

Once the plug is screwed in, the site is covered over and left for weeks or months. Only once it has been allowed to integrate with the bone is the site opened up and the 'stage two' device (known as 'the abutment') is introduced into the body (Fig. 11.5). It is this abutment which passes through the skin and contacts with the outside world. For a single dental implant this abutment is the tooth, but over the years there have been variations. For example, two implants either side of the jaw are used as pegs to mount an entire set of dentures on the jaw. Later refinements included packing bone around the implant, at the stage one insertion, so that a receding jaw can be built up again.

While the first teeth were implanted in 1965 it was not until the mid 1970s that it was accepted as a viable treatment option in the Swedish healthcare system. This was after a number of careful studies had shown its promise. The application to approve the technique had to predict the level of interest that it would enjoy. They suggested roughly a hundred Swedes a year who would use this innovation. By 2005 this prediction was shown to be woefully inadequate, as the numbers had reached 100,000 dental implants in Swedes a year. It was not until 1982 that an international conference accepted Osseointegration as a treatment option for dentists. Since this acceptance its application has grown and by 2015 there was an estimated one million teeth implanted world wide every year with the global numbers running somewhere above twenty million in total [325]. Osseointegration was only slowly adopted taking more than fifteen years to be accepted by the dental community. Per-Ingvar's son, Rickard, notes that even in 2015 writers and journalists still commented on this technique being 'experimental' [325], despite the hundreds of prosthesis users world wide.

Once osseointegration was shown to be viable in the jaw, the team considered many other ideas of application, although it would be sometime before any were attempted. An early example attached a hearing aid to the skull, so that noise vibrations were passed into the head via the implant. One form of hearing loss is when the mechanical connection from the outside world via the ear drum to the cochlea is damaged or lost. Sensors inside the ear are no longer able to receive airborne sounds, but are still be able to pick up the vibrations of the skull. In an osseointegrated device, the tight connection to the bone meant that when the abutment is vibrated, the sound is effectively passed into the skull and some hearing is restored. In 1987 Bo Håkansson and Peder Carlsson were given an industrial award for their bone anchored hearing aid design.

Other aspects of plastic surgery can also benefit from osseointegration. Some procedures might be referred to as 'massive' restorations. Entire body parts, such as ears, can be held on with simple clips to frames and pegs that stick out from the skull. The benefit of this form of fixation is that the prosthesis needs far less effort to attach or remove. It uses no glues or bands, and the edge between the prosthesis and the body can be made thin so that the casual observer cannot see the join. The most extreme of these are the facial reconstructions, where a person has lost a significant portion of their face through accident or disease. Only the face surpasses the hand in its ability to define the person and someone with an incomplete face will find moving around in society always challenging. The osseointegrated attachments for this sort of reconstruction can make an enormous difference to the life of the person who would not wish to go out in public with literally half their face missing [328].

The Brånemark technique has been applied to such injuries. Its most extreme example is a woman who possesses only one eye and the lower part of her jaw. The rest of her face is lost. Before reconstruction, she had always kept her face covered by a scarf. For her, a titanium framework was constructed that fitted into her skull. The frame created an upper and lower jaw, with the dentures held in place with magnets. Then her face was replaced with a high quality silicone sculpture of her original face, which clicks onto the frame using magnets. The result is impressive. Her face will not move or animate, while her personality still shines through. She can move through a modern city with no one really noticing what is different. It is the osseointegration that makes the framework possible. The silicone mask can have edges thin enough to hide the joins. Without the osseointegration any silicone facial reconstruction would have to be laboriously glued on to the person's face and the edges would have to be thicker so a step between the mask and the skin would be apparent.

The idea of using osseointegration in an exo-prosthetic application had been around for years: In 1977 a cartoon of P-I by the chief hand surgeon Göran Lundborg, showed him as a classic amputee pirate brandishing a sword with a bone anchored hook and peg leg [325]. However, as with many other prosthetics stories, it was a potential user who challenged the profession to make the first step [327]. In this case a woman pushed Per-Ingvar into following up the idea. In the mid 1980s an article in a general interest magazine, ran an article about the bone anchored teeth [327]. After reading this a woman wrote to Brånemark. She told him she had lost her legs at the age of sixteen when she was run over by one of Göteborg's many trams. The loss was very high in both legs. Conventional socket prostheses would not function particularly well at her level of loss, hence she used a wheelchair exclusively. She tasked Brånemark to come up with a solution. The magazine article said that a man could practically hang off his own dental implant, and she only wanted to *stand* on hers. Five years later she was the first person to stand on osseointegrated prostheses.

To progress from teeth to legs required the implants to be scaled up. Biomechanical studies showed how much force was needed to allow a person to stand or walk on their legs, and the clinical work showed what forces could be transferred through a given sized implant. This told the team how much contact area the implant should have and an engineering safety factor doubled it. The operation was successful, although today the woman does not walk that much on her prostheses. She mostly wanted to be able to stand up and perhaps move more easily to a chair, which she has been able to do for over a quarter of a century. After twenty-four years Per-Ingvar's son, Rickard, replaced the implants with newer designs, born of the decades of research.

It is Rickard who continues the research into osseointegration. When P-I retired, he continued some research on application of the ideas until his death in December 2014, Rickard pursued the more clinical research.

11.3.1 Rickard Brånemark

Rickard Brånemark was certain was going to be a doctor. At this distance he is less sure why, except perhaps like many a child he wanted to follow in his father's footsteps. Having started on this road he has travelled further and broader than his father, building on his work. Although a project report he wrote at school on the dissection of the human hand suggested that medicine was his destiny, he started at university as an engineer. This was the result of the Swedish medical school system. At the time, it admitted qualified students via a lottery process, so while Rickard had very good marks, he was not guaranteed a place immediately. For the interim period he signed up to read engineering, as he thought the knowledge would be useful in his later career, this has proven to be true. To understand osseointegration requires an understanding of the engineering of the implant and the biology of the bone and tissue around it. Over his career, Rickard has straddled both sides of the divide. The younger Brånemark won his place at medical school after a short time, but he did not stop the engineering degree immediately, for a term he studied both full time. While very proud of this achievement, he then modestly suggests that the early days of a medical degree simply needed an ability to remember material, it was the engineering that needed understanding [325].

In 1995 at the World Congress of the ISPO in Melbourne the Brånemark team revealed the results of the many years of work in Göteborg to a packed house. The entire team was present, therapists, prosthetists and surgeons. The range of results, from the teeth to facial replacements, were presented. Many of us in the audience were impatient for the results of the exo-prosthetics work. Saved to last, videos of the first recipients putting on their prostheses without a socket and being able to reach above and behind themselves with a greater range of motion than a socket prosthesis would allow, was a moment few who were there will ever forget. Yet they left us with as many questions as answers: Who could benefit from this technique? What levels could be it be fitted to? When could we get one for our users? Only over the years have we learned the answers.

Perfectionism is a national trait in Sweden. *"You wouldn't want it until it was right?"* is a question that you often hear, if you work with Swedes. In their defence, Rickard Brånemark suggests they are *"conservative"*. In fact for a technique that cannot be removed easily if you change your mind and where, if it goes wrong, the patient could lose more of their limb, prudence is clearly essential. Osseointegration is not a procedure that should be undertaken lightly. Brånemark admits that it takes a long time to progress the knowledge and get to a clinically acceptable solution.

The challenge he faced was to *"to speed up the rate of research progress, without endangering the patients"*. A lesson from history is the pioneering surgeon, John Charnley. He studied the problem of creating hip arthroplasties. Today these are the most successful surgical technique ever devised. They transform the lives of people with worn natural hip joints who cannot move without extreme pain. They allow them to return to a mobile lifestyle. Some patients with arthroplasties can do everything they did before, others have to slow down a little, but compared with the fate without an artificial hip, it is miraculous. The 'most successful' claim is based on the enormous transformation that is possible to an individual and the huge numbers of individuals who continue to benefit from this technique [325].

When designing the first arthroplasty, Charnley made a ball and socket joint similar to the natural one he was replacing. It consisted of a wedge that fitted in the centre of the femur, with a ball at the top end for a head. This fitted into a socket that was placed in the pelvis as close to the anatomical location as the surgeon could determine. To make this work, Charnley needed to make the ball run in the socket as smoothly as possible. Today high-density polyethylene is used. It is a dense plastic that wears well. Charnley originally chose PTFE (Teflon), as it has extremely low friction. Unfortunately it wears quickly and it produces very small particles of debris that cause inflammation in the joint. The inflammation is painful, it limits the joint motion, and may cause the implant to loosen so much it must be removed. This flaw only became apparent after the first patients had found how much better their lives were with the new hips. The first two thousand recipients of the Charnley hip had to be re-operated and the Teflon replaced [325]. This shows that caution in introducing even something as simple as a joint is essential, if mistakes are to be avoided.

Rickard Brånemark commented that for the dental implants: *"the lead time to come up with new things and get it accepted is at least 15 years and we can see that also with amputees"*. While the first limb implant was fitted in 1990, there were some in 2015 who still do not accept osseointegration as viable.

> "Good and bad things related to that of course... the good is that if it takes some time you learn more and you don't see a lot of disasters... maybe if [Charnley] had waited a little bit longer, there would be less re-operations... however, if you never move on, nothing will happen... If John Charley hadn't moved on, we might not have the hip joints, so it's two sides of the coin, all the time" [325].

This challenge drives the work in Göteborg. After their first few operations with the exo-prostheses, they began to standardise the approach and the technique. The result was the OPRA study [9]. This has produced over one hundred implants with bone anchored limb prostheses spread across the world.

The constant refinement of the technique aims at two important goals: Greater reliability and reduced complexity [327]. The first is simply making sure that every patient has the best chance of a good outcome. The process of fitting includes X-rays and CT scans to establish the correct approach for the implant. Rickard then performs a practice run on a plastic bone model. If the residuum is close to average of the population then he can use a standard plastic skeleton, created for teaching purposes. If there are significant abnormalities then the CT scans can be used to generate a custom model with 3D printing techniques (14.1.4.1). Rickard then practices the operation. On the day, he takes the practice model into the theatre and refers to it constantly [327]. In the theatre there are many small practices that ensure the operation is a success, such as coating the device in marrow blood before it is implanted, to make sure the correct cells are there at the start. These he refers to as *"Swedish Voodoo"*, methods that may have started in the dental community but have died out because they can generally get good results without them [327].

The second goal, is to simplify, is so that the technique can get an even wider global take up. Infections are always a concern and so the cleanliness of the operation and the wound is important. Rickard comments that, if possible, this technique could benefit landmine victims in Laos, Cambodia or Vietnam [327], but the result of a blast is considerable contamination, reducing the chances of a successful osseointegration. The younger Brånemark wants to develop the techniques so that one can just do the surgery and the implant can tolerate the environmental conditions.

> "That is where dental implants are today so why shouldn't we be able to do that? ... If we are as smart and good as dentists we should be able to do it" [327].

He appreciates that the dental surgeons have been at this for more than 50 years and the limb surgeons are about 15 years behind.

> "I am utterly sure that with international, globalised collaboration, lot of good scientists, clinicians and may be new ways of dealing with biological response" [327].

Once a stable abutment is established, there are other innovations that become possible. The goal of making electrical connections directly to the nervous system is long standing and is frustrated by many roadblocks. Getting the signal out of the body and through the skin is equally difficult. It is possible to pass signals through the skin, wirelessly. Indeed it is even possible to pass electrical power into the body to drive the electronics, but the technical problems remain significant for routine provision to prosthesis users. Bypassing the skin is a desirable option. It might seem an obvious step, but the abutment should be able to act as the conduit for wires to sensors into the residual limb. The result can be electrodes that can pick up the EMG signals from individual muscles, making the signals more specific than standard surface EMGs, with the connection unaffected by socket motion. That is not all, electrodes can attach to the nerves so they can trigger sensations of touch or other forms of feedback to the subject. The team investigated the idea for many years before they announced their first user with the implanted electrodes in 2013. As significant was that *any* of the Brånemark implants fitted in the preceding *fifteen years* could also take this upgrade. This is the sort of far sightedness that few innovators in the modern era would ever have the vision or confidence to conceive.

Osseointegration's biggest danger is that of infection. If the implant is well anchored into the bone then there should be little chance of the infection getting in. Skin placed over the end of the screw will attach to bone around the countersunk stage one implant and seal the abutment. In the early days the team needed to learn how important infection control was. Persistent infections were perfused with antibiotics. These days, the care taken ahead of the procedure greatly reduces the chances of infection. The team make sure that the bone and the residuum have no cultivatable bacteria within them before they begin the procedure. This has reduced the later infections to a minimum, even for amputees with blast wounds who certainly have been exposed to many bacteria at the time of the original trauma. Even so infection is the aspect of the early work still remembered by those who are sceptical of the procedure.

As the operation interferes with the conventional use of a socket and takes months to be complete, it is worth remembering that the cautious Swedes excluded many potential recipients. For example, any users who were able to use a socket prosthesis well, were refused an implant. Only wearers who had trouble with conventional prostheses were recipients of an implant. This meant that patient experience in the early days of the procedure was with the cases who had experienced such difficulties. These recipients were therefore atypical and potentially they were persons with problems with their sockets and residua, rather than the simplest cases with the wounds that healed easily and were provided with straightforward prosthetics. This might have meant that many of the cases in the study had troubles, which may or may not have been related to their prostheses. Despite this, once the Brånemarks had shown the way, others were bound to follow. In the mid-teens of the twenty-first century, there were over twenty other groups doing one form of direct bone anchorage or another, showing that it is both possible and practical.

Brånemark created a company, *Integrum* to explore the technology. By the middle the twenty-teens the company had two foci were to conduct clinical trials of the buried electrodes (to prove that they provided a better signal and control for the user), and most importantly, obtaining FDA approval in the United States. The USA had been a closed market to the Swedish technique. With approval to conduct surgeries in the country the technique had, for the first time, access to the largest potential market and a population that will pay for a new technique, if they can see a benefit.

Osseointegration's benefits are many. It removes the need for sockets that limit the range of motion of the residuum. A person with a loss above the knee can sit on the ground or on a chair without the brim of their socket resting on the edge of the chair in an uncomfortable way. If the person has a loss below the elbow, the user can bend their elbow more when not wearing a socket. They can reach behind themselves and scratch their back and they can dress more easily. Without a socket they also have greater surface area of skin to lose heat. The amputation itself takes away a significant area of skin to sweat from, and the socket covers up more skin. Osseointegration gives some of this surface area back. The mass of the prosthesis matters less as the device becomes part of the person, rather than something they wear. The best measure of this integration into the person is that people sleep with their prosthesis on [327]. This is almost unheard of with socket prostheses, much as people rarely wish to sleep with their shoes on, it just feels wrong. Osseointegration changes this practice.

One unexpected side effect of osseointegration is what has been referred to as *osseoperception*. It has been shown that we feel through our bones [329]. This is not something that a physician would have been sure was true, before osseointegration. Those with prosthetic legs feel the floor through their prostheses, and they do not feel that are floating in mid-air as they do with a socket prosthesis. Those with arm prostheses can feel vibrations from their device, such as when they rest it on the gearstick of a car. A result of the implant is that the phantoms change. Over time, following a conventional amputation, some people's phantoms change shape, a foot retracts up to the end of the stump, or the missing arm disappears so the phantom hand floats in space. After osseointegration the phantom leg may extend and the foot may rest back on the floor. Tests with functional MRIs show that the areas of the brain associated with the touch of the missing area are activated when the person has an implant [327].

The entire process to fit an osseointegrated implant can take many months. The initial implant (a threaded section) is placed in the bone and left for at least eight weeks before the stage two abutment is added and the skin is placed on the ends of the bone and implant. In all living systems there is a delicate balance between repair and disuse. If we exercise, we get more muscles and our bones become denser and stronger. If we do not exercise the body will uncompromisingly remove the muscle and bone (7.2.1.1). Exercise encourages growth and addition to bone density. It stimulates repair, a person with a healing leg fracture should exercise, but too little or too much can lead to failure. This is also true for a bone beginning to osseointegrate. Recovery has to be a very careful set of stages increasing the loads placed on the implant, but not overstraining the implant bone interface or submitting the implant to sudden loads. As with a leg fracture, patients who do too much, risk falling and introducing a sudden and large force which can break the new connection. To reduce this happening the newly integrating get a short ‘training prosthesis’ that allows smaller forces to be placed on the implant [327]. The wearers then exercise daily, increasing the loads and repetitions of loading. If the person gets any signs of problem (such as soreness, redness, swelling) they go back to smaller forces and lower numbers of exercises, until the implant is stabilised. Thus it can take some time before the full weight of the person or the prosthesis can be born by the device. For leg implants, the exercise is to stand with partial weight onto a scale for twenty minutes and then to note how painful the implant is when the load is removed. If it is too painful, the loading scheme will be reduced until it can be tolerated. The integration is better in pulling/pushing (tensile) loads than twisting ones, so the connection on the leg prostheses includes a fail safe to release before any twisting becomes too great. For the osseointegration of arms there are similar exercises. The subject exercises the arm with a short prosthesis on.

While the standard Brånemark technique is a two stage process, Rickard will perform a single stage operation when the bone can take the load (i.e. it has not lost its strength due to disuse, age etcetera). Similarly if there is sufficient soft tissue in the right place to cover the implant and make a good joint, the team will routinely undertake single stage operations. For example, a small implant like a peg to hold on the end of the thumb is generally safe to be done as a single procedure. With improvements in the way the soft tissue is handled, more single stage operations may become possible [325].

When asked about the science behind osseointegration, Rickard Brånemark confesses he is still trying to understand it better [325]. The original understanding of the structure of the integration was based on optical microscope images, where the biology and the metal met at the interface. The microscope slides were made by embedding a slice through the limb in plastic and grinding it until it was translucent and could be observed under the microscope. At this resolution the osseointegration looked as if it went smoothly from pure titanium, to titanium oxide and then on to bone. As it became possible to see deeper with electron microscopes, this changed. Where previously they observed continuous transitions with no gaps, they now can see there are places where there were gaps, or other materials in the spaces, as well as places that were smooth transitions from titanium, through titanium oxide to bone. The question becomes; how much direct contact between bone and titanium is enough for osseointegration? This is still being explored [327]. What is certain is that the body has not rejected the prosthesis. It has not thrown up fibrous tissue around the implant to exclude it from the body, as happens with a conventionally implanted hip or knee joint.

11.3.2 The future of osseointegration

Rickard Brånemark remains on the lookout for potential partners that can add to the knowledge of the procedure. This ranges from mathematicians to model the bone and soft tissue, to material scientists and companies to create titanium oxides that are lighter or stronger.

The other implant techniques being offered by the other twenty or so groups are often similar, but with variations in the design of the implant, or occasionally the surface of the device. Some add a coating to the surface of the implant. It is a crystal, hydroxyapatite, which is found in bone, and is there to encourage the bonding of bone with implant. It is used in routine joint arthroplasties, but it does not always work, such as cases where the hydroxyapatite originally bonded to the implant and then bonded with the bone and separate from the metal [330].

How the skin interface is bridged often separates the techniques. One solution comes from the Stanmore Hospital in England [331]. Their Intraosseous Transcutaneous Amputation Prosthesis (ITAP) attempts to mimic the way that some deer antlers grow through the skin of the deer and the skin infiltrates the border of the bone. Most antlers are actually hair, only a few are bone and it is these animals which the ITAP mimics. One of its advantages are that it is a one-shot process, rather than the two of the Brånemark. The hole is bored and the replacement inserted and the wound closed. This allows the technique to be tried with animals, including some paying customers for veterinary solutions. I have seen a video of a Golden Labrador that shows it happily using either leg to stand on to urinate, despite the fact that one leg was artificial and held on with an ITAP. This suggests its value. The pioneering vet, Noel Fitzpatrick, uses a variation of the idea in implants on his patients [332]. The greater flexibility afforded to people experimenting on animals *as patients* allows him an ability to try ideas still waiting trial in humans.

The different approaches may come from the national characters. Brånemark puts the caution down to the Swedish character. While others can go out and try the system until it breaks, Brånemark says they do not yet have the data to approve of that level of exertion. *"We work in a different way... We are from Sweden. We drive Volvos, safety comes first"*. If there is the demand they will need to work on more robust systems. He thinks that people should be able to do anything they like, *"except American football. No implant system will survive that. Not even an intact body will survive that"* [82].

Osseointegration causes engineers and clinicians to ask hard questions of their knowledge and their designs. Not only must the implant integrate within a few months, but it must stay integrated over many decades. The team now has experience with teeth for over four decades, and of limbs, more than thirty years. As a person ages, the bones change. Women especially will tend to lose bone density post menopause, but experience shows that loading through exercise still stimulates bone growth effectively and a person with a well anchored prosthesis is more likely to use the limb and exercise than one with a poorly fitting socket. Indeed the first recipients over seventy are showing *less* bone loss around the abutment than elsewhere [82]. This suggests the additional exercise the prosthesis allows, is helping to protect the subjects.

A changing biological platform is harder for the engineer to cope with. Engineers are used to equipment ageing through fatigue and wear, and not dynamically changing as the device is used more often. Osseointegration is almost certainly one technique pointing to the way bioengineering must progress. As humans become more and more integrated with their technology they need to learn more about how the changing environment of an implant must dictate changes in design.

Rickard clearly feels like he is still at the beginning and there is a great deal yet to study. For example; he is examining how changes in the material surface of the implant effect the response of the growing bone to the presence of the implant (*"nano topography"*) [82]. The aim is to change how the bone reacts to load, adding more density and strength under the right loading regime. He would like to look at changing gene regulation to impact how fast the bone will grow. Local electric fields can also affect the way cells grow and can be used in wound healing and antibacterial control [82].

The insights gained, allow the team to fine tune the designs and match them to the individual. Over the last decade the team have developed the mathematical models of the bone and implant, so an experiment is not about a generic model of any bone and any implant. It is now *the* person's bone and the actual implant. This allows them to create individualised protocols. Rickard referred to the ideal treatment: A CT scan of the bone is made before the surgery. After the implantation the team uses sensors and loggers placed in the bones and implants to record the loading of the bone and implant over days, months or years. With additional CT scans later, united with this very accurate loading recording, they will be able to look at the real bone remodelling: *"we will be able to explain Wolf's law, finally"* [327]. Wolf's law is the cornerstone theory in bone physiology. It explains the bone's response to loading. Too little or too much and the bone weakens. The right range encourages growth and density, so that exercise is the best treatment for osteoporosis, if the person can exercise. This theory is hard to test because it has been impossible to measure in detail what loading is applied or how much bone is laid down. Osseointegration could give the scientists a window to explore this theory.

Rickard has hopes that there will be much progress in the next decade with an explosion in the number of scientists working on the problem: *"In ten years time if we sit down again we will say we didn't expect it to go this way if we had done it ourselves"* [327].

One place Rickard sees avenues to explore is the broader application of surgical techniques to the rehabilitation of the injured person. Following the FDA approval of osseointegration he spent some time on the west coast of the US, before taking up a position as a research scientist at MIT with Hugh Herr. With a stable implant and the ability to get control and feedback signals [220]. Rickard sees the need to take the techniques further: *"To some extent we have solved the anchorage, solved the pass through of signals. Now we need to work on what sort of external signals, how to pick them up..."* [82]. Initial feedback signals felt much like *"odd electrical tingling' not real sensations"*. Others, such as Dustin Tyler of Case Western Reserve University, USA, have attempted to use different forms of modulation of the signal to get different forms of feeling, with some success [221], but this is not a full range of controllable signals.

Brånemark wants to combine the existing surgical techniques to take him further, to create an entirely new form of prosthetic interface. With use of various surgical techniques he believes he could make something akin to the missing limb in miniature which would be the interface to a much more sophisticated prosthetic control system. He hopes it will be possible to assemble a collection of specialist tissues in a geometry that can generate physiologically appropriate signals [82].

The method would be to use the three modern surgical techniques: TMR, RPNI and AMIs. AMIs take existing agonist/antagonist pairs of muscles, now detached through the amputation, and by adding a tendon to link them over the end of the bone recreates the muscle pairing. As a result they have a physiological action once more, one muscle stretching the other as it contracts and vice versa (2.6.1). This is only possible if the appropriate structures remain after the amputation. The RPNI technique relocates small parts of muscles to another area, where they are reinnervated to create small muscles to be used as new EMG sources. A novel concept is to create a biological structure that allows the signals of the missing limb be recreated. It would be an RPNI pair joined to create an AMI and it would be placed it in the bulk of the remnant limb. Finally the appropriate Golgi tendon organ would be reinnervated. The person will be able to feel the muscles as an opposing pair and the muscles would work more similarly to the natural system than two separate reinnervated muscles. The sensations created when the AMIs were stimulated would make more sense to the central nervous system. Rickard wishes to: *"see if that can create a balanced joint perception in the brain... If it works it will seem to the brain like natural proprioceptive feedback"* [82].

He wishes to take this even further. Based on an idea by Hugh Herr: He would take a skin patch where the epithelium has been removed and place it within the remnant limb and reinnervate it. The patch would be arranged so that the stimulated muscles would cause the skin patch to be distorted. This will be interpreted as skin stretch or touch (depending on the geometry) by the brain. As the skin will be generating the right sort of signals the brain should interpret this as contact more easily than artificially stimulated nerves. Similar to Tamar Makin, Rickard Brånemark too is looking to pass basic signals to the brain and leave the complex signal processing to the nervous system (see 3.2.1).

"The brain is fantastic at sensory and control systems... the better the signals we can supply to the brain, the more the brain can do, so (we are) helping the brain sending in good signals... we can pave the way for a totally new way of interacting in a natural way with the brain without having to decode the brain signals. Which no one is up to."

The problem he sees is that one can obtain short term funding, perhaps for a PhD program. But for something like osseointegration he needs to think about it over the next two decades, or an entire career (like his father). He believes he needs to follow the ideas to their conclusion. *"If you only ever have a 2, 3, 4, [or] 5 year perspective you [will only] go to the end of your nose"*.

The Swedish school system no longer needs a lottery to admit the bright students into medical school. The have made the process tougher, so if a student gets the top grades they can immediately choose to do whatever they like. Which means that Rickard, had he been born later would never have studied engineering. *"And that would be a terrible thing"* [327]. He really likes the challenges of the medical arena, but never the rote learning from medicine, preferring the ideas in physics and maths. Engineering gave him a different perspective. *"the training of the brain to be analytical, training is better in Engineering school than the medical school"*, he reflected, and this training has served him well [327].

CHAPTER

Limbs from the Windy City

Prosthetics research in Chicago

12

During the hiatus in prosthetic research in the 1980s and 1990s there were three centres in North America that throughout this period continued to undertake prosthetics research at a meaningful level. To some extent the three centres worked together during this period and have continued their fruitful collaborations since the turn of the century. The two Canadian centres are dealt with together in their own chapter. The third is the one that is based in Chicago.

12.1 The Rehabilitation Institute of Chicago

As with many of the labs in this book the name of this institute has changed over the years, and for simplicity I will use the term that was used for the longest period: 'The Rehabilitation Institute of Chicago' simply *RIC*.

At its inception in 1951 it was housed in a renovated warehouse. In 1974, it moved into a new twenty-storey, state-of-the-art speciality hospital on the campus of the McGaw Medical Center in Chicago, overlooking Lake Michigan, a position that shows it status. On the 25th of March 2017, the RIC officially changed its name to the 'Shirley Ryan AbilityLab', following a donation by Pat and Shirley Ryan who made their money in insurance. This coincided with the new research hospital building, billing itself as *"the first ever translational research hospital"*. The emphasis is on research integrated into the treatment of its patients. This was the natural extension of way that some patients of the RIC had been part of the research process for many years.

The RIC grew through a combination of internal developments and external collaborations. It would expand to cover locations in southern Illinois and north central Indiana, and develop a collaboration with Northwestern University and its Feinberg School of Medicine. A residency program was created in 1967, and in 1968 the Institute became one of six research and training centres in rehabilitation medicine funded by the Rehabilitation Services Administration of the US Department of Health, Education and Welfare. Its move to the McGaw Centre completed its transformation into a state-of-the-art speciality hospital. Being a large centre with a significant catchment area meant that the prosthetists in Chicago saw and treated some of the more involved cases and gained knowledge of ways to help those with more extreme amputations.

The centre was founded with the vision of making a difference to the patients [333]. Prior to founder Paul Magnuson's visit to Germany, the tendency in the USA was to treat disability in residential hospitals, keeping the impaired away from the population. Magnuson observed that the Germans' treatment of their atypical population allowed them to function more easily in mainstream society and so he set up the RIC to have a similar impact. Magnuson's successor, Henry Betts, took this vision and grew the RIC into the seventeen storey hospital building. Betts believed in giving the best treatment to the person to get a significant outcome *irrespective of the cost* [333]. The teams would work with the patients

Making Hands. https://doi.org/10.1016/B978-0-12-820544-0.00021-4

until they had useful function. This gave the RIC a strong reputation with rest of the medical field, but not such a good reputation with suppliers and insurance companies who had to support this effort.

This idea has continued, but the latest director Joanne Smith has aimed to improve the Institute's financial reputation. She is a product of RIC, she went through as a physiatry resident before getting a medical business degree at the University of Chicago, and then returned to the RIC, she became director/CEO when Henry Betts retired. She has worked to maintain the clinical standards while working on the aspects of administration (such as their credit rating) and ensuring ethics paperwork was complete and correct. Her first phase *"tidied up the books"*, while maintaining the clinical focus [333]. Her second phase was to advance the delivery of service. She believes research is the key to clinical success, not only improving the techniques, but also ensuring the patients and researchers share the same space. The idea is that if patients and researchers routinely see each other it will impact on the way research is done and how patients feel as they receive treatment. The new hospital aimed to have the research and the patients on the same floor, so the two groups will be more aware of each other. The concept she strives for is to create a *"research hospital"*, somewhat like a teaching hospital, where all patients know that they are part of the education of the next generation of doctors. At the new RIC they will know that they can be part of the latest research program. One catch is ensuring that the ethical standards are maintained. The team needs to strive to ensure that the patients do not feel *compelled* to be part of the research simply because they are in the centre, but are willing volunteers [333].

The RIC's connection with the Northwestern University and the prosthetics school was always hard to determine from the outside. Visitors to the tower would slip between the RIC and Northwestern without actually realising it. The majority of the staff may have appointments in one and work in the other, or perhaps in both. Papers are published referencing both organisations, but when an outsider, confused by this intermixing, makes an error and refers to the wrong group, they are firmly put in their place. People have studied as a student of Northwestern and worked later in RIC. Early on Northwestern gave the RIC house room in its facilities. In recent years it has been the reverse. In the last few years the Northwestern labs have moved out to their own facility. It is Northwestern that has a prosthetics school and the original prosthetics research lab (NUPRL) [334].

Dudley Childress was the focus of the work in Chicago for a generation, making links with other centres and promoting work in different directions. For example, he met David Simpson and embraced EPP as a control idea and conducted research on it (see 8.1). It is perhaps a measure of Childress's openness to ideas that he not only worked on EPP, but continued to refer to it and Simpson every time he spoke of it (a tendency not always followed in the research community).

Many people who work in the field credit Childress with encouraging them to work in prosthetics. One of the major players of the new century is Todd Kuiken who sees Childress as an early influence. Jon Sensinger, now director of the UNB Institute of Biomedical Engineering, realised he wanted to work on engineering for the disabled from the age of nine. When a teacher pointed him in the direction of prosthetics it was to Northwestern and Childress, that he aimed his sights from the age of thirteen [333].

Todd Kuiken studied at Duke University for a bachelors in Biomedical Engineering (1981-83), which he followed by a PhD and an MD at Northwestern in Chicago (1983-90). He said he loved the technical aspects of study, but in the end he is: *"very much a people person and liked the dynamism of medicine, I liked the idea of taking care of people so I ended up getting an MD/PhD so I didn't even have to make up my mind"* [42]. He easily gravitated to rehabilitation, he said it was the result of *"hanging about at the RIC"*. His PhD was looking at muscle re-innervation of rat muscle. *"That's were*

I got the idea for TMR, from a one liner in a paper by Loeb and Hoffer, in a discussion (section) they were apparently writing late one night". The one line was: *"maybe you could move the nerve from an amputated muscle".* It does seem the sort of throw away line one might put in a paper on a technique, suggesting where else it may go, without any personal intention to follow that route. However, to Todd it laid out the path to improved function for a prosthetic user. It has continued to resonate with him to this day [42].

12.2 Targeted muscle reinnervation

Targeted Muscle Reinnervation (TMR) is a surgical technique to *create* control sites rather than to recover them. It takes a nerve that used to control a muscle that no longer exists and attaches it to muscle that no longer has a body part to move. Conventional reconstruction surgery will try to restore as much of the damaged limb as is practical, but once the muscle is gone, the control and the signals are gone too. In a less damaging accident the limb remains intact but a nerve is severed, it is possible to reinnervate muscles with that nerve. Unattached nerves try to grow and make connections to muscle, any muscle. If a detached or severed nerve is placed back in a muscle, connections are remade, and some control of the muscle is restored. This is a routine procedure for a nonamputated but injured person. This result might be seen as something of a failure, because in some cases less than half of the nerves rejoin the muscle. As described in Section 6.1.3.1, muscles are made of bundles of fibres and each fibre twitches a small contraction. A smooth tension is the result of lots of little twitches being added together to make a larger force. Fine control of muscle tension depends on being able to coordinate all the muscle fibres twitches, making lots of little twitches add up to a smooth increasing level of force. Not all muscles have the same number of nerves per fibre. Put another way, some muscles are made up of a smaller number of bundles of fibres. As the nerve excites a bundle, they all twitch together. Fewer bundles mean more of the muscle turns on at once, so there are bigger steps in the increase of tension. With more bundles of fibres, the output force can increase in smaller steps and so be controlled more precisely. Taken to an extreme, this means that a muscle made up of one hundred bundles could increase its force in one hundred small steps, while a muscle that is all the same bundle could only be fully on or completely off. It is clear that it would be hard to do anything sensitive and precise with the second muscle. The muscles for our hands have many more nerves than the big muscles in our legs. This helps to give us greater control over our hands than our feet and the greater dexterity that can be achieved. If the reinnervation process is less effective than the original innervation of the muscle, then more fibres will fire simultaneously and move the muscle with less finesse than before. This is obviously better than no control at all over the injured muscle, but it might seem to the patient to be less than a total restoration of function for their living arm.

Having read the late night one liner in the paper by Hoffer and Loeb, Kuiken's insight was that while it results in poorer control of a normal muscle, if it is applied to someone with an amputation, it can be used to create completely new control sites that did not exist before. Some control is created where there was none. The more limited control is no longer as big a disadvantage. Anything is very much better than no control sites at all.

The more of the arm that is missing the more motions are needed to bring the hand to the mouth and restore some function. This is the paradox of the conventional controller: The greater the loss, the more degrees of motion the person needs to have restored, but the fewer the sources of control signals for the arm they possess. One option is to intercept the nervous signals going to the muscles and extract control signals from them, then use these signals to control a prosthesis. This has been the goal of researchers for decades. They have tried many ways to attach electrodes to the nerves to sense the signals, but there are many problems with getting a successful match between the biological and the electrical systems. While this might change in the next decades, it remains a limitation for the majority of advanced prostheses.

Since the muscles to control a body part are generally above the joint being controlled there are often muscles in a shortened arm that no longer have a role. For example, the pectoralis muscles. These muscles raise and lower the arm, but if the amputation has removed the upper arm they no longer have a role. Kuiken realised that following the amputation, he could place nerves that used to run to missing muscles from below the loss, in the unused, but still existing, muscles above the amputation. He could employ the unused muscles to amplify the neural signals to the muscles and create new EMG signals. With this he could detect and use the nervous signals more quickly and easily than developing detectors to attach to the nerve. For his PhD, Kuiken looked at 'hyper' reinnervation: To put a bundle of many nerves into a muscle originally controlled by fewer, and observe how the growth distributed between the fibres. He hypothesised that putting a 'large' nerve in a 'small' muscle was *"upping the odds"* on the connections being made. He observed that they distributed well and created more connections even when there was a failure to connect at all by some of the fibres [42].

This is a radical concept: To take a functioning muscle (which no longer has a purpose, but could be used for conventional myoelectric control, and remove its nerves, before reinnervating it with a different nerve (that has nothing to control) to create control signals that are more physiological. Kevin Englehart said that TMR was *"the only thing that seemingly came from nowhere"* [335].

To be allowed to perform this on a living human, Todd first had to establish it in principle with studies of what would happen to the muscle if such a procedure took place. This included mathematical studies followed by TMR surgery on rats. Success gave Todd the confidence to consider performing the operation on a human.

Their first major user was Jesse Sullivan. Jesse had been an electrical linesman. He used to work on the high-tension electrical distribution cables in the US power grid. It is a dangerous job, and one day Sullivan became part of the electrical pathway. The electricity crossed his body through his arms. When this happens very large currents can flow which heats up the skin and bone. The temperature rises rapidly and the flesh burns. The form and extent of the damage depends on what happens to the path and how long it flows. If the flow is through the heart, it tends to be fatal. If the current runs mostly in the arms, the result is often loss of much of the limbs and massive burns to the chest. Such it was with Sullivan. He lost both arms at the shoulder. It would be understating affairs if one said that the prosthetic solutions for such a person were challenging. He had nothing except his feet to undertake some manipulative functions, nothing to hang the prostheses off, and nothing much to control them with.

Kuiken had been planning to perform the first procedure on a person with losses in the mid humerus. It would be a simpler operation. The biceps and triceps could be split into two and three sections simply. The name is a clue to the structure of the muscle. They have two and three 'heads' (places where they start). Each could be separated to create five muscles. The new nerves could then go into the 'extra' heads creating up to four new control sites easily. Then *"Jesse came along, and he was such a remarkable person"* Kuiken said [42]. Sullivan is an easy-going man, he speaks with a relaxed southern drawl and is keen, enthusiastic and ready to try anything that might make a difference.

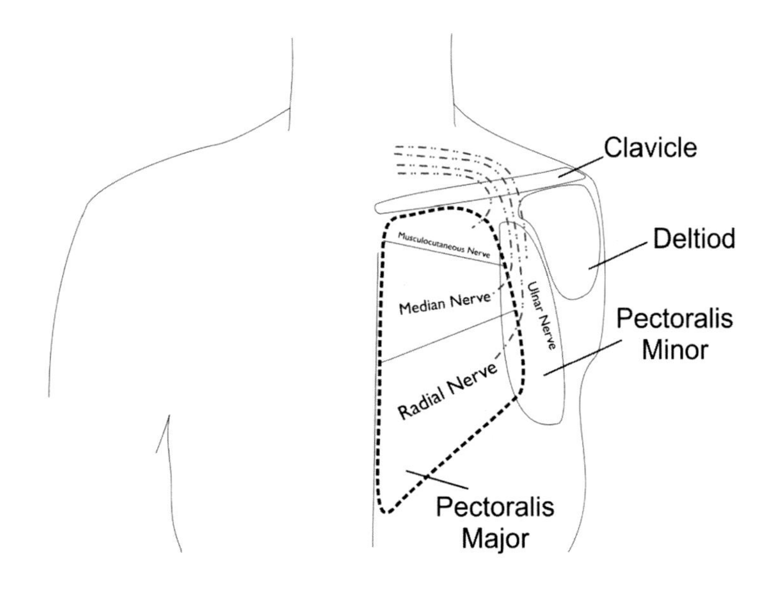

FIGURE 12.1

The muscles reinnervated in Jesse Sullivan's chest. (Adapted from [10].)

All surgical operations carry risk. TMR was an experimental procedure, which means it was ethically impossible to undertake new surgery for just this procedure. Not only might the operation not benefit Sullivan, it might actually cause him more problems. Performing an operation (that could be life threatening), just to fix a problem that is not dangerous, is not an acceptable risk. Limb absence is, not of itself, life threatening, so experimental surgery for prosthetic control alone could not be justified. However, as Jesse's original wounds healed, the injuries were causing him discomfort and so there was a need to perform some surgery on him to alleviate this. Under these circumstances it is possible to undertake other low risk surgery not connected with the main procedure. With this Kuiken and the team got their opportunity.

Sullivan was a willing volunteer and an excellent coinvestigator. Since his operation he has been the subject of numerous studies of his recovered control, and has used a number of the most sophisticated prosthetic arms ever built. He always works with enthusiasm and humour. He remains a friend to the project and to Kuiken in particular.

The surgical procedure on Sullivan's chest was developed by Greg Dumanian, a plastic and hand surgeon at Northwestern Memorial Hospital. He took the larger pectoralis muscle and divided it into three. He electrically isolated the different parts, and moved the smaller pectoralis muscle to the side of the chest to serve as a fourth muscle. The team then took the major nerve running onto the arm and put the musculocutaneous, median, radial and ulnar nerves and placed them in the separated chest muscles [336,337]. They then closed and waited for the healing to progress.

One problem with the procedure, especially in the early days, was that it is hard to know what the result will be. All the preparations cannot remove the element of chance in the outcome. One difficulty they encountered was being sure that the nerve was not damaged in the original accident. What they

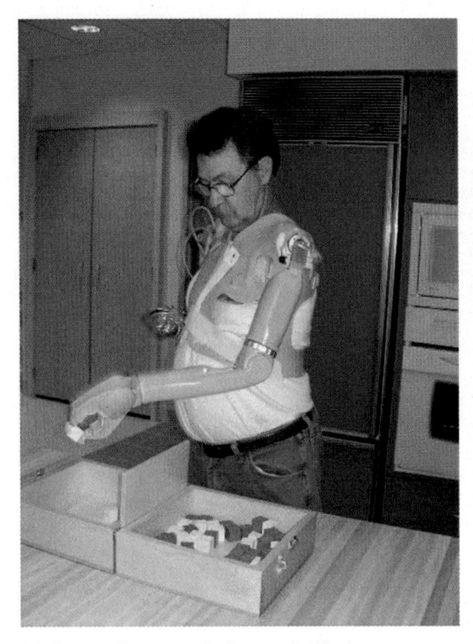

(a) Practical tests with Box and Blocks assessment tool.

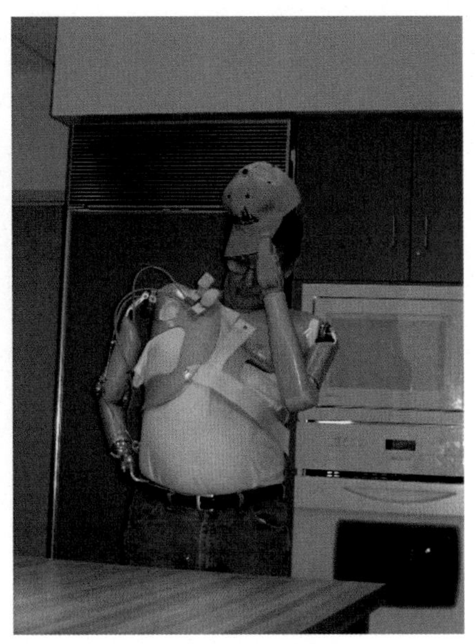

(b) Demonstrating ability to reach up to his head. Compare with Fig. 12.3, with the greater number of degrees of freedom with the research arm he needs to compensate less.

FIGURE 12.2

Jesse Sullivan was one of the first recipients of the TMR technique. These pictures show him operating his routine prosthesis. His dominant arm use body powered devices and the nondominant uses myoelectric systems. It is the left shoulder that has the TMR. (Shirley Ryan Ability Lab.)

found is that with a natural system there can be unexpected results. For Sullivan, the musculocutaneous, median, and radial attached, but the ulnar nerve failed. This would have reduced the number of sites created. However, nature has a way of surprising. Instead, the median nerve naturally split in two and reinnervated the unactivated muscle, so that they still had four control signals. This shows how readily nerves will attach to a muscle, given the chance (Fig. 12.1).

Sullivan's insurance provided funds for a routine arm. He could use a Boston Elbow and Otto Bock wrist with Greifer, which was more than he had been able to control easily before the operation (Fig. 12.2) although the number of degrees of freedom made some actions awkward.

The team also assembled the most sophisticated arm ever built with six degrees of freedom. They wanted to use a powered shoulder. At the time the only such shoulder proven to work clinically was that produced by David Gow in Edinburgh, so they used an EMAS *ProDigits* shoulder (7.1.2.2). They employed a Boston Elbow in its conventional position and a second Boston Elbow drive on its side to create a powered humeral rotator. For the hand they wanted wrist rotation and flexion so they employed an Otto Bock rotator and a Kesheng hand with a built-in wrist flexor (Fig. 12.3). Jesse was able to perform a number of manual tasks with greater flexibility and fewer compensatory movements (Fig. 7.4).

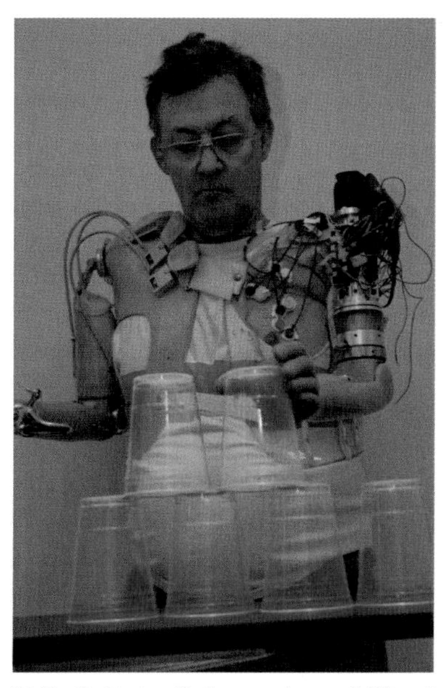

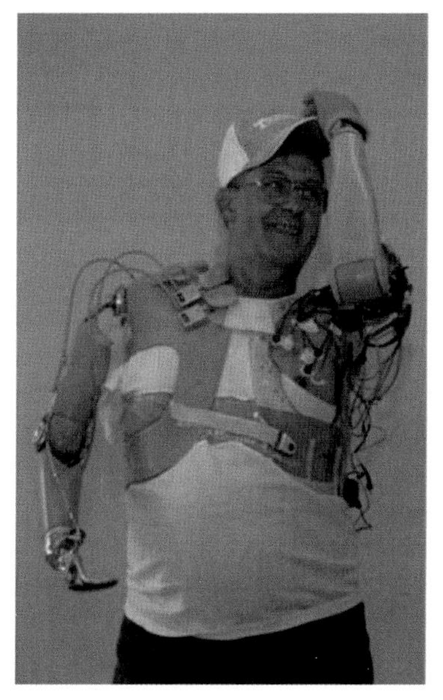

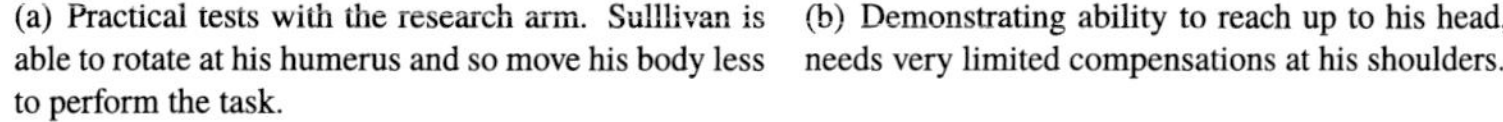

(a) Practical tests with the research arm. Sulllivan is able to rotate at his humerus and so move his body less to perform the task.

(b) Demonstrating ability to reach up to his head, he needs very limited compensations at his shoulders.

FIGURE 12.3

At the time this was largest number of degrees of freedom assembled into a single prosthesis. Scottish shoulder, Boston Elbow and a second used as a humeral rotator, and Otto Bock wrist rotator and Kesheng hand with built-in wrist flexor [5].

Learning to use the 'big arm' with conventional control would have been a daunting task: Move one joint, switch, move the next and so on. Instead they wired each axis with direct control from the muscles so that each joint had its own controller all the time. With this more than one axis could move at once. They tried to match the action of the muscle the nerve used to control to the relevant joint in the arm. It rapidly became obvious to Sullivan he had control of both shoulder and elbow and so he was able to reach forward, moving both in combination so that the hand moved out from his body in a straight line, in a way not possible with almost any other form except the Southampton Arm controller (see 10.2.2). The video of Sullivan reaching the hand forward in almost a straight line by moving his prosthetic shoulder and elbow together is impressive. It looks almost natural and easy. Sullivan clearly has good kinesthetic ability to coordinate his efforts, but it is the TMR that allowed him to channel the same reflexes he used to use with his biological arm [5]. In everyday life Sullivan does not use the full arm. He employs the simpler, lighter, more reliable arm (Fig. 12.2). He was the first of many to benefit from Kuiken's insight.

Luck played an important part in the next part of the process. Little regard had been given to the fate of the sensory nerves in this procedure. The team were not expecting Sullivan to tell them that he could feel his hand within his chest. Again the nerves found a way. In this case it was his sensory

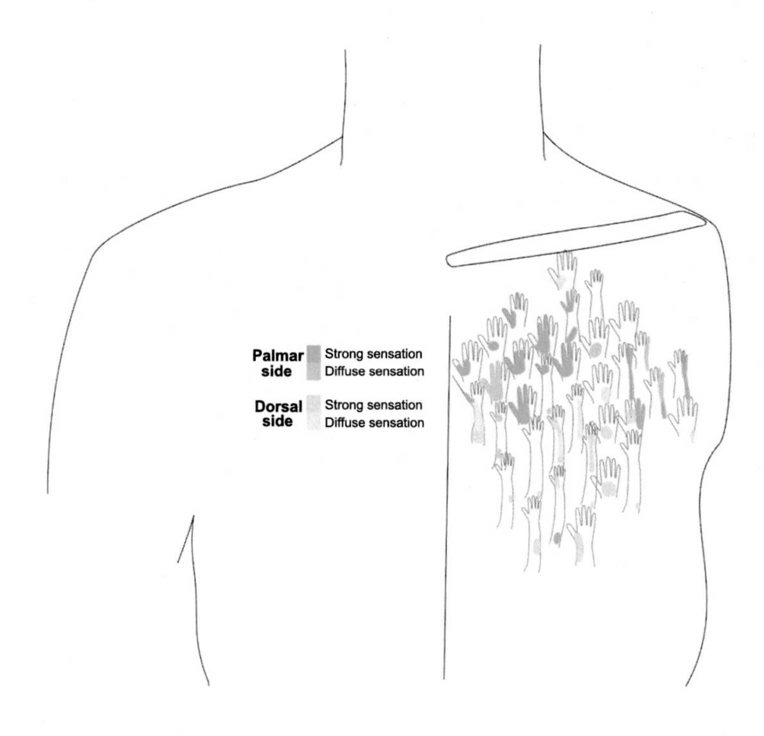

FIGURE 12.4

The sensory reinnervation of Sullivan's chest. The sensations are spread across the muscles in his chest. Different areas have become reinnervated by different nerves, giving distinct sensations. (Adapted from [10].)

nerves. This was investigated by stimulating the skin across his chest and the different locations found were mapped to different places on his missing hand. The different nerves linked to different receptors, so that Jesse could feel sharp sensation and diffuse sensation in distinct areas. The result is a map across his pectoralis muscle that includes fingers, thumb and other parts of his hand. As the index and second finger use a different nerve to the other two fingers, it is not surprising that their sensations are separated on Sullivan's chest (Fig. 12.4).

While the teams have continued to explore these sensations, there is little practical use for them yet. It is necessary to build devices to mechanically stimulate the chest. These 'tactors' remain bulky and it is still part of research to work out what signals will be the most useful to the operator of an arm. Meantime, the TMR sensation does provide a window into how the nervous system works. Animals cannot tell you what they are feeling so there is only so much you can learn from putting implants in animals. Research using implants can make some progress, but there are many problems of matching the electronics to the nerves. This aspect is still only slowly progressing. To test the TMR recipients electrodes are not required, just devices that press on the skin the subjects can describe what they feel. This opens a door to research into nerves and ways to stimulate nerves that is impossible without implanting sensors in the arms. Existing TMR provides a quick route to more research.

When TMR works it gives the user more control sites and more importantly, far greater finesse with their control compared with conventional myoelectric controllers. However, these controllers do not exploit the full potential of TMR. To do this requires the use of pattern recognition of the signals which allow more degrees of freedom to be controlled more intuitively. In fact TMR and PR are made for each other, because it is not clear if it will ever be possible to get sufficient reliable surface signals to fully exploit pattern recognition with conventional muscles. For PR on conventional amputations some of the muscles are usually buried and inaccessible. TMR brings them to the surface and creates patterns computer systems can use very well.

Kuiken suggests that TMR should be used at every opportunity. He has said that it should be considered as part of the original reconstruction of the injury. It would make the process of getting used to the TMR only part of the entire process of getting over the original amputation. However, to be successful the follow up needs to be as extensive as that undertaken at the RIC [338].

Another problem with TMR that might not be welcomed by the recipient is that it takes a long time (up to five months postoperation) for the nerves to connect [295] and during this time the nerves can be very painful until they are fully integrated. Kuiken suggests that the pain occurs because the nerves are disturbed, but they do settle down again. Not performing TMR can also be painful. The severed nerves can scar up or create neuromas, which can be painful for some. Attempting to reduce the pain is one of the common reasons for revision surgery after an amputation. To Kuiken, TMR is *"a healthy way to treat a nerve"* [42]. He has found this in animal studies since Sullivan's operation. Following a cut, if one places nerve into a muscle there is less neuroma and less sprouting of extra nerves trying to find a muscle. It is the unattached nerves that can give rise to pain. *"It just makes sense. Give a nerve a volume of tissue to spread out through and grow into and you are going to have a happier nerve, because the nerve's job is to reinnervate something"* [42]. Retrospective studies in the recipients suggest there is less pain downstream of the TMR [339].

The next stage in the TMR work in Chicago was a multisite randomised trial looking at the treatment of neuroma and phantom limb pain. This aimed to establish if the procedure has real value to prevent long term pain in persons with amputations. Randomisation removes the chances of a placebo effect (3.2.1). Reduction in pain is an important finding and probably a bigger result than using TMR to create control sites. This is because there are more people with neuroma pain than there are users of myoelectric limbs, so the procedure has the chance to improve the lives of many more people. Indeed Kuiken believes that some surgeons have begun to adapt TMR for treating neuromas, but objective trials were needed to ensure the results are meaningful [42]. The team went to considerable effort to ensure the patient and surgeon were blinded to the decisions. The surgeon only knew which treatment they were going to give when an envelope was opened in the operating theatre, to tell them which group the patient was assigned. The results were positive. Indeed the trial had to be stopped because the stories of the effectiveness made recruiting subjects who might not get the TMR ethically problematic. However, Tamar Makin of UCL does not feel the results are good enough to clearly show a benefit [79]. While she felt the technique was the right way to go and made sense, the statistics do not bear it out. They are too equivocal for her to rely on them, especially when the procedure is expensive and invasive and the placebo effect is not clearly demonstrated as being removed (see 3.2.1).

As TMR was not a completely new procedure, simply an adaptation of an existing operation, it has been easily adopted by other centres in the United States, Canada and Europe. Along with Osseointegration, TMR is one of the most innovative approaches that have appeared in prosthetics in half a century. Both can change how a prosthesis is controlled and what they are capable of. Both also require additional surgery. This may not be tolerated by some potential beneficiaries and TMR is not the first surgical solution to address the problem of limited control (2.6.1). TMR opens up a number of possibilities that have been inaccessible previously. As Jesse Sullivan showed, a person with TMR can perform simultaneous, independent control of multiple axes.

Through the US DARPA program, Otto Bock collaborated with Kuiken on TMR. The first upper limb prototype using Otto Bock devices was demonstrated in Vienna in 2007. It had four joints in the arm and a Michelangelo hand. Otto Bock has worked with Todd Kuiken and specialists from the Division of Plastic and Reconstructive Surgery of the Medical University of Vienna to perform TMR on people to be fitted with an Otto Bock upper limb prosthesis.

An important early user was Christian Kandlbauer. He also lost both arms in an electrical accident in 2006. He used Otto Bock devices (DynamicArm) on both arms; conventional control for the right side and a TMR format on the left. Kandlbauer was the first person outside the United States to wear a TMR controlled prosthesis [340], and in common with most prosthetic fittings, the project was not *subject and scientist*, but really a collaboration between designer and user.

According to Otto Bock:

> "The collaboration with him greatly advanced this pioneering technology in the laboratory and created good conditions for enabling future users to make use of the sensing function of this prosthesis" [340].

Kandlbauer said of the TMR.

> "I control my left prosthetic arm with my memories of my previous arm, and when I try to 'move' it, it moves my prosthetic arm. Over time, it's as if it's become a part of my body".

Tragically, in October of 2010 Kandlbauer died as a result of injuries he suffered when the car he was driving veered off the road and crashed into a tree [296]. Andreas Waltensdorfer, a senior physician at the hospital where Kandlbauer had been in intensive care, and the local police said that it was impossible to tell whether the accident was caused by problems with Kandlbauer's prosthetic arms.

Kuiken rode the wave of interest in his technique, being included in both of the DARPA programs that were launched in 2006, and was one of the few voices in the consortia that had direct clinical experience. It was clear that whatever the teams produced, TMR would help their application. It was also obvious that pattern recognition would enable TMR to be used to its greatest extent, so this also began the long and fruitful collaboration between Kuiken's and Englehart's teams (see 13.2). This collaboration has included a two way traffic in staff between the two cities. At one time three UNB graduates were working in Chicago and one wit nominated it 'UNB South'. Later Jon Sensinger moved to Fredericton and eventually became IBME's director.

A significant bar to measuring the growth and impact of TMR is that there is no central tracking of surgical procedures. If a physician does not tell Kuiken that they performed a TMR procedure, Todd cannot know. Hence Kuiken in 2020 he had little idea how many people had TMR. He knew it was *"at least a hundred people... but that's just a guess... [it] could be twice that amount"* [42]. They had trained recipients in at least ten centres around the USA and Europe by 2015.

Additional surgical interventions are not always welcomed by amputees. They can be necessary for reducing the pain and damage caused by the injury, neuromas, heterotopic ossification etcetera. Some patients tolerate the process, but Kuiken believes TMR is becoming better accepted, especially by those patients who have more to gain, such as persons with high level or bilateral loss. *"I would say this isn't for the average user, because you are wanting more than average to do this, because the arms are still too heavy and still a pain"* [42].

Kuiken's next studies moved further down the arm. The idea was to create multiple sites for pattern recognition of multifunction hands. He considered that the forearm has many muscles which in a conventionally innervated arm would not be expected to add much to the richness of the data detected by surface electrodes and used in PR. Todd believed that if these muscles are used for reinnervation to produce new patterns they will create additional functions. However, this was not born out in practice, as when he performed the operations there was no robust improvement in control. An additional challenge was recruiting sufficient volunteers. While with those with higher losses have much to gain from TMR, as they lack the control sites and need to control more joints, those with a trans-radial loss have far less to gain, so fewer were interested in the surgery [341].

12.3 Center for Bionic Medicine

One thing that has come out of the interest in Kuiken's work is the Centre for Bionic Medicine (CBM). Its aim is to translate research going from concept into the field. As a result many of the researchers are post doctoral engineers and scientists. The staff has twice as many professionals as students, who range from mechanical and electrical engineers, to prosthetists and therapists. Todd's model is the time honoured one of hiring: *"a lot of really smart people, when you put them together, it's great"* [42]. Todd placed a strong emphasis on the broad team at Chicago and their ability to find solutions.

The idea extends the remit of INAIL in Italy, Princess Margaret Rose in Scotland, the Oxford Orthopaedic Engineering Centre in England, which were all based on the premise of professional engineers working in a clinical setting. He regards the model as being rare in prosthetics and orthotics. While other multidiscipline centres exist. The key difference is that the CBM is large and well funded. In contrast, the European centres gain funding on the promise of a commercial output. The CBM blends the industrial with the academic, without a need to create money, projects can be higher risk. With a conventional fully academic model projects often end when the students graduate or the money runs out. The CBM is run on money from many different sources, including grants and philanthropic contributions, so there is no immediate need for pay back. The intention is to ultimately license technology to outside firms, so that there might be financial return one day. For now, the team is freed to concentrate on solving problems.

The strong link with the prosthetics and orthotics clinic is crucial to the application of the technology, so the team works closely with it. While the CBM does not rely on students for input, they do host students on the prosthetics program. This allows the students aiming to become prosthetists to conduct research on some project or projects and take a little longer to qualify. This is important to Kuiken. Like others, he is concerned over the future of the field. As the high end prostheses become more sophisticated, the prosthetists need to keep up with the innovation. They can only do this through training and experience of the technology.

Kuiken's focus has shifted in recent years. Having established that surgical intervention can make a difference to how a prosthesis is fitted and how it is controlled, he is keen to extend the idea. When asked what he thought his biggest contribution to the field is he says: *"To open people's minds to the possibility of changing the human body to work with technology"* [42].

During the interview he wanted to share a recent experiment with us: *"that I am really excited about"*. It was not TMR at all. It was a form of plastic surgery, that aims to make the fit of a prosthesis much better. The team took a person with an amputation in their femur (upper leg). The subject was overweight, which is often the result of being made less mobile by the amputation. The team surgically adjusted the remnant limb to make control of the prosthetic leg better. They opened the thigh to remove a lot of fat beneath the skin and then performed liposuction on other parts of the residuum. Liposuction is usually performed for cosmetic reasons. In this case they did it more extensively, removing just over three kilograms from the thigh, and reducing the radius of the thigh by over two centimetres. The result was a much thinner and firmer thigh, making it far easier to place it inside a socket and creating a firmer connection that makes control of the prosthesis easier. This change can improve the person's ability to get about and to exercise and maintain a healthy weight. The person was *"ecstatic"*[42]. Socket fit was not the only benefit. Normally a socket adds bulk to a limb by increasing its girth. Since the residuum was thinner than the intact leg, the new socket on the shortened leg became closer to the size of the intact leg and the two legs look more symmetrical under clothes. There is no need to buy baggy trousers to encompass the bigger socket. Kuiken sees this as more revolutionary than TMR: *"probably going to do more good than almost any prosthetic evolution that is going to come out in the near future. The interface is the most important"* [42].

Kuiken advocates surgeons to: *"change the body to work well with the prosthesis"*. This might be a return to the more significant procedures of the tunnel cineplasty and the Krukenberg (2.6.1). However, the difference between these two procedures and the liposuction is the impact on the body. It is more in sympathy with the person's own self image.

To Kuiken it is not as publically engaging as working on injured soldiers, but as half of the population are overweight enough to make this a useful extra technique in the surgeon's tool kit, *"It's exciting"* [42].

CHAPTER

Prosthetics research in Canada

13

Canada has a population of 35 million spread across roughly 10 million square kilometres, it is gifted with a wealth of resources that make it an affluent country. Its geography, weather and distributed population create challenges that other countries do not have to face. During the lean years of prosthetic development in the 1980s there were two active centres in Canada that contributed to the development of prosthetic arms. One was known as the Ontario Crippled Children's Centre (OCCC), and the other is the Institute of Biomedical Engineering (IBME) at the University of New Brunswick (UNB). The two centres worked closely together from early in the 1960s to the 1990s. As their foci changed they looked to other partners so that the IBME's strongest clinical partner shifted from the OCCC to the Rehabilitation Institute of Chicago.

13.1 Ontario Crippled Children's Centre

The history of the Ontario Crippled Children's Centre's contribution to myoelectric control is complicated by its changing name. *The Ontario Crippled Children's Centre*, was established in 1957. It was renamed in 1985 to the *Hugh MacMillan Medical Centre*, after its first administrator. In 1990 it became the *Hugh MacMillan Rehabilitation Centre*, clarifying its focus. The Centre was subsequently amalgamated with the Bloorview Children's Hospital, and this resulted in the *Bloorview MacMillan Centre*. In line with the trend to informal names for children's hospitals, it later became known as *Bloorview Kids Rehab*. The most recent change occurred in 2010 following a donation from the Holland family of Toronto. It is now called *Holland Bloorview Kids Rehabilitation Hospital*. For the sake of simplicity, the OCCC successor and renamed organisations are collectively referred to here as the *Centre*. The other organisation, the IBME, changed its name from the Technical Assistance and Research Group for Physical Rehabilitation and so it is referred to as the *Institute* or the *IBME* (see 13.2).

From the early 1960s to the early 1990s the Centre's rehabilitation group were pioneers in the development of myoelectrically controlled hands, wrists, and elbows for children. These new prostheses were created by an interdisciplinary team of researchers, occupational therapists, electrical engineers, mechanical engineers, and product managers from the Centre, IBME, and a local technology transfer company. This activity was supported by the Variety Club of Toronto, a chapter of an international charitable organisation. In 1970, the Variety Club later incorporated a subsidiary, *Variety Ability Systems Incorporated* or VASI, which produced the devices [285]. The early years, from the 1960s to the mid-1970s, saw the creation of the technologies and institutional structures that would underlie the development of VASI's commercial products. The coming together of the interdisciplinary team and development of these legacy products occurred between the late 1970s and the early 1990s. By the late

Making Hands. https://doi.org/10.1016/B978-0-12-820544-0.00022-6

1990s the novel development activities had begun to wind down, although VASI's production business continued until 2005, when it was acquired by Otto Bock.

13.1.1 Origins

In 1963 the Canadian response to the impact of the drug Thalidomide (4.4) was to found three Prosthetic Research and Training Centres, in Toronto, Winnipeg and Montreal funded by a collective $200,000 National Health Grant from the Department of National Health and Welfare (UNB was a recipient of an additional $50,000 annual grant).

The first leader of the OCCC's Prosthetic Research and Training Unit (PRTU) was Colin A. McLaurin. He is credited by Dudley Childress as being one of the founding fathers of the field of rehabilitation engineering, and is *"perhaps the most prolific, practical, and broad-based designer/innovator the field has ever known"* [342]. Born in 1922, McLaurin was educated in aeronautical engineering and served with the Royal Canadian Air Force in World War II. His work building prostheses began in 1949, when he was working at the Sunnybrook Hospital in Toronto, a facility of Canada's Department of Veterans Affairs. The early work investigated the use of new materials. It explored the introduction of suction socket fitting, and a new mechanical hand with cosmetic glove [343]. James Foort, who would subsequently lead the Winnipeg Unit, described the laboratory facilities they had in 1951 as a *"broom closet made available when the janitors moved to better quarters"* [342].

McLaurin left the Sunnybrook Hospital and Canada in 1957 to be the founding director of the Prosthetic Research Center at the Rehabilitation Institute of Chicago (see 12.1). There he developed the Michigan Feeder Arm, in collaboration with physician Dr. George Aitken. This arm for children with bilateral absences was used in several centres (4.4). David Simpson in Edinburgh would develop a gas powered variant (8.1). The arm was one of the first electric-powered assistive devices in the US that was used in daily activities. In 1963 McLaurin returned to Toronto to direct the OCCC. By 1964 the Toronto PRTU personnel consisted of McLaurin, William Sauter (chief prosthetist) and Karre Lind (machinist and electronics technician, who was responsible for developing and producing prototype models). The Toronto PRTU had new equipment, a prosthetists' training program, an out-patient clinic, research projects, and materials application activities [344]. Included in the 600 children registered with the institution's clinic were twenty-one thalidomiders [345].

The first prosthesis developed by the group was an electric hook. The hook was deemed to be functionally essential and was challenging to design [344]. The group determined that electric drive had advantages over compressed gas. The project aimed to optimise the size, weight, and reduce noise generation. Commercial motors were bench tested. All were suffered from wear and noise. The motor from Globe Motor Company of New Jersey, an aircraft and missile manufacturer, was chosen because it provided motors for free for trial purposes [344]. The Committee on Prosthetics Research and Development (CPRD) in the US recommended that electric devices should use a twelve-volt power supply derived from ten nickel-cadmium button cell type batteries.

They were unaware of Reinhold Reiter's work (4.2.1) in Toronto [346,347]. The Russian Hand suggested that a clinically useful myoelectric arm could be designed and manufactured (4.2.1.1). However, it was deemed to be too heavy, and motor torque too low. The Toronto hook weighed less and had twice the force of the Soviet hand [344].

McLaurin knew from the beginning that a critical element of any electric hand was the control system. He monitored developments at other research centres and his assessment in the mid-1960s was that although there were many claims that EMG could be used for proportional control, no practical examples had been demonstrated. Nevertheless, they were keen to see the technology develop and planned to adapt these controllers to their devices. In 1965 McLaurin had found the UNB myoelectric control system showed enough promise for use in their program. On this basis, Scott from UNB was brought into the Toronto PRTU project as its electronics consultant [348].

Two years later twelve child sized electric elbows had been developed by the Toronto PRTU and fitted to a variety of patients. They were found to be 'particularly acceptable', with the most common complaints being noise, wire breaking, and occasional clutch slipping [349]. The problem of masking gear noise was identified as one of the major improvements to be undertaken in 1968 [349]. By this time they used five different types of controls on their elbows, including the UNB controller.

The project grew to include the operation of the clinic, and manufacturing and distribution of prosthetic components [349]. At the same time Sheila Hubbard was hired to work at the PRTU. She was first employed as a physiotherapist at the University of Toronto Institute for Biomedical Engineering and the PRTU. She would eventually become manager of clinical technology and adjunct clinician investigator, running research projects under the supervision of academic staff. Even then, she would keep her hand in clinical practice by filling in when someone was away. Hubbard was qualified as both a PT and an OT, while this was not uncommon then, Sheila crossed a divide that is rarely straddled today.

Hubbard's interest in prosthetics was piqued while working at her first job after qualifying as a therapist. One of the children in her case load was a young Thalidomider who had been fitted in Toronto and then passed to the London Ontario, clinic nearer to where he lived. Sheila worked closely with Karl Ruder, the prosthetist, who helped her become more interested in prosthetics. A final spur to her interest was that one of her contemporaries on the therapist's course was now working in Toronto where a larger number of Thalidomiders were being treated and the experimental nature of the fittings interested her. She left the Centre in 1971 to have her first daughter and she continued part-time until her children went to school in 1979. When another OT who was working on a research program on myoelectric hands for children left, Micky Milner (then director of the Centre) asked her to come on to the project and finish. The role of a research OT in Canada was rare, then as now. She found the flexibility of the clinical team stimulating and regarded the ability to *"bounce around"* between specific projects as what kept the job fresh and the atmosphere of enquiry she described as *"heady"*. She felt it was rare for a therapist to be sent abroad to present their own work [21].

In 1968, five years into the program, they had fitted a number of electric hooks, wrists, elbows and arms, but with a few exceptions, the external power had little clinical impact [350]. The available systems were expensive, mechanically complex, and little used. Again, the option of bottled compressed gas as power source was ruled out because of the complexity in recharging or replacing bottles, even though they were used in Montreal and Edinburgh. The advisory board identified that electric activation had two advantages over bottled gas: For a given mass, batteries could store much more energy than carbon dioxide. Also, greater future technological advances could be foreseen for batteries, which remains true. In 1968, fast charging battery systems were newly introduced to the market, replacing overnight charging. According to the 1968 report, battery technology was *"progressing continuously"* and it predicted eventually silver zinc or other cells would be able to produce four times the energy for the same mass [351]. Later, other technologies would surpass even this prediction.

For McLaurin the solution to the power problem would be solved in time. The critical technical challenge for effective devices was to develop better control systems. A goal was to produce electric controls that rivalled the system from Edinburgh. Simpson had achieved far greater acceptance of gas-powered arms in his clinic than other groups had achieved with electric systems. The key to this was the feedback method, Extended Physiological Proprioception (EPP) (see 8.1), not the power source.

Robert Scott of UNB developed four models for control of terminal devices. In 1970 initial fittings and testing were undertaken at a clinical evaluation program in Los Angeles under the auspices of the CRPD, with the UNB control system included in the production designs [351]. That year production began at a facility in Toronto, organised by the charity, the Variety Club of Toronto. While the Toronto chapter operated a vocational training school for boys with physical handicaps. This new venture was intended to be different, a manufacturing facility that, *"will fulfil a needed function in making available components that are much needed but cannot otherwise be obtained"* [352].

Ultimately the Centre did research, development testing and fitting [345], and VASI manufactured prostheses. This included the North Electric Hand developed by Northern Electric Corporation (subsequently Nortel) [353]. Childress called VASI *"one of the early, and one of the few really successful, ventures to transfer technology from the development laboratory to enterprise"* [342].

William Sauter fitted the first child sized Northern Electric hand in 1971 to an eight-year-old, establishing the Centre as one of the leaders in the movement to show that children of preschool age could successfully operate myoelectric controls. Later, in 1976, the Systemteknik hand (see 10.1.1.1) was made available for preschool children.

In 1975 the PRTU annual grant from the federal Department of Health was cancelled. In the same year McLaurin left to lead a newly founded rehabilitation engineering centre at the University of Virginia. The departure of McLaurin was a loss for the Toronto group. Replacing McLaurin was Morris (Mickey) Milner. Born and educated in Johannesburg, South Africa, Milner received B.Sc. (Eng) and Ph.D. degrees, respectively in 1957 and 1968 for his work in the department of electrical engineering of the University of the Witwatersrand. His PhD was on the physiology of peripheral nerves. He then worked at various bioengineering centres, before in 1978, becoming the second director of the Centre.

Shortly thereafter, Issan El Timmen was hired as a mechanical engineer, with an undergraduate degree in mechanical engineering from the University of Sheffield in the UK, and seven years experience in product development at National Cash Register Corporation. El Timmen would eventually work twenty-six years for the Centre and VASI. El Timmen was a design engineer. His first major project from 1979 to 1981, was the development of an electric hand for children aged two to six years. Following successful testing of a design, it was passed to VASI for production.

According to Milner, the work to develop powered prosthetic products was interdisciplinary and highly collaborative [354]. Although the interdisciplinary approach was practised well before Milner, it intensified during his tenure. Milner's appreciation of this approach had roots in personal experience; his first child was born with cerebral palsy, which opened his eyes to the problems in rehabilitation medicine [355]. In interacting with various medical professionals he came to see that interdisciplinary approaches had much to offer the field [356].

Collaborative projects involving the Centre, VASI, and initially UNB personnel, involved weekly meetings involving therapists, prosthetists, VASI representatives, and researchers. According to Milner: *"Everything was discussed and there was a spirit of camaraderie. I thought it was the most potent model"*.

Sheila Hubbard agreed, saying that the meetings included researchers, clinicians, occupational therapists, prosthetists, engineers, and VASI representatives and the atmosphere of collaboration was very stimulating. Included in the larger group was Robert Scott from UNB, who Hubbard said was a mentor, intimately involved in designing systems and almost like a clinical colleague [181]. The group discussed progress on everything they were working on. It was a forum for discussing what clinical ideas might be taken into research. Most important for Hubbard was the clinical testing of the products.

The team's effectiveness can be observed from an occasion when Hubbard was trying to teach a child to use a bicycle. The wrist on the child's prosthesis only had wrist rotation. This made it almost impossible to hold the handle bars. Frustrated, she went into the next meeting and insisted on putting this need to the top of the agenda. Within two weeks she received the first prototype of what became the VASI product, the 'Omniwrist' [21]. The design is based round a ball and socket and allows the hand to be placed in almost any orientation. According to Hubbard, Martin Mifsud, the product manager at VASI, was a catalyst in this innovative design work. He challenged people at the meetings, asking questions such as the anticipated case load for devices and potential sales.

13.1.2 Paediatric hand development in Toronto

The Centre was influenced by Swedish experience in fitting children with powered prosthetics. The influence began when Rolf Sörbye presented at a meeting in Grand Rapids in 1978. A group from Toronto (including Hubbard) attended the meeting and were excited by the presentation [21]. As a result they got together and tried three experimental fittings between 1978 and 1980. This ultimately led to the research project that brought Hubbard into the Centre. The drive was towards increasing the range of devices available and to push the age at which they could fit a child. This goal required the evolution of equipment that was light and compact enough to be used by smaller children. Starting with a few children, the population grew as children came to the Centre from all over the province. The first children had below elbow absences but later the fittings were above elbow as smaller joints became available. The job was always interesting to Hubbard, there was always something new, she liked that they were pushing the boundaries [21]. She did not find it was difficult to teach the children how to control their devices. Over time the standard fitting for a mid-humeral user evolved, with a myoelectric hand and a body powered elbow (if there was enough musculature and remnant limb to accommodate it). By 1981 they had fitted sixteen three-year-old children with myoelectric prostheses using commercial versions of the UNB three-state myoelectric control (see 13.2.2). Most were built with the Systemteknik/Steeper hand [357].

One result of this drive was the production of myoelectric devices for infants. This needed a compact control circuit to permit one-muscle, voluntary opening control of infant electric hands. The key was in making it simple enough for an infant to use. Hubbard credits the work of Thomas Haslam, a prosthetist at the Saint Anthony Center in Houston, for designing what became the only practical circuit for infants. In Hubbard's words it *"revolutionized"* their approach, allowing for fittings and training down to ten months of age, instead of three years, and increased the odds that myoelectric devices would be used [181]. This pushed commercial companies into developing their own prosthetic products for children.

In 1983 Milner became director of research for the OCCC and associate director of the Institute of Biomedical Engineering at the University of Toronto. He was succeeded as chair of the research meetings in the Centre Steve Naumann who was also adjunct assistant professor in the University of Toronto Department of Rehabilitation Science, and associate professor in the Institute of Biomaterials and Biomedical Engineering.

Their projects were user-focused. The 1987 annual report wrote: *"Close proximity of the clinical service to the design process promotes ongoing interactions between engineering and clinical staff and communication with children and their families. This aids in understanding users' needs and expectations while examining the developmental feasibility in terms of a practical outcomes"* [358]. It was thus not surprising that clinically trained Hubbard had, by the mid-1980s, moved into a role of coordinating research and development projects on the use of myoelectric prostheses aided by microcomputers, with supervision from professional research staff including Naumann and Milner [358].

This collaborative and interdisciplinary approach did not last. Milner felt that the pressure to publish research results increased as they became more research intensive. As academic positions became dependent on publications and track record, the pressure was on more on research based projects and not necessarily clinical outcomes. Hubbard agreed, she said this collaborative and interdisciplinary approach lasted for about fifteen years from the late 1970s to the early 1990s. Among the reasons for the change, she cited the formation of a new research institute (creating physical and mental distances between the clinics and research), the loss of regular attendance by the VASI personnel, and the change in the Centre's myoelectric provider. Hubbard said: *"UNB in the early days was both a clinical collaborator and control system supplier. The UNB system was replaced by Otto Bock. Otto Bock was a supplier to the hospital"*. Otto Bock did not replace UNB as a product development collaborator [181]. The original group worked together so that any member could easily go to a colleague's office and ask a question and get a response quickly; the new institute removed this close proximity. Timmen also concurred with Milner and Hubbard's view about the decline of this interdisciplinary approach at the Centre, although Timmen attributed the change more to the loss of its leader. *"It changed after Mickey Milner left. He was a people person and knew how to bring people together. He believed in teamwork and sharing. He was more consensus than directive"* [359].

During the 1990s and early 2000s VASI continued to develop, improve, and sell upper limb prosthetic products, but it was no longer building novel products to prove that myoelectric upper limb prosthetics could be used by infants and children, nor did it have the same kind of contributions from the Bloorview Centre. As a result, the fortunes of VASI began to change. Development of myoelectric prostheses were now no longer a top priority for the Bloorview Centre. Autism and head injuries were demanding more of the Centre's time and resources, and creating new research funding opportunities. In 2005 the Centre found a buyer in Otto Bock [285] (see 9.1).

13.2 The Institute of Biomedical Engineering

Toronto is a modern city alongside Lake Ontario. It has towers and parks that stretch along the lakeshore for miles. Toronto was known to the original European settlers as York and it featured as one of the centres of fighting during the War of 1812 (when the newly formed American Republic tried to take parts of British North America away from the empire, while the UK and the rest of Europe were more focused on fighting Napoleon). It is cosmopolitan, busy and modern. In contrast, Fredericton is the capital of the province of New Brunswick. It is the sleepy home of the administration of the province. Fredericton has retained a small size and low rise buildings. Fredericton possesses two universities, a cathedral and the legislative assembly. The province's heavy industry is elsewhere. Fredericton is small, it more has the feeling of a village than the city it professes to be. One of the two universities is the University of New Brunswick (UNB) the home of the Institute of Biomedical Engineering (IBME).

13.2.1 Origins

In 1961 Dr. Lynn Bashfield, the medical director of the Forest Hill Rehabilitation Centre (later called the Stan Cassidy Centre for Rehabilitation) in Fredericton was looking for technological advances that would help with the work of the centre. Dr. Bashfield was aware of the research at UCLA on new prosthetic devices, and he asked the UNB dean of science if he knew of anyone who might be interested in the technological challenge presented by two people with quadriplegia. At a meeting convened by the dean, one of the UNB faculty members who attended was a young assistant professor of electrical engineering, Robert N. Scott.

Bob Scott began his bachelor of science degree at UNB in 1950, graduating with twelve other electrical engineers in 1955. According to Scott, education in electrical engineering at the time was mostly practical, with very little theory in the curriculum. To a student of electrical engineering today the course would be unrecognisable. There was no choice on what to study, only eight mandatory courses per term [360]. The curriculum did not include anything on solid-state electronics (the transistor was only invented in 1947), and nothing on computer/digital systems or control systems theory as they were very much in their infancy.

The education of undergraduate electrical engineering students was still largely grounded in a craft tradition. For example, at the start of his studies Scott was instructed to bring a good pair of pliers and an electrician's knife to his courses. A newer applied science model was emerging from US institutions like MIT, and it was embodied in places like Oxford University that taught *Engineering Science*. Scott joined the electrical engineering department as a lecturer in 1959. At that time no one in UNB's electrical engineering department had a doctoral degree. There were two faculty members with master's degrees but, according to Scott, these were almost honorary, based on teaching experience and one paper [360]. The faculty focus was on teaching, not research. However, the Dean of Engineering, James Dineen (an electrical engineer) saw the direction the profession was heading and encouraged Scott to do research. This would not be easy, as he lacked available mentors and possessed meagre equipment and budgets.

The meeting that happened following the request from Dr. Bashfield led to the formation of a new group, called *'The Technical Assistance and Research Group for Physical Rehabilitation'* (TARGPR), which uncharacteristically put 'assistance' ahead of 'research'. The meeting also led to the development of low-tech assistance for wheelchair control.

To investigate possibilities, Scott contacted researchers at UCLA, who were building a powered splint that was controlled by a tongue-operated switch. Users were able to feed themselves with it. This was the seed of the idea, that not just hands but other parts of the body could be used to control an assistive device. This would influence Scott and the Institute throughout its history. For the two patients at the Forest Hill Rehabilitation Centre, Scott built a tongue-operated controller for their electric wheelchairs. It was, Scott claimed, the most useful thing he ever did at UNB. According to Scott, it drew their: *"research was too clinical... Indeed, the complaint was made from time to time that the work was too practical to constitute academically respectable research"* [360].

This picture of engineers undertaking projects for the needs of a small group of patients changed with the arrival of Thalidomide, and the subsequent formation of the PRTUs. According to the Institute's next director, Ed Biden, it was a once in a life-time opportunity: A brand new field with substantial long-term funding [361]. Indeed, it was a novel step to include an electrical engineering group from New Brunswick in the work at the three other research and training units associated with teaching hospitals. New Brunswick would have been considered a backwater province and the clinical centres they collaborated with were in major cities across the country. As the authors of the 1963 *The Report of the Expert Committee of the Habilitation of Congenital Anomalies Associated with Thalidomide* wrote: *"...limb abnormalities... can usually be met by existing paediatric facilities, particularly within university centres"* [362]. The expert committee also foresaw that: *"The use of external power in artificial limbs is in its infancy, and will undoubtedly be required in the long-germ [sic] management of severely involved phocomelic children"* [362]. The expert committee's recommendations only called for the development of three centres at Winnipeg, Montreal and Toronto. There was no mention of electrical engineers among the list of critical professionals to be associated with these units [362].

The location of the other centres had political relevance. Winnipeg would have been regarded as to the west of the very large country, providing support to the neglected 'pioneer' half of the country. Toronto and Montreal are both in the central part of Canada where the majority of the population live (the provinces of Ontario and Quebec). They also represent the two specific populations within the Canadian political sphere. Which may not be visible to outsiders, but their representation is essential for any agreement within Canada, whether local or national. They represent the Anglophone and Francophone populations. Both must be served and seen to be served, equally and independently, irrespective of population size.

Over the years the centres specialised on different aspects of the problem: The Montreal centre focused on the electronics design. Winnipeg was more concerned with the lower extremity problems. UNB focused almost entirely on myoelectrics and only the OCCC covered both extremities.

The annual award to UNB from Canada's Department of National Health and Welfare was $50,000 per year (beginning in 1963), was enough to hire technical professional staff. UNB was unique in this approach. The university of Toronto had research staff, not practising engineers and technicians. The research staff approach was consistent with the funding, which was to be used for research and training of prosthetists. In progress reports, Scott sought to have the word 'research' in every paragraph, to make clear the distinction between research on the one hand, and patient care on the other (a provincial responsibility). According to Scott, *"That is a lot easier to define on paper than in the hospital"* [360].

The four centres were required to meet twice a year to exchange information and work together. Some PTRU meetings were less about collaboration than forums for presentation of success stories, with little discussion of the many issues that were yet to be addressed in developing useful devices for clinical practice [360]. The experience of failure, a great source of learning, was not spoken of at these meetings. A difference between these centres and those in the USA was that children were the primary focus of the funding from the department of health.

Another external influence was the publicity generated by the Russian Hand (see 4.2.1.1). The impact was twofold: It generated a lot of media attention. It raised both questions and expectations. Scott thought it was also *"a good example of the idea being right, but the technology not being there, as the Russian Hand had a slight closing force, one size only (adult male), enormous power requirements, and quality issues with batteries (sometimes exploding when recharging)"* [360].

The Canadian licensee of the Soviet technology, the Rehabilitation Institute of Montreal (RIM), made electronics that were better in 1964, but were still not good enough for clinical practice [363]. It was this version Scott saw and evaluated. Scott said that the publication of the concept of a myoelectric hand at the Expo 1958 in Brussels, opened up the imagination of engineers at UNB and elsewhere to the potential of the basic design concept.

Scott decided to focus on myoelectric controls as it seemed the most promising technology. Of the technical program that developed he said: *"We initially defined the objectives in terms of the clinical education, and not in terms of academically respectable research. We did not try not to do good science. We were not at this very long before we needed research to support the application we were working on"* [360].

13.2.2 Myoelectric control at the University of New Brunswick

An early project engineer was UNB graduate student Phil Parker, he was originally an electrical engineering graduate student of Scott's in the early 1960s. Subsequently he was a faculty member for twenty-nine years in UNB's department of Electrical Engineering. Both men were interested in research to assist in developing a system to meet clinical requirements. It made for a good fit. Scott directed the Institute, Parker developed the signal processing algorithms. A third engineer, Dow Dorcas designed the control systems. Parker was the research leader, with a focus on understanding the human neuromuscular system and control of prosthetics limbs. Instead of importing systems from elsewhere to produce makeshift solutions, (such as hardware from UCLA to make the wheelchair controller), UNB now had the people to develop concepts and implement them.

Philip Parker was born in New Brunswick on an island called *Grand Manan*. It is just off the coast of the mainland. It is a small island of *"Hard drinking, hard living, fishermen"* [364] with their own unique accent and an independent spirit, attributes common to many an island community. Although he has lived 'away' for most of his life, when asked where he's from, he hesitates; it depends in where you are from and if you have heard of the island; he's a Grand Mananer first. Parker went through UNB's undergraduate program a few years after Scott. While the course was not quite so craft-based, it was still very practical. It now included Control Engineering. Signals, Communications and Information Theory were not yet taught. Scott was Parker's project supervisor: *"All my work has been myoelectric signal processing, even then we were looking at the relationship between myoelectric firing frequency and force"* [364]. That is, how the signal varies with force. This is one of the places that has totally changed with advances in technology. Today a simple piece of equipment or a line of computer code would render the complex signal into the individual frequencies. In the 1960s if Parker wanted this sort of information he had to measure each frequency in turn, a long laborious process.

Parker went to Scotland for his master's, he was at the Dundee campus of St Andrews University. His project was on signal processing. It was early work on character recognition, computers reading text. Like EMG pattern recognition, character recognition requires the computer to detect the underlying character, despite any variations in the layout that a human would not even realise was different. Any change of size or orientation of a character would have defeated the software at the time. Parker's work considered distortions such as rotations, and scaling. He then went to Ottawa to work for the Canadian *National Research Council* (NRC) on national defence projects. He still believes that if the telecoms giant, Bell had offered him a job when he worked for them one summer, he might have joined them and stayed away. However, when passing through Fredericton on a visit, he spoke to dean Jim Dineen, and mentioned that he would like to come back one day, not considering how long that might be. It was a matter of months before Bob Scott wrote to him and told him there was a post in Electrical Engineering he could apply for.

Parker was employed as a research associate, working on instrumentation, and the analysis of myoelectric signals. The work was with kineseologist Barry Thompson, on studies of human gait and the muscle's function during walking. To investigate this required putting small wires (acting as electrodes) into the muscles and recording the activity in the muscles. There was no supply of fine wire electrodes,

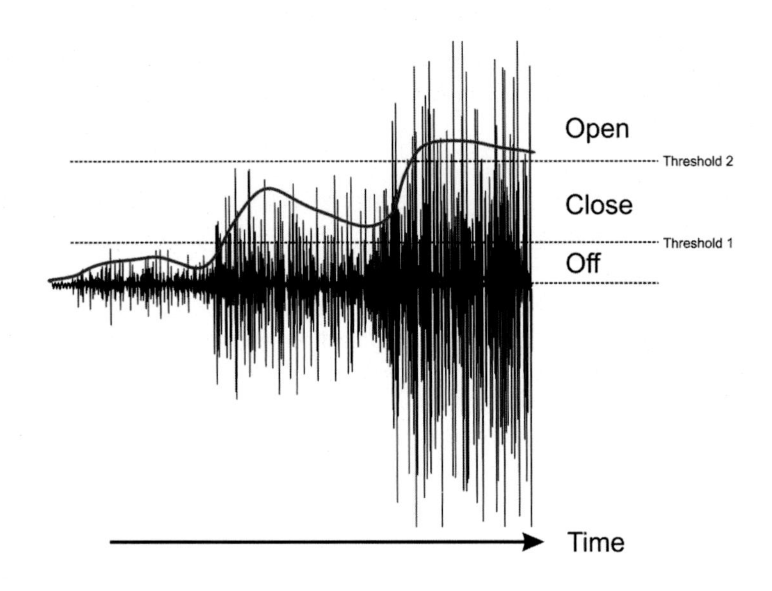

FIGURE 13.1

The UNB three state controller. Used when the wearer could only generate one clear EMG signal. Small activity caused the hand to close, larger to open. The important aspect for successful operation was selecting the thresholds, based on the noise in the signal.

so they had to make their own. This consisted of winding hair-like wires into spirals, removing the insulation at the tips and then sterilising them.

Later his PhD looked at a multistate myoelectric controller.

This took the activity of one muscle and divided it up into bands so that if the muscle was tensioned in the range for long enough it activated the 'state'. This had grown out of the need to find a way for a single muscle to control two directions to open and close a hand in cases where the user could only control one muscle easily. For a three state controller the lowest band is off (Fig. 13.1). This band has to include more than just no signal at all. An EMG signal is noisy even when the muscle is still, and if there was no band at the bottom of the range that corresponded to zero the noise would occasionally trigger something to happen. Making the hand unpredictable. The next control band up required a small level of activity that triggered the hand to close and lastly a larger signal level was used for the hand to open. They attempted to extend the three states to five levels to switch between an elbow and a hand. This goal was harder to achieve, unless the subject was looking at a display of the signal. This was clearly not convenient for a prosthesis.

The critical aspect of the design was where to put the bands so that the users could achieve them simply, with few mistakes. If they were set too low it would be difficult to close the hand as the user would overshoot into the 'open' band. If the threshold was set too high the user would rarely achieve the open level without much effort. The EMG is a noisy and variable signal, this variability gets greater with bigger signals. The size of the signal is related to how hard the muscle is contracted. To divide the range of the muscle up into bands of increasing tension the same width bands would mean that the higher bands would be harder to target precisely. Instead, the bands must get bigger as the tension increases. Parker's work looked at how much information and how much 'noise' was in the signal. He could determine how many bands were possible, given a level of noise in the signal, and where the bands should go. The result of this work was tabulated, and in the clinic the prosthetist could look at the signal and dial in the bands based on the signal and the contents of the table. A simplicity of application that remains essential for clinical use of technology. The original ideas for the three state controller came from Dow Dorcas and Vaughan Dunfield in the early 1970s. Dunfield built the electronics, which would do the switching without use of a computer. At the time it was utterly impossible to use a computer clinically.

While on sabbatical at the Winnipeg PRTU in 1967, Scott saw work being conducted on implants to detect the EMG signal. He came back to the IBME keen to continue the research at UNB. To create enough space for the electronics package, the device would reside in the cut end of the bone of an amputated limb. A wireless link was used for the signals and to power the electronics. According to Parker, the electrodes were *"rough looking"*, with wires running around the end of the bone [364].

By 1965, Bob Scott was spending most of his time at UNB supervising a staff of about a dozen involved in development, some clinical practice and research. TARGPR was informal, unincorporated and unaffiliated. Dean Dineen did not object to this arrangement, but he saw it was not fair to the rest of the department of Electrical Engineering, as it was taking Scott away from the more routine duties of teaching. Sensing a lightening and openness of the mood of the UNB president (Colin B. Mackay) at one of their regularly scheduled meetings, Dineen produced a proposal for a new bioengineering institute. It had no business plan or models presented, no supervisory group or structure, yet Mckay liked it and at the next senate meeting in 1965 the Bioengineering Institute was approved. The Institute was created to undertake interdisciplinary research involving more than one faculty, and was operationally and financially responsible to the senior UNB administration.

One of UNB's major contributions to the advancement of myoelectric hand technology was biological signal processing. Electronics can capture the biological signal and make it useful to the circuits used to control a prosthesis. In the early days this was performed using fixed electronic circuits, but once it is possible to use a computer to process the signal more elegant ways could be devised. It was this that the Institute began to excel at, under Scott and Parker.

Throughout the early to mid-1960s the team at UNB were committed to the idea of prosthetics as primarily functional devices, the equivalent of *"pliers on wires"* [360]. In 1968, John Hall, head of paediatric orthopaedics at the University of Toronto, introduced the IBME group to a fifteen-year-old girl missing her arm below the elbow. She was highly functional with her prosthesis and had been wearing a conventional body powered prosthesis, including the harnessing and straps to control and retain it. Although she agreed to a trial of an electric powered hand, in the fitting she told Hall she would not use the electric hand as it neither looked good nor fit well [360]. The message from Hall and the user was that they wanted something comfortable, good looking, and last but not least, *functional*. At the time the IBME team had been focused on making an electric pincer work better. This was a turning point for Scott. UNB's focus would have to change to listening to the user, and fitting the technology to the requirements of a comfortable and attractive prosthesis. This allowed the Bioengineering Institute and the OCCC to advance in the development of myoelectric controls for integration in a natural looking and comfortable prosthetic device. In other words, the electronics and myoelectric controls would be built *into* the arm, not worn outside the prosthesis.

A second novel feature of the UNB myoelectric controls is that it was not a highly intelligent device requiring minimal input from the user. The UNB group found that such a design posed a psychological issue for users. They did not get the same sense of satisfaction from a device that was highly automated, users preferred a level of control. Scott recalled that quadriplegic users at the local rehabilitation centre wanted a wheelchair they could drive, and did not want to be driven or pushed [360]. This may have changed over the years, given the level of automatic action incorporated into a modern smart phone or computer. Auto selection/completion of words and the use of artificial intelligence to generate choice for the consumer have changed what is thought of as independent thought and action. It is likely that the right sort of automatic control will be tolerated.

In prosthetic control, Bob Scott also looked at supplying feedback to the user. The idea was to feed an electrical stimulation to the forearm of the operator that reflected the force the hand was imparting on the held object. The team worked on a design of EMG electrode that was also an electrical stimulator, where the feedback stimulation did not drown out the detected electromyogram [189]. Unfortunately it was not useful to any but one of their wearers. He employed feedback at work, where he carried drawings in his prosthesis. He found the information useful to be able to tell if he had dropped the plans [365].

Few attempts to supply feedback have met with any success. They can work in the lab, but they seldom give the user any real functional benefits. This may be because the level of information is not enough to be useful to the operator. Finger tips are capable of detecting a substantial amount of information. They can reveal much about the weight, texture, temperature, stability of the object. In this case of the UNB feedback, all the signal told the user was if the hand was closed round an object. This is so little information that it is likely to be ignored by the average user. Users get more information by looking at their hand. There are other forms of incidental feedback to the prosthesis operator which they do use, such as the sound and vibrations of the motors or the weight shift of the prosthesis as it picks up an object.

13.2.3 Myoelectric controls symposium

UNB's largest and most long lasting contribution to the field, and to the industry, has been the organisation of meetings known as the 'MEC (MyoElectric Controls) Symposium'. First held in 1972, the initial focus of the symposiums was to educate the profession on the many technical aspects of providing myoelectric solutions for users. Delegates and presenters include manufacturers and suppliers.

In the early 1970s there were few prosthetists with any experience of electrical systems at all. Externally powered prostheses were only then being introduced. A myoelectric limb requires a different approach to casting, fitting, adjustment and user training. As no prosthetist or therapist would have been trained in these differences, training was necessary. MEC ran annually for years, as a course teaching professionals to fit and train users. Soon it became apparent that the market was beginning to saturate, with many practitioners now being aware of the technology. It was also becoming clear that there was a need for a forum for the profession to share the latest developments in the field. At this point MEC began the transformation into a conference where delegates present papers on novel aspects of powered, upper limb prosthetics. The training aspect was mostly taken on by the manufacturers, informing the field about their latest products.

MEC still is a multidisciplinary event, delegates and presenters include engineers, developers, prosthetists, prosthetic technicians, occupational and physical therapists, social workers, psychologists, physicians, surgeons, and manufacturers. Originally focused on myoelectric controls, this made MEC a conference exclusively on upper limb prosthetics, which (at the time) made it unique in the prosthetics field. In conventional prosthetics conferences the level of interest from the profession reflects the smaller number of patients who need a prosthetic arm. Thus, in a major prosthetics and orthotics conference, groups of upper limb specialists are in a minority. Having a conference focused exclusively on this area allows for a more concentrated experience for the delegates who can attend any of the sessions and be sure that there is something relevant for them. The result is that MEC ceased to be just about myoelectrics but upper limb prosthetics in general. In the 1990s (for instance), one of the keynote speakers was Dick Plettenburg from Delft, the long-term enthusiast for body powered solutions (see 5.4).

In the 1980s a second upper limb only conference grew up in the UK, the *Trent Prosthetics Symposium* (TPS). It was a more national gathering and generally kept to a more clinical focus, keeping research papers to a minimum. Like MEC it was unusual for being based at a single location (Nottingham, UK), rather than moving to different locations as the majority of conferences do. The local nature of both of the upper limb conferences was because they were both the creation of local groups. At the end of the 1990s the three organisers of TPS retired and the National ISPO organisation took on the conference and determined to make it fully international (changing its title to the 'Trent *International* Prosthetics Symposium' - TIPS). This took TIPS closer to MEC's traditional research and development territory. Since 2008 MEC's focus has begun to change in response to the changing research environment. Myoelectric signals have always been used as input for other forms of device or system, but it was rare for any person working in any other field to present at MEC. By 2010 some groups were beginning to use myoelectric inputs to control prosthetic legs, and so there has been a shift for MEC towards prosthetic legs (if controlled this way). Attendance figures in both conferences reflect the history of the subject with rising numbers in the 1970s, followed by reduced interest in the 1980s and 1990s, and then a resurgence in the 2000s with changes in technology, interests and funding for the field. The most recent MECs reflected the newer research lines into neural signal detection and signal processing. This has moved the centre of the conference well away from devices and clinical fittings towards the understanding of the underlying processes. Meanwhile TPS gained an international dimension and has begun to move around the UK to different venues. It continues to focus more on the clinical aspects of the field.

13.2.4 Consolidation

As far back as 1972 Scott was diversifying the role of the Institute. In that year they won a contract with the provincial department of health to deliver the clinical engineering program for New Brunswick hospitals. At the time the Institute was running on a budget of about $500,000 per year, and it was all external funding. To keep the Institute together the UNB vice president, finance and administration (Jim O'Sullivan), gave the institute funding, encouraged them to continue as they were, while the development office sought other funding. When he was interviewed Bob Scott could not recall a time from the early 1960s until his retirement in 1990 when his requests to the University for support were denied. *"There was an incredible level of support from the administration"* [360]. In about four months new funding was secured from the Canadian Imperial Bank of Commerce (CIBC). The funding was generous, in the millions of dollars, with no strings attached except to oblige UNB to prepare press releases with the bank's public relations. CIBC funding ended abruptly resulting in one of only two occasions when staff have had to be laid off. With the CIBC funding it was possible to extend the manufacturing capabilities of the Institute. After his retirement Scott maintained an interest in the fate of the Institute, staying on the board and attending the MEC symposia. This continued up to his death in late 2014.

With changes in funding came changes in focus. Research shifted from clinically targeted research and development to more blue sky research funded by national granting councils [360]. There were also collaborative projects with companies such as LTI and Steeper in the United Kingdom. Scott's role changed as well. Although the myoelectric control system development was still not theoretically oriented, he began writing research grant applications. This required him to frame technical issues in terms of theories from the academic literature, but according to Scott he was not sure he believed any of them. Scott wanted answers to questions that arose in developing new systems and products, and did not care where they came from [360].

A long standing criticism of many ideas that appear in the literature or the media, is that the work is not grounded in any clinical experience. Designers need to think broadly, but in the end real people have to use the results and they provide the spur and the limits on the technology. As the IBME did not have direct access to patients, this was always a problem. In 1981 the Institute developed a regional limb fitting clinic and service, a myoelectric controls manufacturing group and a distribution group. It was originally housed off-campus in a suburban neighbourhood, a short distance from the university at a rehabilitation centre. Later the limb fitting service was moved to an extended IBME building making the research and the clinic closely tied. A manufacturing and distribution business (also conceived as an IBME activity) was relatively short lived.

The concept behind the manufacturing and distribution business was that it would make and sell products developed by Institute researchers as well as distribute products in Canada for LTI and Steeper, with profits returned to the Institute to further the research agenda. Mark-up on UNB-based products were unsustainably modest. For example, in 1980 a myoelectric trainer that had a cost of production of $3300 was sold for $3500 [360], but to Scott making a profit on the products or the conference, when they were directed at helping people, was *"inappropriate"* [172]. It was an unconventional approach. The convention was (and still is) for universities not to sell biomedical products, but rather to license technology to firms that can afford the cost of commercial or clinical product development and accept the liabilities that go along with product sales and service. However, these were early days for university technology transfer. The very small mark up on products, combined with low sales numbers (in the biggest year, twenty-two systems were sold) meant that the Institute was loosing money [361]. The decision to wind-up the manufacturing business occurred in 1989. Initially, the advisory board aimed to divest themselves of all three aspects of the service: limb fitting, manufacturing and distribution. However, one member persuaded them to keep the limb fitting service on the basis that the Institute's reputation relied as much on the fitting centre as it did on research. In 1989, the rights to the locally manufactured products were sold to LTI for one dollar. Some Institute technology found application in LTI products, whereas others were shelved. In the end, keeping the fitting service and moving it to an on campus location gave them greater ability to integrate research into clinical deployment.

Bob Scott's successor was Edmund Biden, who was also a mechanical engineering graduate from UNB engineering. Biden was a Rhodes scholar and gained a DPhil (PhD from the University of Oxford) working on the biomechanics of human knees. This work contributed to the design of an internal knee prosthesis (arthroplasty) by his supervisor John J. O'Connor. Scott had started by focusing resources on myoelectric upper limb prosthetics. Towards the end of his tenure he had begun to broaden the remit. Biden extended this, reaching out to a variety of groups on and off campus. He led strategic planning exercises and broadened the Institute's activities to include more mechanical related projects, motion capture-based research on gait analysis, and subsequently upper extremity movement analysis. Previously, the reputation of IBME was based on Scott alone. With the change in leadership the reputation rested on the work of a number of researchers, including Parker, Biden, and two of Parker's students, Bernie Hudgins and Kevin Englehart (the third and fourth directors of the Institute). But there was also continuity, both in the staff and in Scott's model of having technology development activities and limb fitting work performed by professional staff, with student efforts informing and feeding into this work. Biden also built on the relationships with LTI and Steeper to develop collaborative research and development projects.

Biden was director from 1990 to 1999. Like Parker, on a return trip to his home town of Fredericton, Ed had made it known to David Bonham, the chair of mechanical engineering, he would be happy to return to UNB, so when a faculty position in Mechanical Engineering became vacant Biden was encouraged to apply. When asked for his contribution to the Institute, Biden is very succinct: *"Survival"*. The 1990s were a particularly difficult time for finding money to do research. During his tenure there were times when they saved money to pay salaries by restricting use of the telephones, and saved on the bill for stationery by hoarding the promotional pens given away at conferences. He sees keeping the Institute together, and the fact that it is still in existence, as a measure of his success. As the Institute had historically relied on single large grants, the withdrawal of one large grant threatened the continuation of the Institute. Under Biden they ensured that the sources of funding and the topics of research were diversified, so no one loss could potentially be fatal. Success with larger grants creates boom and bust.

The next director, Hudgins, (who took over from Biden in 2000), also received his undergraduate and doctoral engineering degree from UNB. His earliest publications were coauthored with Scott, and Parker (his dissertation supervisor). He began work as a research engineer at the Institute in 1980 after his undergraduate degree. His master's work addressed the basic problem of understanding a signal from a muscle contraction. The doctoral program advanced on the master's thesis, seeking to understand myoelectric signals and examining how they arose, as well as control systems. It was funded in part with a grant from the Natural Science and Engineering Research council of Canada (NSERC), under its Collaborative Research and Development Program, and support from both Steeper and LTI.

The extension of existing lines of research, such as adding more states to the three state controller, proved to be a dead end. Instead, another path opened up in the late 1980s and early 1990s. It was directed toward the application of pattern recognition for control of myoelectric prostheses. According to Hudgins *"We had gone through twenty years of a lull in the field since the introduction of myoelectric systems by UNB and others in the 1960s"* [366]. Indeed, the control systems for single function electronic hands for users with below elbow absences had changed little since the late 1960s.

According to Kevin Englehart, the program's focus was:

> "based on the conviction that the most urgent need for externally powered prostheses was in high-level amputation. The efforts put forth to meet the intrinsic technical challenge of this complex control problem have produced anatomical and physiological models and have drawn upon statistical signal processing methods to maximally extract information from the myoelectric signal" [335].

As part of his doctoral research in the late 1980s and early 1990s, Bernie Hudgins observed a 'motor plan' in the myoelectric signals accompanying the onset of arm or hand movement [367,368].

The question he asked was; why do these structured patterns occur? His idea was that motor plans were the outcome of learning simple, ballistic contractions, which, once a movement had been learned, became stable for a given task. It is like artillery preplanned or 'predicted fire' plans in at least two respects. First, the individual guns in artillery units are trained to fire according to specific and detailed sequences. Like human movement, there may be improvisation on the field of battle, but the effectiveness of predicted firing by artillery is often in the ability of individual guns to execute highly coordinated plans worked out well in advance. Second, artillery firing plans may be discerned by the sound of the guns, but this provides only a very gross understanding of the intended movement of shells. It is no substitute for knowing gun position, elevation of the target, wind speed and direction, barometric pressure, gun barrel wear and even propellant batch and temperature. Likewise, surface myoelectric signals provide only approximate neuromuscular information. They are the by-product of complex temporal and spatial muscular signals, which are a subset of the signals delivered to the muscles by the motor neurons. The trick is to be able to get enough information out of the signals to be able to predict roughly where the shells will land or the hand will end up.

13.2.5 Pattern recognition

This work resurrected the field of pattern recognition (PR) within prosthetics. Pattern recognition techniques are not unique to myoelectrics, and are applied in many places. For instance; it is used to allow computers to recognise faces in video images. Most computer programs are told directly how to respond to input information. When information is complex or vague it is much harder to draw up a rule that a computer can follow, while humans easily sort through vague imprecise data and come up with a definite interpretation. In a hope to deal with messy or imprecise data, Artificial Neural Networks (ANNs) are electronic versions of the sort of wiring that takes place in the brain. Brains are made of neurons (brain cells) linked together. Strong links between cells mean the information flows between the cells and on to the output, weaker links have less traffic and may be removed. The ANNs can be a computer program simulating the network, or a physical web of real connections.

ANNs can be taught what input is the desired, and which needs to be rejected by showing examples to the network. It learns a task through repetition. The links between neurons then become strengthened or weakened based on repeated exposure. The advantage, and downfall, of these systems is that there is no need to understand the complexity of the input. The network becomes a 'black box' that just gives you the answer. A danger of this is that one cannot be sure what it is about the input that the program is using to distinguish between the two categories, or if it has been given sufficient information to work reliably.

For myoelectric control, the first challenge was to determine the different characteristics of the signal that would be used to differentiate between the muscular signals for different uses (e.g. opening and closing the hand). These characteristics are referred to as 'features' and are simply different mathematical qualities that every signal has. For a myoelectric signal, a most obvious feature is the amplitude of the signal (how big one muscle signal is compared to others). This was the quality that the Sven hand processor used in the 1970s (see 11.2.1). Since the amplitude is related to the level of muscular effort, it is something one can easily relate to. If you hold your forearm and wiggle your fingers you will feel a pattern of bulges that relate to the pattern of contractions in your arm, you can intuitively sense that there is a pattern to be detected and the resulting motion is probably related to the size of the different contractions.

Alone, amplitude may not be enough to get as many separable actions as one might wish to reliably control a prosthetic hand. Another challenge is to work out how many different signals one needs. It is easy to see that if you want to tell two things apart, you need two distinct inputs (e.g. left/right, up/down). For a greater number aimed at the control of a hand *and* wrist, you need to gather information from a greater number of muscles and look at a greater number of features. Hudgins' major contribution was to identify a set of features that work. These features evolved during his PhD, he chose features he could reliably extract with limited technology, they were as simple as it was possible to do with existing computing resources in the early 1990s.

Any signal can be looked at in terms of changes over time (the 'time domain') or over the frequencies that make up the signal (the 'frequency domain'). One way that to analyse signals is to use the Fourier transform. This breaks any signal, no matter how complex, into individual frequencies (such as separate notes in a musical analogy). In the early 1990s the transformations would have taken far too much computer processing. Bernie chose to use features of the signal one could observe changing over time. Looking in the time domain meant things in the signal that happen over time that could be easily obtained through addition, subtraction and moving information about in the computer. Any other mathematical operations would have taken too long with the equipment that he had, and naturally, a prosthetic controller has to be fast or the hand will move noticeably after the command, which was and is totally unacceptable.

Hudgins chose:

- **Mean absolute value** - the average size of the signal.
- **Zero crossing** - the number of times the signal went from positive to negative or vice versa, this is a measure of frequency.
- **Slope sign changes** - the number of times the line goes from trending up to going down, another measure of frequency so not used as widely afterwards.
- **Waveform length** - this was the running sum of the input values that gives you the length of the signal in some form and is also related to the frequency.

The last feature was found to be very powerful as it contains both frequency and time features. The derivation of the features are shown in Fig. 13.2. The drawback was that it was tricky to set up the system as the operator had to know the data. For example, 'zero crossings', these are shown centre right of Fig. 13.2. There are many, if the signal has noise on it. This would give far too large a number of zero crossings to be useful. To make them work the results had to be tweaked in response to the signal and what the engineer had seen before. To address this Hudgins added a deadband (top right). This meant that when the signal was big enough and exceeded some minimum threshold the zero crossing count increases, otherwise it was taken as silent. In the figure, two potential thresholds can be seen. Without a threshold there would be hundreds of zero crossings. The count for a smaller threshold than indictated would exceed one hundred. The inner threhold shown, reduces it to ten, and the outer threshold has zero count.

The problem with this approach is that those values which are useful are based on the data but not on any theory that predicts it. To get any system to work reliably, this led to 'magic' numbers for these limits and for the boundaries. As this was all based on experience of the engineer, it was difficult to transfer to the clinic. In the clinic the prosthetist wanted to put the device on and easily adjust the settings, or tell the user how to change what they are doing to get a better result. For a simple system such as the three state controller, the prosthetist could find the numbers easily. If the user was

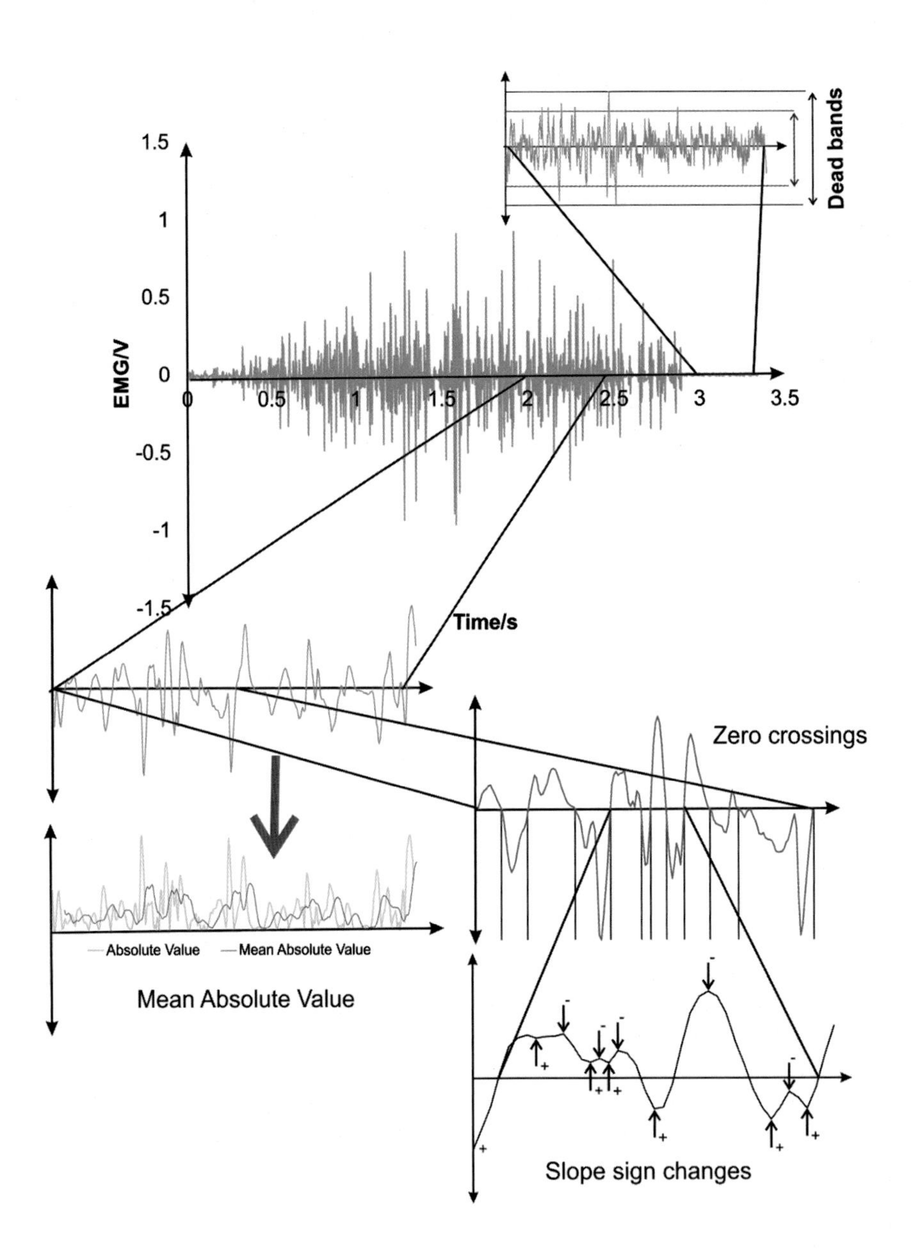

FIGURE 13.2

The Hudgins feature set. Different parts of an EMG signal are emphasised for clarity. Absolute value takes all the signal with a negative value and inverts it. The mean is a mean over a moving window of time, giving a trace that follows the general trend of the signal rather than its detailed signal. Zero crossings can be every time the signal changes sign, but this will lead to a very high number if there is underlying noise in the signal, hence use of a deadband.

not generating a big enough signal, the prosthetist could ask the user to exert more effort, or turn up the gain of the amplifier or adjust down the thresholds. In a multidimensional system this was far too complex to make it easy. This made any PR system difficult to transfer to the clinic.

Despite its simplicity (or perhaps because of it), most of Hudgins' features have become something of a standard in the pattern recognition world [368] and the paper based on his dissertation has been cited over 1100 times in the ten years 2009 to 2019, while it was only referenced 149 times in its first ten years (according to Google Scholar on 28th October 2019).

The ultimate aim of the work was to determine what was unique about a set of contractions. This allows a computer program to easily determine different patterns being generated to control a prosthesis, such as the difference between opening and closing the hand and flexing and extending the wrist. With each new contraction of a muscle, the pattern recognition program would analyse the new data, and select a predefined movement (e.g. open hand) to send instructions to the prosthesis.

A final problem related to PR is the blocks of time over which you look at the data, this is called 'segmentation'. The data is observed over a series of windows in time. By looking at it this way one can see the changes over time (signal getting bigger or smaller). The length of the window having an impact on the speed of response of the controller. A long window might be more sure of the users intent, but means the arm will only react slowly. Shorter windows might mistake the user's intention.

For any pattern recognition system the set of contractions are unique to the person and the set up. The size of the signal and its frequency content depend on the electrodes' position on the arm of the user, the relative strengths of the muscles, the level of fatigue of the muscles and the electrical environment (level of dead skin, sweat etc). Most of these qualities can also depend on the orientation of the arm (such as if it is positioned above the head or below the waist). This means that the program has to be customised to the user through exposing it to a number of examples of each contraction. This is referred to as 'training' of the program. Before a program can be used it has to be trained. Since the number of factors that need to be taken into account are many and can change by the minute as the arm is moved about or the person tires, this set up can become useless very quickly. The first portable system developed at the Institute was better at coping with changes, even different users, but it was never used in the field and Englehart moved on to newer ideas on signal processing.

The method of training a prosthesis system for pattern recognition has remained the same. A person is asked to perform a series of example contractions for each of the operations that they want to separate. For example, performing a fist, then a lateral grip, then a pinch grip. It is necessary to give the program more than one example of each motion in order for it to get some idea about how the signal will vary. This is the crux of the problem, if the definitions of a particular motion are made too broadly, there is a strong possibility that the system will misclassify one action as another. Taking facial recognition as an example, this would be like accepting anyone with dark hair as one individual. Conversely, if the definitions are too restrictive then most correct actions will not be classified at all. This it would be like assuming the hair style (and the absence of a beard) must always the same, so on days when the target did not shave he would not be classified. The number of times the target is correctly identified is the 'recognition rate'. The problem is that if the system makes too many mistakes (either misses a contraction or picks the wrong one) then the user will become frustrated as the arm performs the wrong action (or takes no action at all).

In the community of scientists and engineers who are working on pattern recognition in other fields, ninety-five percent recognition is seen as a very good rate, indeed for a career researcher it may well be enough to prove their system works and it is time to move to a new topic. For prosthetic PR it is the point at which the work begins. Ninety-five percent success means five percent failure, or one in twenty actions/commands is wrong. This is far too high a failure rate for a real prosthesis controller.

The Achilles heel of pattern recognition has always been its susceptibility to the variability of the signal. In the human body things change all the time, but the person will rapidly adapt to fatigue or injury of one muscle and still control their body well (although perhaps not with the same skill level as when fresh and uninjured). For PR, one command in training might be then interpreted as the opposite once the user is tired. Such variability in a prosthesis is less well tolerated, if the prosthesis does something unexpected. Predictability is key. Users can adapt the way they control their prostheses in response to changes of socket, muscular strength or fatigue if the changes are consistent. From early in the new century the team at UNB, and many others, began to look at ways of improving the success rate beyond ninety percent.

One problem is that the muscle activity is different depending on the orientation of the arm. When it is above the head the muscles will pull in slightly different ways than when the hand is by the side. This might make the hand open inadvertently, or refuse to open when the person is reaching for a high shelf. To remove this chance, the UNB team combined training the prosthesis in different arm orientations (above the head or down beside the leg) with a wider range of signals [369].

A second source of misclassification is transitory signals. A muscle may slightly change its action briefly during a contraction and in the natural system, this brief disturbance is hidden in the inertia of the system. A digital computer reads the patterns regularly (such as twenty times a second), so it can spot these changes and react to them if the computer algorithms did not take into account what has been seen in previous readings. For example, if the signal has been continuously closing the hand for half a second and then one brief 'open hand' command is observed, is likely to be wrong and the one open command is a misclassification. The majority is followed and the hand continues to close. While this makes the controller less prone to those sorts of failures, it does add a little delay to every action, as the controller needs to be 'convinced' a new command is real.

To be successful a controller needs to be trained thoroughly in all the potential cases. This requires many training examples for each case. This might mean many more actions as part of the training session, which would make the session longer and more tiring. Another problem is that the user must be consistent about how they perform the training actions; if they are not then the computer becomes so permissive of variations of pattern that it cannot tell one pattern from another.

In contrast, if the muscle signals of an intact limb change as it is moved (if perhaps we are tired or picking up an unexpectedly heavy object), we will rapidly see and feel the difference and change what we do. This is not so easy to do with a prosthetic controller. One thing is clear from all of the above: The way that conventional myoelectric hands are set up in the clinic, with the user going home, and bringing the prosthesis back for further adjustment months later, (the usual provision of 'fit and forget'), could never succeed with a PR device. The signals will certainly be different the next time the user puts on the prosthesis. It may well be different ten minutes after they put on the socket because it has shifted when they picked up a heavy object, or they have become fatigued or have taken it off and put it back on.

It was Blair Lock, one of Englehart's graduate students working at the Rehabilitation Institute of Chicago, whose insight was to give pattern recognition any real chance of use in the field. He conceived the idea of letting the user retrain the prosthesis any time that they felt it was not working to their satisfaction. He did not suggest plugging the device back into a computer and getting the computer to show what the person needed to do. Instead, the prosthesis would go through a set of preprogrammed actions that the user must follow. *Prosthesis Guided Training* (PGT) was a real breakthrough. It allowed the user to decide when they wanted to retrain. It meant that any changes in the distribution of mass within the prosthesis, as it moves that might affect the signal were incorporated into the signal [370]. To initiate the PGT, the user can press a button, there is a pause, a signal from the controller to warn it is starting and then it goes through the sequence: Open hand, close hand, flex and extend wrist and so on. The user can trigger this training any time they wish. Experience shows that the user gets better at controlling their muscles, the retraining becomes less frequent as time goes on [371]. Because they can retrain it, the user remains in control and so is more likely to persist until they are better at using the device, rather than give up early in frustration. These ideas have been embodied in the products from Coapt. By 2019 hundreds of units had been made and about 500 potential users had tried them (see 14.1.2). They reported a 90% acceptance rate and only 0.8% of the products were returned in five years [372]. The one drawback is the system requires the commands to each axis to be sequential. Parallel recognition of two actions at once remains an elusive goal.

Pattern recognition of a signal requires quite modest processing requirements, once the training has been completed. This was why it was possible for Hudgins and Englehart to build a microprocessor pattern recognition system in the 1990s. The PGT training, on the other hand, needs quite a lot of storage of the data while the training is taking place, so that it was not possible to build a cheap and compact system with PGT much before 2010.

Hudgins recognised that the biggest problem UNB encountered with pattern recognition was the training of the program. As with the basic myoelectric systems, it took more than two decades to apply the concept in clinical devices. By the time he retired in 2012, this was changing with Otto Bock, LTI, and Touch Bionics planning to include these systems in their future myoelectric products [335]. For Hudgins and Englehart this has meant persevering towards a very long-term research goal of increasing the number of devices under control of the myoelectric signal, and finding ways to extract and exploit more information from the signal more reliably. This has meant continuously applying improvements in signal processing techniques to enhance their pattern recognition system.

The fourth director of the IBME was Kevin Englehart, he was also Hudgins' associate director (2000-2013). Like Hudgins, all of Englehart's electrical engineering degrees are from UNB. After his first degree he worked for the telecommunications giant Bell for a year, before starting a master's in telephony. He rapidly changed target when he took Parker's 'Biosignals' course. Following his master's he wasn't sure he wanted to do a PhD, but the application of the severe conservative market driven policies of Ronald Reagan south of the border, was having a significant impact on the job market in Canada. Englehart stayed to finish a PhD, which included the project to produce a stand alone pattern recognition system. Over the space of twenty years he became one of the leaders in the application of digital signal processing, especially for myoelectric signals.

According to Englehart, advances in the field have been bedevilled by challenges in developing systems and getting them to work with users. As a result, the manufacturers have had to be very conservative in the adoption of new technologies. Nevertheless, there was progress. To work together, physicians, surgeons and engineers needed to develop a common language. There was a move from *"a very use oriented field, focused on how to make prostheses work in the 1960s, to interest in the 1970s in the concept of a man-machine interface"* [335]. For Englehart, the prosthetics conferences in Dubrovnik were the most important symposiums for the research that came out of that era. In the late 1970s there was a focus on commercialisation of devices and use-oriented application of EMG signal processing [335]. For fifteen or so years there were no fundamental changes in myoelectric control.

According to Englehart, what brought the field out of this hiatus was Hudgins's dissertation in 1991. It created an interest in the application of pattern recognition to EMG control, and was followed by hundreds of papers written thereafter. For Englehart, it was here that he saw a role and potential for science. The relevant questions were; how to make it work and how to measure it, in order to improve the user experience.

Englehart combined this influence from Hudgins' work with experience he gained while working in telecoms on special purpose computers. These are not stand-alone computers but ones built into another device, the 'embedded systems' which are now part of every smart device on the market (and many apparently dumb ones too). His doctoral dissertation presented a language for using pattern recognition in the field. A series of papers he published in the 1990s demonstrated that it could be done with modest computer systems with low power requirements. The objective of his subsequent research has been to deliver clinically robust, dexterous control to myoelectric prostheses. This means taking pattern recognition methods from the laboratory setting to clinical application in users' homes and at work. It also required not just a new signal processing paradigm, but also developing new training programs for prosthetists, occupational therapists and users. These concepts have found their way into prototype upper limb devices built for Steeper, RIC, Otto Bock, Johns Hopkins, and UNB. According to Englehart, *"I did not see where the work was going, and was lucky to have met the right people to take the work from scientific bench-top curiosity to work and be relevant to the field"* [335].

For Englehart, this experience of developing new systems in collaboration with other groups, often in international partnerships, has been part of a larger trend: *"Teams that worked in isolation now work together"* [335]. UNB became a subcontractor to two major DARPA funded projects under its 'Revolutionizing prosthetics program'. Both projects focused on the application of TMR and use of a pattern recognition control system derived from the work in UNB. The inroad for UNB was its long standing expertise in pattern recognition, and the partnership with the Rehabilitation Institute of Chicago. The partnership focused on the advancement of targeted muscle reinnervation.

Kevin captured the tension in research that is oriented to both fundamental knowledge and customers.

> "This institute has a very strong culture in myoelectrics and does it very well. We have a clinic that grounds the research and provides a daily reminder of where the research needs to go, but that does not mean it guides the research. We do EMG and do very little outside of the prosthetics field (which is a minor part of the EMG). We have a well defined niche" [335].

According to Englehart, the Institute has changed from Scott's focus on keeping the technology simple and trying to implement what works, to an emphasis on clinicalover commercial activities during Biden's tenure. Hudgins and Englehart, under the mentorship of Phil Parker, began their careers by publishing important work and establishing reputations in the academic realm, and then moved into applied work and international collaborations when the new innovation funding was there to support it. The changing landscape of research makes sticking with only prosthetics control a liability where funding may once again dry up. Diversification is the option being taken by Scott's successors. The interest in human computer interfaces for a wide range of conditions and circumstances makes the skills in processing of biological signals that the IBME has developed very marketable, and it is in other areas of rehabilitation that the latest director, Jon Sensinger (12.1) and his team are now looking. Sensinger is the first director who has not studied at UNB. He is a graduate of the University of Illinois, Chicago.

Before coming to UNB as Englehart's deputy he worked at the RIC, his appointment strengthened links between the centres still further.

IBME is housed in a single stone and brick building that was once at the top of the small UNB campus. Above it on the hillside were woods. The building had been the home of a number of different roles before in 1982 Scott's team moved in. The structure started as a smaller building further down the hill, but in common with other buildings in North America, when it was in the way of a major construction, they simply moved it. The building was then extended for the first time and played home to the entomology department. (Tunnels linked the building to green houses, so that later when Prince Charles visited the IBME in 1996 an RCMP police cruiser was parked over the external exit to prevent anyone getting in or out of the building unscrutinised). Later still the building was the annex to the Geology department. Legends of it at that time are that: *"it was a wild and crazy place... they had a coke machine and they filled it with beer"* [172].

When the IBME moved in, they filled every room in the building, packing the students in when their numbers boomed. Eventually extra space was provided by the addition of a trailer in the car park at the back. The trailer's original aim was to allow more animal studies to be conducted in a custom-built theatre suite, but the researchers with the interest in conducting the studies, and the environment around such work, changed. As its primary role faded it became rooms for the IBME's graduate students. When the building was extended again, the trailer was sold on. Over the years the campus grew, the provincial archives extended their building below the Institute and the University Bookstore took up residence above. In 1996 the building itself was extended a second time to allow for more offices and the prosthetic clinic to be housed on the main floor, and in 2019 a further extension was announced that will house a more extensive motion analysis laboratory. The IBME sits unobtrusively between the Provincial Archives and the University Bookstore, and many, walking up the steep hill, do not even know what goes on in 25 Dineen Drive.

When the building was extended in 1996 it was given a name. Many buildings on campus have been named after famous New Brunswickers: Leonard Tilley was a father of confederation. Marshal D'Avray a local politician. The Halls of residence and departments are named after family (UNB graduates or faculty): FJ Toole, was founder of the Chemistry department, Edmund Walker Head (Engineering), Lieutenant Governor, Sir Howard Douglas (Old Arts), Andrew Bonar Law (Fredericton born, UK Prime Minister). Deans and VPs have roads named after them (Mackay, McCaulay and Dineen). For a building to be named after a person, the proposer has to make sure between a third and a quarter of the price of the building comes from an external source to 'sponsor' the construction. At this point, members of the Institute eschewed the notion of finding a sponsor for the name of the building or the institute, and persuaded UNB that they had a candidate who had contributed far more to the Institute than any monetary input. When offered the choice, Bob preferred the building to the organisation. So they were allowed to name the building after their founder: Bob Scott.

CHAPTER

Prosthetics in the twenty first century

14

At the start of the new century technological progress in the prosthetics industry was very much an extension of existing social, cultural, technological and industry trends. As the first decade unfolded the rise of compact low power technology moved from the computer industry to the domestic market, and the technology to develop affordable advanced prosthetics became readily available. David Gow had been developing his technologies in the 1990s, and during the first decade of the new century Mike Hunter sold his hand design to Steepers and Otto Bock released the Michaelangelo. As a result multifunction hands began to be seen as possible. After these pioneers, the market had changed, new companies and new designers found interest and backing which created a wider range of designs. Turn over of new ideas and new players, take overs, mergers and innovations have been part of the story of exo-prosthetics from the beginning (see 2.7). Since the turn of the century this trend of innovation and change has been very much apparent. This chapter considers some of these newer players in the field and where prosthetics might be going in the future.

14.1 New activity in the prosthetics market

All fields of endeavour have periods of rapid growth and times of slow progress or stagnation. After the excitement of the 1970s interest in upper limb prosthetics waned. During the dip I joined the area and was able to observe the following decades when it went from a small number of specialists to a topic that excites interest from diverse groups. By the turn of the twenty first century an increase in technology coincided with a broader public interest in disability and rehabilitation. The Middle East conflicts made the injuries of the soldiers more visible, disabled sport achieved a higher profile. It might have been thought that this boost would wane as the available resources to pay for the work and devices was reduced by the slow down in the global economy in 2008. This did not seem to happen. Instead, enabled by the internet and cheap manufacturing capabilities (see 14.1.4.1), society moved into the era of the *citizen engineer* where members of the public who wished to contribute to business, medicine or science, directed their energies and abilities in new ways. It is very common to see in the media, a group or an individual who have decided to help people who have some disadvantage. The premise is that the skills that are needed to solve the problem exist within the population and motivated citizens can make a difference to one person or many people's lives. This range of areas of interest for the citizen engineers is wide. They wish to help with the effects of poverty, or the barriers imposed by physical or mental limitations at home or abroad. One area that seemed to attract more attention than others is upper limb prosthetics. Many of these innovators state they wish to make their devices commercially available.

Making Hands. https://doi.org/10.1016/B978-0-12-820544-0.00023-8

Small companies are able to introduce innovation much faster than the established players. For example, the introduction of the first new technology batteries (lithium ion) came to the market by a small Norwegian company, Cypromed in 1997, against resistance from all the major manufacturers. This resistance Hans Dietl of Otto Bock sees as a reflection of the commercial liability. The energy density in modern batteries is considerable and there have been many stories reporting spectacular failures of the newer batteries [373]. A small company risks far less than a large one and much progress has been made through innovation from below. This changes the market, which has to respond: within a few years of Cypromed's formation all the major players were shipping lithium ion batteries of their own and Cypromed had not suffered any catastrophic failures of their batteries.

What follows is some of the more significant entries in the prosthetics market since the start of the millennium.

14.1.1 The Vincent Hand

In 1998, in the city of Karlesruhe Germany, Stefan Schultz and Christian Pylatiuk were looking for a new project to explore, they chose upper limb prosthetics. Their first design addressed the problem of the mismatch between the high speed motion of the motors and the low speed reciprocal motion of a hand. They used a motor as a pump. The motor could be run at its most efficient speed to create gas pressure and then the joints of the hand were driven hydraulically. The actuators mimicked the action of spiders' legs which are driven by inflating bags at their joints [207,374]. The bags press against levers and flex the legs. This idea was taken as far as an entire arm [375], although the flexibility of the joints made the motion of the arm somewhat unsteady. The flexibility in the fingers was a distinct advantage. It made them compliant and less prone to damage (see 14.1.3), while also adaptable to unevenly shaped objects. A number of different models were made and tried by users [375]. They were mounted on the standard QDW and operated like a standard single degree of freedom myoelectric hand. While they were not true multifunction hands with separate finger control, they achieved more grip flexibility than any previous nonmultifunction hand had previously obtained.

The German rehabilitation physician, Michael Schäfer spotted what he considered a flaw in the Touch Bionics ProDigits design. As ProDigits were self-contained fingers they could be used by someone with a loss of the entire finger. Anyone with the remains of the proximal joints of the finger could not use the Touch Bionics finger without making their hand longer. Michael encouraged Schultz to look at shorter finger designs, pointing out there were many people to be catered for. The result was a smaller finger that went through two distinct iterations, the first being too close to the Touch Bionics patent. Like the ProDigits, once the fingers were made they could also be built into complete hands. However the fingers were shorter the hands could be made much smaller than the other multifunction hands. While Touch Bionics, RSL Steeper and Otto Bock were making hands that fit the fifty percentile male hand, the new hand (marketed as the Vincent Hand), could cater for the 50% female dimension.

The size is a critical determination. Fifty percentile size is the size you get if you lined up every hand in increasing size, and picked one half way down the line. A hand this size can cover many members of the population without being too big or small. It is one of the features with human perception that an artificial hand looks bigger than it is. Thus fifty percentile hands still look big. If the hand is at the 50% female range, then not only can it cater for many more females but it can easily fit teenage males as well. The market's response to Vincent was not long in coming. Touch Bionics adjusted the design of the proximal joints to make their fingers a little shorter, and so the entire hand too, Steeper had more scope to reduce the size of the bebionic hand to have a female size range too.

The Vincent hand evolved slowly. Schultz did not seek external funding to accelerate the development. What has resulted is a very elegant mechanism with powered fingers and thumb [376]. The hand does not use position feedback. The flexion of the fingers is controlled through the electronics adjusting the current to the motor, but this control is achieved more tightly than the Touch Bionics fingers. The digits of a Vincent Hand remain in their relative places as they close and open more readily. Faced with the different grips the hand can achieve, Schultz chose a form of state machine to switch between the different grips [377]. The final elegant twist to his design is that with a few modifications Schultz could make a child's size hand. The fingers and thumb have fewer joints, but still more than any other small hand [377]. The hand cannot be fitted to infants or preschool children, but it has added to an under served market and perhaps may create loyal users who will move from the child's hand to the adult multifunction device as they age.

14.1.2 Coapt, the first commercial application of pattern recognition

One of the people who bridge between Canada's IBME and the RIC is Blair Lock. Blair was a graduate student of the IBME. He says his upbringing in rural Nova Scotia in Canada's Maritimes allowed him to develop his skills as a *"builder"* [371]. His father was an engineer and growing up he was allowed room to play and explore. By the time he reached UNB in the late 1990s to study electronic engineering he was clearly set on a path to be a professional engineer, not an academic. A successful undergraduate project (which he thought up himself) leads to a master's studentship under Kevin Englehart. The project was a noise cancelling system for public address systems. Blair's system spotted a feedback howl as it formed and suppressed that particular frequency without impacting on the rest of the signal. He was able to demonstrate this live at the project examination and with it he demonstrated his ability to deliver on a task. Englehart noticed this skill and for a master's project, Kevin asked Blair to build a system that integrated all the different experimental elements of a pattern recognition system his students already had produced. This combined data acquisition, feature extraction, classification and post processing into one *"quasi functional system"* [371]. Before Lock, any experimenter doing PR research at UNB would have had to run a series of individual programs, passing the output file of one program into a subsequent program which would perform separate parts of the process. Now the single program took the data, performed the training and the result was used to recognise patterns of signals from the subject. This is much closer to a real working system. The program could be given to other researchers who could conduct tests and generate results without needing to be able to program themselves.

At this time Todd Kuiken was looking to develop the idea of TMR from an interesting experiment into a clinically significant procedure. He needed someone to see through all of the elements of the work. The job specification was an *"engineer"* to do *"stuff"* [371]. Little more specific than that, except Kuiken specified it must be for a minimum of two years, which felt like an eternity at the time. i.e. leaving *"the safety of eastern Canada"* for the big city. Lock was hired for the job, in 2019 after fifteen years in Chicago, Lock remarked: *"Turns out I made the two years, and I am still here"* [371]. Blair was the first full time employee of the new lab. Over the next nine years at RIC, he became the director of operations for the research lab, keeping his finger on the pulse of the research. The role ranged from public relations to legal and marketing considerations as well as research dissemination, subject recruitment and data analysis. *"It was one of those dream jobs"* [371].

Perhaps his most significant contribution to the field is Prosthesis Guided Training, which came from the simple observation that during standard training, all eyes were on the computer screen, with

FIGURE 14.1

A Coapt pattern recognition system. The additional electronics required to be mounted inside the socket are shown. The multiple electrodes use for the pattern recognition system can be seen on the inside and outside of the inner socket. (Copyright: Coapt, LLC.)

the prosthesis hanging idle (see 13.2.5). Once the prosthesis was enabled all eyes went to the device. Blair wanted to keep the focus where it belonged; on the device. It remains a builder's response, to want to simplify things. He first questioned his own thoughts. Was it too simple? Why had it not been done before? The truth is that most research in pattern recognition was (and to a certain extent still is) done on simulations and with able bodied subjects. Only when a real user is wearing a real prosthesis is this sort of thought likely to occur. He agrees that it saved pattern recognition. Without the simplification it might never have been adopted by clinicians and wearers. *"Knowing what I know now about the real world of practitioners and users, these are not research subjects. These are not researchers that are employing it every day, and if it was one iota more complicated than it currently is, boy, I don't know"* [371].

This also led to the setting up of the company Coapt. Initially, he waited for one of the existing prosthetic companies to take on pattern recognition. No one did. It is not a surprise, the major players resisted the idea of a complex hand for decades. It took David Gow to set up his own company to prove to the industry that was possible (see 8.2.2). Lock and his good friend and fellow IBME graduate, Levi Hargrove, were both working at RIC and after one of the first meetings promoting surgery and TMR, the pair thought that since no one else was going to do this *"I bet we can"* [371]. Chats became meetings. Licencing discussions became legal conversations and in 2012 Coapt was launched.

Coapt considered both the licencing model, of creating computer code to work with the individual company's hardware, and the hardware model (developing their own hardware to control the device). This was the route they went for (Fig. 14.1). It had some precedence, in the way that one company's devices worked with other company's electrodes. Motion Control only made an elbow for many years but they ensured that Otto Bock and other's hands could work when mounted on their arm.

Coapt has given Blair a perspective on the market and how it fits together. The cost of the hardware is a very small part of the overall price. For the money, the users get training and support over a long time. Clinicians and users need to be trained to fit and use the devices. Lock's aim is to make the Coapt system more reliable and easier to use. He sees an irony that while the new generation of clinicians are not scared of technology, they are also not prepared to spend much time fiddling with it to set it up. Now much of the latest domestic technology sets itself up. *"Setting the clock on your video is not acceptable any more. It's a bit of a paradox, but it's what we see"* [371].

As the book was going to press the latest development in the Coapt story was that in second summer of the pandemic, in 2021, Coapt acquired LTI from College Park.

14.1.3 TASKA prosthetics

New Zealand is a large landmass but possesses a small population. It was one of the last placed to be settled by homo sapiens, far later than Australia, and only a relatively short time before European explorers reached it. This is because it is a long way away from any other landmass. Its isolation means that almost every item produced by industry has to be imported. All products are expensive. This leads to people preserving complex and expensive devices like cars for far longer than it would be financially viable in Europe or North America. Prosthetic components have to pass through at least one extra distributor before they get to the island. This pushes up the price of an already expensive device [118]. It is against this background that Mat Jury decided that he wanted to make advanced prosthetic hands.

Mat is another motivated citizen who wanted to help and make a difference. He was a successful engineer with a career in IT. After a mountain bike accident in his early thirties where he broke both elbows and both wrists, he was significantly disabled for several weeks. The insight of not being able to look after himself or pick up his children was a spur to moving into this area. His first thought in 2010 was to become a prosthetist. This was impractical. There was no training in New Zealand meaning he would have to study in Australia for three years. As he had a family, this was clearly not an option. Instead he started building hands in his garage. He was trained as an electronics technician so he was comfortable in building circuits and programming microprocessor controllers. He reached out to the profession and talked to Americans John Billock, Troy Farnsworth and Richard Weir. He asked them what was his best course of action and they all replied that he should make the hands more robust. This was at a time when the initial Touch Bionics hands were delivered with additional fingers to make up for any breakages during the fitting process.

Mat's breakthrough was realising that making the hand compliant was a way to make it more robust. Touch Bionics had been devoting considerable effort into making their hand stronger by using titanium in the fingers. *"It was the fundamental recognition that actually compliance makes it more robust, but hang on, it also gives a bunch of other advantages too... It's nicer to shake hands, you can push things between fingers and they splay and you can grab things in different ways"* [118]. It gave Mat confidence that it was a feature that put his hand ahead of the others. It gave him a way to *"clamber up the investment process"*. The financial market in New Zealand was immature. The process was a gamble and he obtained backing mostly from friends who are now shareholders in the company [118].

Jury sees not being the first to market as an advantage: *"We got the luxury of being the second generation of product to market"*. He was able to talk to users about the problems they had with their hands and learn what had gone wrong. He could see that contact with the outside world was damaging to the hands. The Touch Bionics hand had compliance in the top of the fingers and the bebionic hand further down the digit. Both can survive blows to the back of the hand, but the idea of the *entire* hand having compliance seemed to Jury a good way to go. It allows the hand to grasp objects from many angles without aligning the grip precisely to the shape as the standard single degree of freedom requires [253].

FIGURE 14.2

TASKA prosthetic hand. The first upper limb system on the market that can be routinely immersed in water. (Copyright: TASKA Prosthetics.)

Following the award of his patent on a compliant palm, Mat felt able to discuss it with the rest of the industry, so he took the hand to the MEC conference in Canada in 2017 (13.2.3). It was here he met representatives of Motion Control. The aim of creating a robust waterproof hand fitted well within Motion Control's own aims and they began negotiating a distribution deal for North America. This gave TASKA a new and potentially big market. Jury saw the waterproofing as *"the icing on the cake"* [118]. It is the compliance he is most pleased with, but he admits it does not sell the hands as effectively as waterproofing does. They are always happy to show off the hand's abilities: *"we still whack the hand on its side a hundred times a week at trade shows"*. Even so he still has clear aspirations to make the hand even more robust. *"We want to see people using our hands all the time and not having that cognitive loading about whether they can or can't use it for washing the car, or whatever"* [118] (Fig. 14.2).

14.1.4 Open Bionics

Over the first decade of the new century, the one technology that was being talked about as the future of manufacture moved from very expensive machines for the specialist to the home market, and with it an entirely new form of citizen engineer emerged. Home engineering had a long history of people making items from model boats to timepieces, but all of this required a high level of skill and a range of expensive machines at the makers' disposal. The arrival of 3D printing, with other aspects of compact electronics and computers, changed who could be a maker and what they could build.

14.1.4.1 3D printing

Following the launch of multifunction hands and the reduction in price of Additive Manufacture (AM, also known as. '3D printing'), it seemed to a very large number of engineers that prosthetics was a good application of the technology. Unfortunately most designs still lacked important elements: Often the materials used were simply not strong enough for prostheses to survive in the real world. Additionally, the resulting product was close to being the same as a prosthesis made by conventional materials and methods, it did not exploit the advantages an AM hand *should* have. An exception to this was the approach taken by Joel Gibbard of Open Bionics.

Throughout the history of homo sapiens the production of *complex* tools has been a mark of their abilities. Many items are made by removing material from the basic block, be it carving wood or shaping metal. It generally requires skill and considerable training to be able to produce a complex item. *Additive Manufacturing* works by fusing materials together into a new shape. An idea drawn on a computer can be directly printed in three dimensions in a matter of hours. For this technology to be realised it required the development of powerful computers and technology to control the printing of the material. Once the technology was cheap enough it was then useful to the small company or the citizen engineer.

A 3D printed part is built up a layer at a time. Generally, the thinner the layer the smoother the form, but the slower the printing, and the more expensive the printer. How the layers are created is the major differences in manufacturing methods. One printing method uses a laser to melt a powder so that it fuses the grains of the powder together. This is known as Selective Laser Sintering and some nylons and metals can be printed this way. The laser simply 'draws' the shape in the powder one layer at a time, melting the powder and fusing it to the layer beneath. When a layer is complete, the printer moves the job down into the powder and sweeps a new layer of powder over the top, before drawing the next layer.

A second method uses a coil of a filament of building material, the coil is heated and the hot filament is placed against the existing part so it fuses to it. The filament is scanned across the layer to create the form, before that layer is lowered and another layer is added on top. Controlling this is complex, the material has to be heated up enough to fuse on the layer below, without melting it. To not to lose the shape, the new material then must cool fast enough to harden. With heated printing techniques, one side of the piece is hot enough to fuse and the other side is cold. This sets up enormous internal forces within the job which would tend to bend the piece if it was not held down firmly. This means that mounting the job is part of the problems of designing an effective printer. These factors need to be considered in the design of the part to be made.

A final additive technique uses a jet of liquid material sprayed onto the job to create a layer, which sets before the spray head moves up. This is used with concrete to create complex construction elements in buildings.

The arrival of cheap 3D printers has been seen as the dawn of a new age where everyone can make anything they can think of. Indeed it is possible to make things using additive manufacturing that *cannot be made* with conventional production techniques. However, the majority of the items created using the cheaper 3D printing techniques are not strong enough and shatter easily. Additionally, there are many designs, especially in upper limb prosthetics, that are not novel and could be made as quickly, cheaply and as easily using more conventional techniques, *and* the resulting product will be stronger than the 3D printed part. A value of 3D printing is to make one-off devices to match the needs or interests of the customer, and to do it quickly and more cheaply than employing highly skilled craftsmen and women. While the 3D printing of a part does not need skilled technicians to make the devices, it still requires good engineering to make this possible. To realise the design, it is not as simple as drawing something and sending it to the machine. This is where the talent lies.

14.1.4.2 Joel Gibbard

Joel Gibbard was a student of robotics at Plymouth University, when the bebionic hand was launched in 2009. The press interest generated by it occurred at a formative time in his life. During his degree he was looking at what he might do for a career. The opportunity to undertake something as exciting as designing hands was a clear choice. His impression of the existing technology was that it was heavily overpriced and he believed he could do better. He could see a gap between the numbers of people who could use an advanced hand and the few that actually were able to have one. He felt the cost of the advanced hands created a *"social injustice"*. He also believed there was no technical reason why this was true. He admits he was naïve in this analysis [378]. Joel undertook a series of projects as an undergraduate to work on the idea. Initially, as he worked, he expected that someone else would appear and fill this gap, but by the time he graduated he could see nothing happening in this direction, so he set out to fill it himself.

The story of Open Bionics is very much a contemporary tale. Joel got his initial stake to work in his ideas via a crowd funding project. The *Open Hand* project gave him time to work on the ideas for a *"3D printable robotic hand"*. It was during this time that he met Samantha Payne, a journalist whose ideas expanded his. She brought *"a storytelling perspective"* to the project [378]. They founded Open Bionics in 2014. What brought them together was that they had the same vision of a cost effective hand *"with bionic functionality"*. It was Payne who changed the focus towards *"social empowerment that people can get from a really effective and inspiring prosthetic limb"* [378]. The path to this goal was complex, needing both technical and business innovation.

The basic concept was to make a hand that is as cheap as is practical. Gibbard saw the existing market set up as one to create expensive products. It encouraged the suppliers to price hands as high as the market will stand. To get reasonable levels of recompense, a clinic must ask for reimbursement at the highest feasible sum and then contest the decision when it is rejected by a medical insurance company. Prosthetics is a small market and the ability of consumers to push down the price by taking their custom elsewhere is small. After a long negotiation between manufacturer and insurance payer, the eventual prosthetic product remains expensive. Thus the manufacturer has the choice of making a few expensive hands or getting the same profit from many more lower priced devices. The result is there is no middle ground between the most basic body powered terminal devices and the top of the range multifunction hands. It is this niche the cofounders aimed to pursue [378].

Their hand is made using different materials and different 3D printing techniques, matched to the needs within the design. The socket is made from a soft urethane which is flexible, while the main body is made of SLS nylon which is rigid and much stronger. The decorative covers are made from a Polylactic Acid (PLA) plastic which is not as strong, but does not need to be. PLA can be almost any colour which improves the long term appearance of the hand as it does not need to be sprayed and so it will not chip during use. The plastics chosen are the result of constant scanning of the market and testing new candidate materials. Gibbard points out that 3D printing has been used for production prosthetics before, and this is not new. What is novel is they have not been used for structural components before [378].

The 'Hero Arm' is their product. It is a complete arm system consisting of hand, socket, and electrodes all produced to the user's specification. An arm consists of the following elements:

1. Lightweight prosthesis
2. Multigrip hand

3. Electrodes
4. Changeable covers
5. BOA fasteners
6. Liner with vents
7. Soft liner material

The last three combine to make a liner that allows the socket size to change to match the user dimension fluctuation, thus increase comfort. The BOA fastener was developed to simplify the tightening of snowboards boots. It is a capstan which tightens a cable loop that threads like a lace in the boot. It has a one-way ratchet so it is tightened by turning and is released by pulling on the capstan. It is easy to use, designed to be operated through thick ski gloves, it makes it perfect to be used in other applications such as this. One handed, the socket can be cinched up to a level of tightness that is comfortable and is repeatedly adjustable by the user.

The drive to make the hand affordable has meant that Open Bionics had to develop their own electrodes. The cost of any of their rivals would make the resulting prosthesis far too expensive. They have created electrodes that can be placed in a 'daisy chain'. The digitised EMG signal can run along a common communications bus (see 9.2.1), which cuts down on the number of wires in the socket. For two electrodes this is a small saving, but they have built in the capability to allow many more electrodes to be included so that they can use pattern recognition to control their hands. Included in the housing of the electrode is a small buzzer, this can be used to feed information about the hand back to the user. So far they have not had time to explore this feature.

Appearance is one of the significant features of the arm. It can be made to match a range of high-profile characters from the films *Iron Man, Frozen, Star Wars* (Fig. 14.3). That they are able to offer these is a significant development that allows them to gather a level of positive publicity that is unprecedented. Gibbard said that during his Open Hand project he wanted to use the *Iron Man* look and wrote to the copyright holders, and received a standard rejection [378]. The breakthrough came after he entered a Walt Disney accelerator run by Techstars in 2016. This program aims to help new companies to get licences with the Disney corporation. Through the program, Disney picks some companies to invest in and then fast tracks them through the process of getting the licencing deals. By doing this Disney makes money on the deals and is able to control their intellectual property.

Joel Gibbard is a British entrepreneur and his modesty is deeply ingrained. While he actively promotes his product he does not overstate his contribution or the work it took to get to where he is. He wishes the rest of the company and especially his co founder Samantha to get a share of the credit. When asked about getting on to the Techstars program he is modest: *"We definitely brought something unique that nobody else did"*. Open Bionics are not a traditional company for Disney. Most of the other partners are toy manufacturers. *"I don't know. I guess they saw something in us and wanted to have something on the program that was a little bit unique and not what everyone would expect"* [378]. Joel considers himself lucky to get on the program. He suspects that Disney probably did not think they would make money from the deal. Open Bionics was about *"making children happy"* which fitted Disney's overall aims [378].

This simplifies what is likely to be a difficult relationship that needed significant effort to make work successfully. The Disney corporation jealously guards their copyright. To protect their intellectual property they prevent organisations large and tiny from using unlicensed versions of their characters through strong legal representation. It is also true that fewer than one percent of the applicants to the Techstars program get funded. This suggests a great deal of care went into the Techstars application

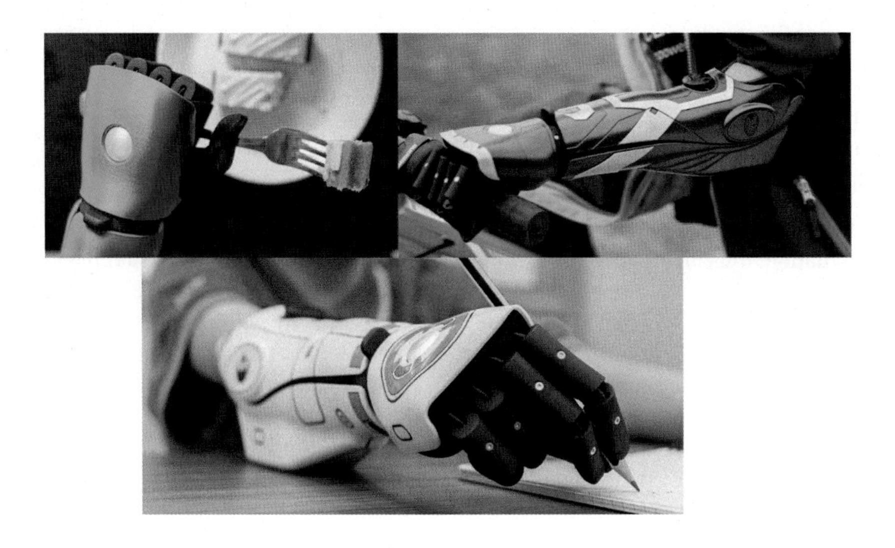

FIGURE 14.3

The Hero Arm. Commercially available 3D printed multifunction hand capable of different grips. One aspect that appeals to users is the company's licence to use Disney Corporation graphics on their arms. An innovation is the use of the BOA fastening system to make the socket adjustable. (Copyright: Open Bionics.)

and much effort into making the deal work for Open Bionics. The benefit of the deal is that they have access to some of the significant heroes from modern films. Its appeal to the users is clear. It attracts a particular group of users; young persons with congenital absences.

The process to become fitted with an Open Bionics arm generally begins with the potential user approaching the company. Open Bionics will in turn will refer them to a clinical team. The clinic will either make a cast of their residuum and modify it to match the needs of the user, or they will optically scan the remnant limb. The company supplies a kit to enable the arm to be mocked up, dimensions established and the location of the EMG amplifiers to be determined. At their facility in Bristol, the team at Open Bionics then scan the cast or use the residuum scan from the clinic. This digitised shape is then used to create the prosthesis. The aim is to automate the process, taking the basic parameters about the user plus the scan and creating the computer files to drive the 3D printer. In 2020, they were a little way away from fully automating the process. It is part of Gibbard's aim to bring the price of the arm down still further so that he continues to work towards this. Once the arm is manufactured and assembled it is returned to the clinic for fitting and training. Joel sees that this is the only difference from the standard relationship between the clinic and the manufacturer; the definitive prosthesis is made in Open Bionic's workshops not the clinics. In fact this is a trend in prosthetics, in some regions manufacturers like Otto Bock have a central fabrication service for their prosthetic clinics, where they make the sockets and custom liners.

One of the problems that has plagued the launch of the Hero Arm is inaccurate information about Open Bionics' aims or methods. This might be seen as a surprise for a company with an effective marketing presence. Members of the prosthetics profession may have seen the novel methods used by Open Bionics, the rhetoric from the press and the high profile of the advertising and have misinterpreted intentions. Alternatively, professionals might have received miscommunications about the methods and motivations of the company. Stories abound of the company only supplying to users and cutting out the clinical services. Added to the impressions gained about the poor quality of other 3D printed hands, this may have created a resistant reception from the profession. When asked, Gibbard says that the idea of taking the role from the prosthetist *"could not be further from the truth"*. The Open Bionics aim is to broaden the access to technology as far as they can. *"[We] firmly believe that the clinician is incredibly important in the fitting of a prosthesis and should always be the one to be fitting it, and are keen to do what we can to make their jobs easier and make the most out of any budget they can get"* [378].

This reveals the underlying philosophy of the company: *"Our mission is to create and democratise technology that enhances the human body"* [378]. Even though the Hero Arm is cheaper than any of his rivals Gibbard wishes it to become cheaper still. He would like the hands to cost nothing to the user. If a potential customer approaches the company, Joel says there are ways to gain additional financial support so that it may cost the user little or nothing at all. His wishes echo Jim Henniquin of the Creature Shop who worked on affordable aids for those in wheelchairs in the 1990s. He suggested it was immoral to show a starving man a meal and then tell him he is too poor to afford to eat.

The business model for the company is to supply the hand to the user via the clinic, rather than *"business to business"* as offered by the majority of prosthetic suppliers. In reality, the supply of other prostheses to many clinics is more complex than that: Often, the prosthetic manufacturers are also the contractors that provide the clinical service. In these systems, keeping the clinics at arm's length makes for a complex relationship between supplier and user.

The next stage for the hand is to attempt to gain independent validation of its use. Open Bionics is therefore involved in a clinical trial. It is being held in Bristol and Manchester and it is being independently run to ensure it is an objective assessment. The aim is to have eight children fitted with an arm and allow an *equivalency study* to be conducted. This sort of study aims to show that for an equal cost it has equivalent function. Since there are currently no multifunction children's hands, Open Bionics is aiming to be as effective as any of the standard products. Joel is confident that it will prove *more* effective. If it is counted as equal it will open up the market for his product more widely. The largest previous independent study of a multifunction hand in children was six users with the Reach hands (see 8.2.1) in the 1990s [257]. There have been few published independent studies assessing groups of users of the same hand. One barrier to such a trial is the finance to buy and fit all the hands. It is likely that previously, the hands tested were so expensive it was hard to finance a pool of users easily. The price of the Hero Arm enables this advance to be made.

Open Bionics' next target is to move away from their current user base. The arm currently appeals to younger people. It is perhaps not hard to see why these hands might not appeal as much to some older users in their current form, but the beauty of the manufacturing method is that Open Bionics should be able to easily change the appearance of the hand to attract a different audience.

The Hero Arm was unique and it is at least due in part to Gibbard and team's persistence and ingenuity. The arm is not just a standard design, made cheaply, it does exploit the advantages of 3D printing by aiming to reduce the cost of production through automation. Others who propose cheaper prostheses have yet to show similar skills. Anyone who can handle a computer mouse can design and manufacture, but that does not guarantee the results are new or useful. The rapid prototyping boom might well open up possibilities in prosthetics and the Hero Arm may not be the only such prosthesis for long. To succeed others need to recognise that robust prostheses are essential.

A future for rapid manufacture in general, is that new designers must be able to draw up a design simply and have it made, so that everyone can customise a tool to match the next job. A prosthesis will just be an extension of this, with devices matching the person and their occupations. It is worth remembering that it has *always* been possible to make tools specific to a task. The trade of *Toolmaker* used to be one of the most respected jobs in the field of manufacturing. Few people were toolmakers, because the job was hard and required especial skill and innovation. What is going to be key to making rapid prototyping more than the democratising of mediocrity will be if the software and manufacturing can be made to unlock people's ingenuity so that they use the technology for more than making the same thing badly again and again.

14.1.5 Norwegian prosthetic innovations and Hy5

Norway has had a smaller impact on clinical prosthetics than other wealthy countries. In the 1990s the company Cypromed was founded in Ottestad outside of Hamar. It was designed to take the spin offs from a collaboration between researchers at the Norwegian University of Science and Technology (NTNU) in Trondheim and an existing cluster of clinical and technical personnel and companies in Ottestad. One of the products they produced were packs of lithium ion batteries designed to replace the standard externally mounted six volt electric hand batteries. The system included the charge circuitry which ensured the batteries were charged correctly and they did not get too hot during the charging process. Once the batteries were being used their endurance and reduced mass were recognised, it subsequently became was necessary for all manufacturers to extend their product line to incorporate each emerging battery technology

In 2015 a new prosthetic hand from Norway appeared on the market. It did not fit simply into the standard classification of single axis or multifunction hands. It was a three fingered anthropomorphic hand that used hydraulic drives. The fingers and thumb are separately actuated, but the fingers are also linked compliantly allowing them to adapt to the shape of the object (Fig. 14.4).

The company was the creation of Jos Poirters a Dutch engineer working in Norway [379]. Poirters background is not dissimilar to Mark Hunter in that he made animatronic puppets, in this case for fairground rides. Like many others he considered the challenge of building a prosthetic hand a worthy goal. Poirters made puppet hands, which did not need to grasp anything successfully, but they did have to work repeatedly for a very long time without needing constant servicing, features that all successful prosthetic hands need. Jos began thinking about the designs for robot hands and arms in 2002 and worked on the ideas for about thirteen years [379].

The first design took him two years, by then he had he knew he could make a hand lighter, more powerful and more functional. It was after about five years, in 2007 that he knew he wanted his design to be a prosthesis and to help people. His target was for the hand to be the same mass as a human hand and have more function than existing prostheses. In fact a prosthesis with the same mass as a natural hand will seem heavy to the user. He gained experience with users through making links with *Norwegian Technical Orthopaedics AS* (NTO) which facilitated user experience with his prototypes in 2008. This also gave him exposure to conventional hand designs. The users he met through this contact were not satisfied with the current designs which gave him confidence to develop the hand further.

Further development meant he needed additional funding. He considered selling the hand to commercial partners (such as Otto Bock) but he did not pursue this as he also wanted to see it through *"I did not want to give it up"* he thought it had *"huge potential"* [379]. In 2015 the first Hy5 Pro was fitted to users.

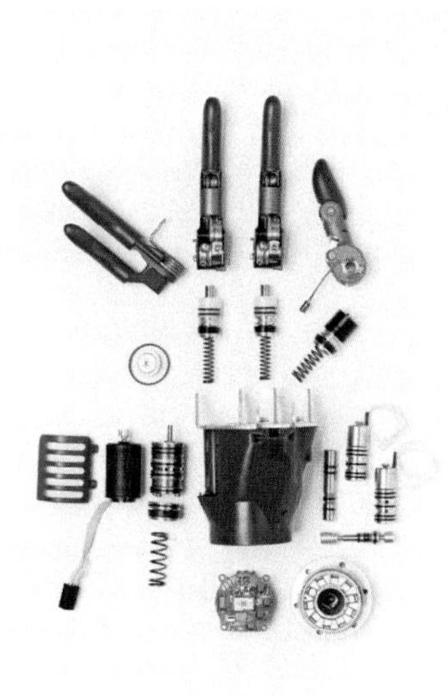

(a) All the hydraulics are retained in the palm section to reduce weaknesses introduced by hoses. The pressure is created by a single electric motor and separate hydraulic pumps.

(b) The hand is capable of a range of grips through the use of separate hydraulic pumps for each finger. It also makes the hand more robust and water resistant than earlier generations of hands.

FIGURE 14.4

The MyHand1. A novel design of a hand which uses hydraulics to drive two fingers and thumb. (Copyright: Paul Paiewonsky.)

The first commercial device was the *MyHand 1*. It is hydraulically driven rather than electric. He chose hydraulics because he felt he could build a hand that was strong, compact, reliable and be easier to build with few moving parts [379]. For this he had to think about the system in a different way, with the palm section containing all the hydraulics. This is because hoses and connections are the *"Achilles heel of hydraulics"* [379]. So he designed a hand without any of these elements. He believes this design makes it more reliable. The block is also 3D printed in plastic making it waterproof. The fingers are 3D printed from titanium which is very expensive but also makes them very tough. So far no one has broken any of his fingers [379]. The three fingers have a single common axis similar, to the majority of conventional single axis hands.

The MyHand1's performance:

1. One size 7 $\frac{1}{2}$ inches equivalent
2. Pinch grip 55 N
3. Power grip 110 N
4. Closing speed 0.8 s from extended to pinch.

The design incorporates one electric motor with two hydraulic pumps on the same shaft. It allows for synergetic prehension. The low pressure moves the fingers quickly, but when the hand contacts an object, the motor changes to drive only the high force pump. The pressure changes from ten to thirty-five bars. Each digit is made of three phalanxes which are controlled from one pump each. The phalanxes have a cylinder each and are linked, but the flow of hydraulic fluid allows them to adapt to the grip shape. The system is 'open' which allows the phalanxes to take up any position relative to each other. As a result they require a mechanical brake on the middle joint to stiffen the fingers when they take up a load, such as in precision grip (Fig. 14.4). If they did not the joints would keep closing inward past the distal phalanx. Simplicity and reliability mean this automatic brake mechanism is purely mechanical system (it is also patented). The overwhelming majority of wearers use a silicone glove with the hand and it gives the device a better grip [379]. Jos has built his own version of the QDW. It is simpler than the Otto Bock design and the rotational resistance can be adjusted without an expensive and complex specialist tool. The hand can form an 'autograsp' which allows the hand to maintain a light touch on an object as the grip is adjusted. This is the result of the hydraulic mechanism, the only other hands that have been able to achieve this before, required sensors on the fingertips and a microprocessor controller (see 10.2.1).

14.1.5.1 MyHand 2

Poirters has continued to develop the hand and a new version was due for release in 2020. The changes include the option of 3D printed aluminium fingers or titanium fingers. Aluminium is half the price of titanium, but also half the strength. The user will be able to choose what material they want, and swap in or out the different fingers as they wish. He expanded the range to four sizes from small to a large equivalent of 7 $\frac{3}{4}$ inches. Importantly, the thumb orientation is changed to a more anthropomorphic direction and will be controlled separately. He employs shells to change the size. The palm is the most complex and expensive, and he saves costs by using the same size palm for all hands.

The hands are generally controlled in a similar way to conventional single degree of freedom hands. However, like the early Touch Bionics hands the user creates a grip by intercepting the fingers in motion and framing the response. The difference is the MyHand 2 will have separated thumb control as a step towards less dependency on the use of the other hand or placing the prosthesis in a specific position to inhibit the motion of the fingers.

14.2 Trends in the field of prosthetics

In the course of researching this book I spoke to people in the field, or now retired from it. They were asked to try and see into the future. This was greeted with mixed reactions, many were reticent in making predictions that will likely prove wrong.

One thing that often seems true about the way people predict the future is that they tend to underestimate the near term, but exaggerate the long term. Before Kennedy's challenge to NASA, only the very well informed, or brave, predicted men landing on the moon before the end of the century, yet at the same time in the 1950s, flying cars and robot butlers seemed just round the corner. Landing on the moon took a staggeringly short handful of years to achieve, neither of the latter innovations have yet come to pass. Predictions of the increase in computer power have been more in line with the facts. However, what this massive increase in available computation has meant for many nontechnical arenas, has left all but the most prescient astounded. Computer technology's role in the current and next generation of advanced prostheses is undeniable, but what that role will be is less clear.

The progress of prosthetics depends on the rest of society and how they prioritise support for those with impairments, as much as it depends on the advances of the technology or ideas. The field suffered a downturn from the early 1980s through to the turn of the century, as much due to the lack of funds as the need for the technology to catch up with the ideas. In recent years the promises of many years (or even decades) past have come close to fruition. More significantly, with a few exceptions, the same list of ideas to advance prosthetics that are being investigated by the profession today could have been made thirty five years ago. The difference today is that these ideas are not in the lab but are in products. If not yet a product they are on platforms that allow more than two users to try them, and these users might get to take the prostheses home for extended periods of time.

14.2.1 Technological potential

The technical aspects include the surgical and the engineering. Both have moved forwards in the last two decades. Ideas about implants being used to mount the prosthesis on to the body, or employed to detect the control signals, go back to the 1960s, but it took time for them to become clinically significant. Osseointegration is now routine and neural implants are able to survive in the body for extended periods of time. One advance is that the signals being acquired are of greater range and quality than before, thus effective control looks more possible.

The design of the prosthesis has changed radically in the last two decades. It is not clear if there will be the same level of innovation over the next decade. It might be that there will not be space in the market for additional multifunction hand designs. There seems no slowing of the emergence of new companies producing new hands. Hans Dietl (Otto Bock) does not believe there is space for more than a couple of major companies who can afford to do all the quality assurance work on their designs [281]. Others (such as Paul Steeper and Mat Jury) are more enthusiastic about the competition with Jury seeing compliance and waterproofing being the trend in the designs over the next five years [118]. Paul Steeper believes there is a place for the new manufacturers, *"but I doubt that any one will dominate"* [298]. In contrast, Harold Sears was concerned that the focus remains away from the user and that the industry is held back. It could make greater progress *"if it paid more attention to actual outcomes and actual needs of the consumers"* [34].

Multiaxis hands with curling fingers are by no means new. With some hands being on the market for more than a decade, multiaxis hands are clearly more than a passing interest. There are more than four products on the market, people can now try them for themselves. Many of those questioned did see a future for the multifunction hands and celebrated the increased levels of competition.

Iversen, Jury and Poirters believe in making tougher hands. Poirters said the future will be simpler/cheaper hands with *"ok function that can take a beating"* [379], Jury and Motion Control are adding waterproofing, and more robust hands have been the goal of Motion Control for many years.

The majority of respondents hoped that pattern recognition would be the next significant change to the control of prosthetic hands. The desire is a simple and intuitive control format for the multifunction hands. Paul Steeper commented that it needs to have *"a platform in place where it's workable for the patient... and become instinctive to operate"* [298].

One question concerns the way that pattern recognition is added as a feature in a prosthetic limb. Currently the hand systems are sold much like cars or computers with additional features being added to a basic model, thus a PR system is a bolt-on for an already expensive hand. Joel Gibbard of Open Bionics does not regard this as an acceptable solution; *"I feel like the control system should be part of the product and shouldn't cost extra"* [378]. A difference in perspective comes from how the hand is supplied; if the healthcare is state based or from private funding. It is a truth that a nation's healthcare system is an important part of the way a country sees itself and to compare it unfavourably is a quick way to make enemies. In his re-election campaign in 2004 George W Bush stated that the USA *"had the best healthcare system in the world"*. It is undoubtable that some of the best facilities and surgeons on the planet are in the USA. However, it is only for those who can afford them. Other countries would regard this as unacceptable and many have a form of centralised or state provision. Health insurance rarely covers existing conditions, so that even in countries with some level of centralised provision, such as Canada, children with congenital absences do not get provincial funding for their prostheses and have to depend on charity to provide a solution. Prosthetics companies operate in an international market, the insurance based system tends to drive the way that prosthetics are priced. This creates the market where a bolt-on is a way to gain greater financial advantage. For example, the additional grasping feature from Motion Control (FLAG, see 7.2.2) is a bolt-on that adds $700 to the wholesale price of the device, but this is not the same as the cost to the user. Sears of Motion Control commented: *"If Motion Control gives away FLAG (any manufacturer occasionally gives a feature away to build the market) the prosthetist still wants/needs to charge normally for it"* [196]. Clearly in a market based healthcare system, it would still cost the end user for the increased functionality. Gibbard started in the UK where the National Health Service provides for all citizens. The clinics have a limited budget for all their potential users, thus his aim to reduce the price of the device is a powerful one if they want to benefit as wider population as possible. Regrettably, these cheaper prosthetics may not penetrate an insurance based market. In that market the price of the finished and delivered prosthesis, can simply be a multiplier of the cost of the componentry. If this is the case then a cheaper hand might not provide the practitioner with sufficient funds to fit the hand economically.

Feedback of information from the hand to the user has been the touchstone for many researchers. Closed loop control is the only way to make any complex system work well and it is how we control our natural limbs. All but a very few attempts to use feedback in prosthetic arms have so far been limited and inconclusive. Now we have enough technology implanted in enough volunteers, we might be able to work out how much feedback is enough, and what sort of feedback is useful. Although the warning is that if the designers wish to use EMG as a control input, experiments have shown that EMG is too slow to be used to control the hand successfully with feedback [380].

In contrast with the introduction of the newer surgical techniques, manufacturers now face the challenge of developing technology that maximises TMR and Osseointegration's prosthetic potential. This would mean prosthetic arms and hands that are capable of much greater control input from the user, but also are able to transmit information from the prosthesis back to the user.

There are still those who suggest that people will want to 'upgrade' to advanced prosthetic limbs and be able to perform more, greater and for longer. It is always possible to find a few people who might want these changes. The simple truth is, however, that the majority of people want a quiet life with fewer inconveniences, not superpowers. The ability to crush a rock or run faster than a car, might seem like a good idea, but the domestic world is made of objects that were produced to be grasped by the human hand and picked up by the human arm. The objects have a maximum mass that fits the power of an average person and a strength to resist crushing by humans. Genetic variations throw up exceptional people who are taller, stronger, quicker than the average, but they find working within normal society awkward or tedious as they have to make changes to accommodate their differences. Taller people need larger doors, but most of the time they have to duck. They are a lesson and tell us that being superhuman is not as good as some people claim. Generally, persons with hand amputations just want their arms back, not to excel in some new endeavour. To choose an over powerful prosthesis that deliberately makes their lives harder suggests a contrary nature. A decade ago it was clear that most people would not want the inconvenience of surgery. Milos Popovic, when speaking about stimulation of the nerves to make paralysed limbs move, suggested that while implanted electrodes had all the technical advantages, few persons with paralysis (who would benefit from this surgery) actually wanted it, or asked for it. They preferred something less invasive. These are people with far more to gain than simply replacing a missing limb. It might have seemed that they would be prepared to put up with the surgery, effort and discomfort to improve their lives. They did not. Even the 1970s iconic program the *Six Million Dollar Man* had a second bionic superman who turned down the opportunity to be *"better than he was"* and had his super limbs down tuned to 'normal'.

Surgical solutions to amputation have many advantages, but also some drawbacks. Hand allografts are now more established as a technique, but the recipient has to invest significant effort to gain control of a grafted arm. In my experience, persons missing one arm below the elbow generally do not consider themselves 'disabled'; it would be more appropriate to describe their attitude of themselves as being 'inconvenienced'. To go to a lot of effort to overcome something that unimportant seems to be improbable. It is not very likely that a person who had already had an amputation would want the extra hassle of more surgery, especially if they already have had more than one surgery following their initial trauma. For example. Todd Kuiken found it harder to recruit subjects for TMR in the forearm, compared with those with higher losses as the majority of potential subjects were content. The one exception is if the surgery will clearly have a significant positive impact on their lives, such as someone missing *both* hands. If we can be more confident that advanced prostheses and surgical techniques will make peoples' lives easier, better, simpler, potential users might agree to the interventions.

Certainly, a younger generation of people seem in general to be more open to radically changing their circumstances, whether it's their clothes or their bodies. This means there might be more demand for surgical 'readjustment' of the limb different than in the past. Todd Kuiken led the way to reintroduce heroic surgery in prosthetics with TMR. He would very much like to see the procedure become a routine part of post amputation care in his lifetime [381]. His latest ventures are no less radical, but more focused on the routine. He can make a remnant limb able to be fit more stably within a socket through the use of liposuction. This is not cosmetic and it is clearly very functional and gives the recipient huge advantages. This and some of the other techniques like AMI (see 2.6.1) promise more function for those who can tolerate the surgery. Perhaps there will be a class of prosthesis user in the future that will want all the modifications technology and surgery can deliver and will genuinely become a cyborg.

14.2.2 Societal attitudes

Many societies have become more tolerant of difference (diversity) in the last decade. This has included sexual orientation, and some forms of disability, but even within these societies there remains a minority who object to change, race and difference. The attitudes of the changing society can be, to some extent, gauged by their response to difference in main stream media. Whereas once a nonwhite face on a major broadcasting channel in the UK or North America was a rarity, the level of tolerance has increased. Now more noncaucasian persons are seen introducing television shows, so much so that it is not worthy of comment any longer, reflecting a positive change in society. However, this improved state does not yet extend to some aboriginal populations who remain in special interest shows. Similarly, persons with impairments on television are still worthy of comment. It was only in 2009 that the choice of anchor for a BBC TV's flagship children's program elicited comment. Kerry Bonnel was congenitally missing her forearm, and she did not hide it or wear a prosthesis. Sadly, it was suggested by some commentators that this would 'scare' the children, rather than praising such diversity, making it more acceptable [382].

The coverage of recent Para-Olympics is also a reflection of this. The Paralympics is *in parallel* with the primary games. In North America its coverage was still a long way second to the primary games. In Europe the last few championships have seen broadcasting companies in bidding wars for the rights to broadcast the paralympics, and they do so with similar levels of hype and airtime as the older games. Channel 4 in the UK emphasised the sporting background and the impairment of each of their specialist presenters, who were the primary commentators on their sport, much as retired footballers and athletes have commentated on mainstream sports for years. This has made a difference, the paralympics now have a higher profile and they show the competitors' ability to compete is just as compelling as any 'conventional' competition. I hope that this trend continues, so that people with differences will be tolerated more and the need to conform to one form of acceptable 'lifelike' prosthetic solution will lessen.

In the past, surveys of users of 'cosmetic' hands reported that wearers used them for the comfort of others and not specifically for their own peace of mind [106]. Perhaps the acceptance of the different designs is leading the way for users who celebrate their difference with their devices. If this is the case, then the design and appearance of prosthetic terminal devices will be able to flourish, matching more the manipulative task than the visual one. This might also allow greater acceptance of the surgical solutions that are more functional and less cosmetic.

Talking to manufacturers such as Bill Hanson (LTI) and Bob Radocy (TRS), they comment that there are some innovations that might be possible, but they will never be commercially viable. At one time that would have been said about multifunction hands, and some may still maintain that it is true for the overwhelming majority of potential prosthesis users. As society and its needs change, then so might the technology that prosthetic limbs are made from.

Many of the interviewees saw that implant technology would open up the control and use of prosthetic arms (Meek, Hansen, Kuiken, Zahedi, Hudgins), although fewer were confident about the use of neural implants, suspecting that the peripheral interfaces were easier to do and safer (Hansen and Hudgins, for example). Makin and Brånemark believe that we need to use the processing power of the existing central nervous system with peripheral interfaces as the correct place to aim improvements that will lead to enhanced control.

This is the crux of the discussion, given that the target hasn't changed: an independent life for the limb absent. How much progress will be made in the next ten to twenty five years, and how different would a book written a generation from now be? What will the differences be and what will the designers of the next generation trying to solve? A need for normality, or a need for function? Will people continue to embrace the technical, or will simpler plainer prostheses still be the most affordable solution? While the technology shifts, the challenge remains as great and as interesting. Few predictions made twenty-five years ago about what society would look like now would have seen what is in place today, so trying to see that far ahead might always be in vain.

One of the predictions most likely to be true is that there will still be engineers, clinicians and users to whom one of the biggest and most rewarding challenges they can think of, is to make a better prosthetic hand.

Interviewees

Christian Almström, Retired,
Chalmers University of Technology, Göteborg, Västra Götaland, Sweden

Mike Amrich, Clinical Specialist,
Liberating Technologies, Inc., Boston, Massachusetts, USA

Ed Biden, Former director,
Institute of Biomedical Engineering, University of New Brunswick, Fredericton, NB, Canada

Rickard Brånemark,
University of Göteborg and University of California, Department of Orthopaedics, Göteborg, Västra Götaland, Sweden

Dave Brown, Retired,
Lecturer, University of Portsmouth, Portsmouth, Hampshire, UK

Alix Chadwell, Engineer,
School of Health and Society, University of Salford, Salford, Greater Manchester, UK

Robin Cooper, Retired,
Formerly Hugh Steeper, Leeds, Yorkshire, UK

Hans Dietl, Chief Technology Officer,
Ottobock HealthCare, Dudestadt, Göttingen, Lower Saxony, Germany

Kevin Englehart, Former director,
Institute of Biomedical Engineering, University of New Brunswick, Fredericton, NB, Canada

Joel Gibbard, CEO and Co-Founder,
Open Bionics, Bristol, UK

David Gow, Retired,
Formerly Head of SMART Centre, Astley Ainslie Hospital, Edinburgh, Scotland

William Hanson, Retired,
Former CEO, Liberating Technologies, Inc., Boston, Massachusetts, USA

Peter Herberts, Professor emeritus,
Department of Orthopaedics, University of Göteborg, Göteborg, Västra Götaland, Sweden

Liselotte Hermansson, Adjunct clinical professor of occupational therapy,
School of Health Sciences, Örebro University, Örebro, Sweden

Shiela Hubbard, Retired,
Clinical and Research Physical and Occupational Therapist, Hugh MacMillan Rehabilitation Centre, Toronto, Ontario, Canada

Bernie Hudgins, Former director,
Institute of Biomedical Engineering, University of New Brunswick, Fredericton, NB, Canada

Mark Hunter, Founder,
Epic Inventing, Los Angeles, California, USA

Ed Iversen, V.P. Research and Development,
Motion Control, Inc., Salt Lake City, Utah, USA

Steve Jacobsen, Formerly,
Professor of Engineering, University of Utah, Salt Lake City, Utah, USA

Matt Jury, Product director,
TASKA Prosthetics, Wellington, New Zealand

Todd Kuiken, Director Emeritus,
Shirley Ryan Ability Lab, Chicago, Illinois, USA

Debra Latour, Assistant Professor of Occupational Therapy,
College of Pharmacy and Health Sciences, Western New England University, Springfield, Massachusetts, USA

Blair Lock, CEO,
Coapt, LLC, Chicago, Illinois, USA

Tamar Makin, Professor in Cognitive Neuroscience,
Institute of Cognitive Neuroscience, University College London, London, UK

Gordon McLeary, Principal Mechanical Engineer,
TouchBionics, Livingstone, West Lothian, Scotland

Stuart Mead, Former director,
TouchBionics, Livingstone, West Lothian, Scotland

Sanford Meek, Associate Professor,
Mechanical Engineering Department, University of Utah, Salt Lake City, Utah, USA

Debra Latour, Assistant Professor of Occupational Therapy,
Division of Occupational Therapy, Western New England University, Main, USA

Philip Parker, Retired,
Institute of Biomedical Engineering, University of New Brunswick, Fredericton, NB, Canada

Dick Plettenburg, Retired,
Director, Delft Institute of Prosthetics and Orthotics, Delft, Netherlands

Jos Poirters, Co-Founder / Senior R&D Engineer,
Hy5, Oppland County, Norway

Bob Radocy, CEO,
TRS Inc., Boulder, Colorado, USA

Robert Reiner, Professor in sensory-motor systems,
Department of Health Sciences and Technology, ETH Zurich., Zurich, Switzerland

Robert Scott, Former director,
Institute of Biomedical Engineering, University of New Brunswick, Fredericton, NB, Canada

Harold Sears, Retired,
President, Motion Control, Salt Lake City, Utah, USA

Jon Sensinger, Director,
Institute of Biomedical Engineering, University of New Brunswick, Fredericton, NB, Canada

Adam Spiers, Lecturer,
Department of Electrical and Electronic Engineering, Imperial College, London, UK

Paul Steeper, CEO,
Steeper Group, Leeds, Yorkshire, UK

Niel Storey, Retired,
University of Warwick, Warwick, Warwickshire, UK

Ian Swain, Retired,
Director of Clinical Science and Engineering, Clinical Director of Odstock Medical, Salisbury, Wiltshire, UK

Bob Todd, Retired,
Engineer, Centre for Alternative Technology, Machynlleth, Powys, Wales

T Walley Williams III, Retired,
Liberating Technologies, Inc., Boston, Massachusetts, USA

Kerstin Cain-Winterberg, Occupational Therapist,
Rehabilitation Centre for Upper Limb Prosthetics, Sahlgrenska University Hospital, Göteborg, Västra Götaland, Sweden

Saeed Zahedi, Technical Director,
Chas A Blatchford & Sons Ltd, Baisingstoke, Hampshire, UK

Bibliography

[1] S. Kikkert, J. Kolasinski, A. Jbabdi, I. Tracey, Christian F.C.F. Beckmann, H. Johansen-Berg, T.R. Makin, Revealing the neural fingerprints of a missing hand, eLife 5 (2016) e15292.

[2] M. Borchardt, K. Hartmann, L. Radike, S. Schlesinger, Ersatzglieder und Arbeitshilfen für Kriegsbeschädigte und Unfallverletzte (Artificial Limbs and Aids for Disabled Veterans and Accident Victims), Springer Verlag, Berlin, 1919.

[3] G. Smit, D.H. Plettenburg, Efficiency of voluntary closing hand and hook prostheses, Prosthetics and Orthotics International 34 (4) (December 2010) 411–427.

[4] G. Smit, Natural Grasping: Design and evaluation of a voluntary closing adaptive hand prosthesis, PhD thesis, TU Delft, 2013.

[5] L.A. Miller, R.D. Lipschutz, K.A. Stubblefield, B.A. Lock, H. Huang, T.W. Williams III, R. Weir, T.A. Kuiken, Control of a six degree of freedom prosthetic arm after targeted muscle reinnervation surgery, Archives of Physical Medicine and Rehabilitation 89 (11) (2008) 2057–2065.

[6] J.M. Nightingale, Microprocessor control of an artificial arm, Journal of Microcomputer Applications 8 (1985) 167–173.

[7] P. Herberts, C. Almström, R. Kadefors, P. Lawrence, Hand prosthesis control via myoelectric patterns, Acta Orthopaedica Scandinavica 44 (1973) 389–409.

[8] P. Herberts, C. Almström, K. Caine, Clinical application study of multifunctional prosthetic hands, Journal of Bone and Joint Surgery. British Volume 60 (4) (1978) 552–560.

[9] K. Hagberg, R. Brånemark, One hundred patients treated with osseointegrated transfemoral amputation prostheses - rehabilitation perspective, Journal of Rehabilitation Research and Development 46 (3) (2009) 331–344.

[10] T.A. Kuiken, P.D. Marasco, L.B. Lock, R. Norman Harden, J.P.A. Dewald, Redirection of cutaneous sensation from the hand to the chest skin of human amputees with targeted reinnervation, Proceedings of the National Academy of Sciences 104 (50) (2007) 20061–20066.

[11] N. Shubin, Your Inner Fish, Penguin Books, 2008.

[12] H. Focillon, Éloge de la main, Vie des Formes, 1934.

[13] M. Fox, Talking Hands, Simon and Schuster, New York, 2007.

[14] H. Sears, telephone interview, 26/11/14.

[15] N. Vignier, J-F. Ravaud, M. Winance, F-X. Lepoutre, I. Ville, Demographics of wheelchair users in France: results of national community-based handicaps-incapacités-dépendance surveys, Journal of Rehabilitation Medicine 40 (3) (2008) 231–239.

[16] P.J. Kyberd, D.J. Beard, J.D. Morrison, The population of users of upper limb prostheses in Oxfordshire, Prosthetics and Orthotics International 21 (2) (1997) 85–91.

[17] R.A. Roeschlein, E. Domholdt, Factors related to successful upper extremity prosthetic use, Prosthetics and Orthotics International 13 (1) (1989) 14–18.

[18] B. Pilditch, Gestural Armware, 2010.

[19] A. Hess, Upper limb body-powered components, in: J.I. Krajbich, M.S. Pinzur, B.J. Potter, P.M. Stevens (Eds.), Atlas of Amputations and Limb Deficiencies - 4th Edition, Vol. 1: General Topics and Upper Limb, American Academy of Orthopaedic Surgeons, Rosemont IL 60018, 2016, pp. 139–158, Chapter 11.

[20] W. Hanson, telephone interview, 25/2/15.

[21] S. Hubbard, telephone interview, 24/2/15.

[22] L. Hermansson, telephone interview, 17/2/15.

[23] J.T. Makley, K.G. Heiple, Scoliosis associated with congenital deficiencies of the upper extremity, JBJS 52 (2) (1970) 279–287.
[24] A. Zinck, Investigation of compensatory movements in prosthesis users and the design of a novel wrist, Master's thesis, Electrical and Computer Science Department, University of New Brunswick, December 2008.
[25] R. Ingle, A struggle against rapid decline, Irish Times, October 2008.
[26] R. Porter, Blood and Guts, Penguin, 2003.
[27] S.D. Taffler, P.J. Kyberd, Differences in the activity of the muscles in the forearm of individuals with a congenital absence of the hand, IEEE Transactions on Biomedical Engineering 54 (8) (August 2007) 1514–1519.
[28] B. Radocy, telephone interview, 21/1/15.
[29] G. McGimpsey, T.C. Bradford, Limb prosthetics services and devices critical unmet need: Market analysis white paper, Technical report, Polytechnic Institution, Bioengineering Institute Center for Neuroprosthetics, Worcester, Mass, 2011.
[30] A. Esquenazi, Amputation rehabilitation and prosthetic restoration. From surgery to community reintegration, Disability and Rehabilitation 26 (14–15) (August 2004) 831–836.
[31] P.H. Chappell, Arm amputation statistics for England 1958-88: an exploratory statistical analysis, International Journal of Rehabilitation Research 15 (1992) 9–14.
[32] G.H. Kejlaa, The social and economic outcome after upper limb amputation, Prosthetics and Orthotics International 16 (1) (April 1992) 25–31.
[33] S. Godfrey, Workers with prostheses, Journal of Hand Therapy 3 (3) (April–June 1990) 101–110.
[34] H. Sears, telephone interview, 7/10/19.
[35] S.G. Milstien, H. Heger, G.A. Hunter, Prosthetic use in adult upper-limb amputees: a comparison of body powered and electrically powered prostheses, Prosthetics and Orthotics International 10 (1986) 27–34.
[36] C. Martin, Upper limb prostheses: a review of the literature with a focus on myoelectric hands, Technical Report 3, WorkSafeBC Evidence-Based Practice Group, Victoria, BC, 2011.
[37] A. Hashim, N.A. Abd Razak, N.A. Abu Osman, H. Gholizadeh, Improvement on upper-limb body-powered prosthesis (1988-2015): systematic review, Proceedings of the Institution of Mechanical Engineers. Part H, Journal of Engineering in Medicine 232 (1) (2017) 1–9.
[38] E. Marquardt, G. Neff, The angulation osteotomy of above-elbow stumps, Clinical Orthopaedics and Related Research 104 (1974) 232–238.
[39] T.P. Schonhowd, T. Kristensen, S. Sivertsen, E. Witsø, New technology for the suspension of trans-humeral prothesis - SISA (Subfascial Implant Supported Attachment), in: "MEC '05 Integrating Prosthetics and Medicine", Proceedings of the 2005 MyoElectric Controls/Powered Prosthetics, Symposium, Institute of Biomedical Engineering, Fredericton, New Brunswick, Canada, August 2005, pp. 17–19.
[40] R.J. Garst, The Krukenberg hand, Journal of Bone and Joint Surgery. British Volume 73–B (1991) 385–388.
[41] Brückner, Lutz, Sauerbruch-Lebsche-Vanghetti Cineplasty: the surgical procedure, Orthopaedics and Traumatology 1 (2) (June 1992) 90–99.
[42] T. Kuiken, telephone interview, 15/1/15.
[43] T.R. Clites, H.M. Herr, S.S. Srinivasan, A.N. Zorzos, M.J. Carty, The Ewing Amputation: the first human implementation of the agonist-antagonist myoneural interface, Plastic Reconstruction Surgery Global Open 6 (2018) e1997, Published online 16 November 2018.
[44] B. Radocy, telephone interview, 25/5/19.
[45] S. Schneeberger, Transplantation of the hand, Transplantation Reviews 19 (2) (April 2005) 100–107.
[46] M. Ninkovic, Hand transplantation, in: K. Ott, D. Serlin, S. Mihm (Eds.), Technical and Surgical Details of Hand Transplantation, Springer, 2007, pp. 197–203.
[47] N.S. Hakim, The history of arm transplantation, in: Composite Tissue Allograft, Imperial College Press, 2006, pp. 1–9.

[48] J. Slatman, G. Widdershoven, Hand transplants and bodily integrity, Body & Society 16 (3) (2010) 69–92.
[49] Surgeon says first hand-transplant now reversed, Irish Times, Saturday, February 3 2001.
[50] F.A.K. Campbell, The case of Clint Hallam's Wayward hand, in: Culture and the State: Disability Studies & Indigenous Studies, CRC Humanities Studio, 2003.
[51] M. Ninkovic, A. Weissenbacher, M. Gabl, G. Pierer, J. Pratschke, R. Margreiter, G. Brandacher, S. Schneeberger, Functional outcome after hand and forearm transplantation: what can be achieved?, Hand Clinics 27 (2011) 455–465.
[52] D. Shehab, A.H. Elgazzar, B.D. Collier, Heterotopic ossification, Journal of Nuclear Medicine 43 (3) (2002) 346–353.
[53] Chad A. Krueger, Joseph C. Wenke, James R. Ficke, Ten years at war: comprehensive analysis of amputation trends, Journal of Trauma and Acute Care Surgery 73 (6) (2012) S438–S444.
[54] G. Phillips, Best Foot Forward: Chas. A. Blatchford and Sons Ltd. (Artificial Limb Specialists) 1890-1990, Granta Editions, 47 Norfolk Street, Cambridge, CB1 2LE, 1990.
[55] R. Cooper, private correspondence, August 2020.
[56] P.E. Klopsteg, P.D. Wilson, Human Limbs and Their Substitutes, Hafner Publishing Co, 1954.
[57] M. Wernke, R. Schroeder, C. Kelley, J. Denune, J. Colvin, SmartTemp prosthetic liner significantly reduces residual limb temperature and perspiration, Journal of Prosthetics and Orthotics 27 (4) (October 2015) 134–139.
[58] International Society of Prosthetics and Orthotics, Education standards, https://www.ispoint.org/page/EducationStandards2.
[59] H.R. Perry, Re-arming the disabled veteran, in: K. Ott, D. Serlin, S. Mihm (Eds.), Artificial Parts, Practical Lives: Modern History of Prosthetics, New York University Press, New York, 2002, pp. 75–101.
[60] H. Takechi, History of prostheses and orthoses in Japan, Prosthetics and Orthotics International 16 (2) (1992) 98–103.
[61] L. Abernethy, J.C. Duncan, W.L. Childers, A content analysis on the media portrayal of characters with limb loss, Journal of Prosthetics and Orthotics 29 (4) (October 2017) 170–176.
[62] British Broadcasting Company, Abu Hamza profile, https://www.bbc.co.uk/news/uk-11701269.
[63] Stella Young, I'm not your inspiration, thank you very much, TEDxSydney, April, 2014.
[64] J.R. Napier, The prehensile movements of the human hand, Journal of Bone and Joint Surgery. British Volume 38 (4) (November 1956) 902–913.
[65] M.R. Cutkosky, On grasp choice, grasp models, and the design of hands for manufacturing tasks, IEEE Transactions on Robotics and Automation 5 (3) (1989) 269–279.
[66] T. Feix, J. Romero, H-B. Schmiedmayer, A. Dollar, D. Kragic, The GRASP taxonomy of human grasp types, IEEE Transactions on Human-Machine Systems 46 (1) (2015) 66–77.
[67] T. Iberall, Human prehension and dexterous robot hands, The International Journal of Robotics Research 16 (3) (June 1997) 285–299.
[68] M.A. Arbib, T. Iberall, D.M. Lyons, Hand Function and the Neocortex, Springer-Verlag, Berlin, 1985.
[69] C. Sollerman, A. Ejeskär, Sollerman hand function test: a standardised method and its use in tetraplegic patients, Scandinavian Journal of Plastic and Reconstructive Surgery and Hand Surgery 29 (2) (1995) 167–176.
[70] A.M. Wing, A. Turton, C. Fraser, Grasp size and accuracy of approach in reaching, Journal of Motor Behaviour 18 (3) (1986) 245–260.
[71] S. Aglioti, J.F.X. DeSouza, M.A. Goodale, Size-contrast illusions deceive the eye but not the hand, Current Biology 5 (5) (1995) 679–685.
[72] J.A. George, D.T. Kluger, T.S. Davis, S.M. Wendelken, E.V. Okorokova, Q. He, C.C. Duncan, D.T. Hutchinson, Z.C. Thumser, D.T. Beckler, P.D. Marasco, S.J. Bensmaia, G.A. Clark, Biomimetic sensory feedback through peripheral nerve stimulation improves dexterous use of a bionic hand, Science Robotics 4 (July 2019) 385.

[73] M.O. Ernst, M.S. Banks, Humans integrate visual and haptic information in a statistically optimal fashion, Nature 415 (24 January 2002) 429.
[74] J.A. Doubler, D.S. Childress, An analysis of Extended Physiological Proprioception as a prosthesis control technique, Journal of Rehabilitation Research and Development 21 (1) (1984) 5–18.
[75] A.M. Wing, C. Fraser, The contribution of the thumb to reaching movements, Quarterly Journal of Experimental Psychology 35A (1983) 297–309.
[76] C. Fraser, A.M. Wing, A case study of reaching by a user of a manually-operated artificial hand, Prosthetics and Orthotics International 5 (1981) 151–156.
[77] C. Fraser, Fact and fiction: a clarification of phantom limb phenomena, British Journal of Occupational Therapy 65 (6) (June 1993) 256–260.
[78] V.S. Ramachandran, S. Blakeslee, Phantoms in the Brain, Fourth Estate, 1998.
[79] T. Makin, telephone interview, 22/11/19.
[80] T. Makin, et al., Phantom limbs and brain plasticity in amputees, in: Oxford Research Encyclopedia of Neuroscience, Oxford University Press, 2020.
[81] T.R. Makin, A.O. Cramer, J. Scholz, A. Hahamy, D. Henderson-Slater, I. Tracey, H. Johansen-Berg, Deprivation-related and use-dependent plasticity go hand in hand, eLife 2 (2013) e01273.
[82] R. Brånemark, telephone interview, 25/10/19.
[83] R.O. Maimon Mor, T.R. Makin, Is an artificial limb embodied as a hand? Brain decoding in prosthetic limb users, PLoS Biology 18 (6) (2020) e3000729.
[84] S.J. Tanenbaum, The Boston Elbow, in: Health Technology Case Study 29, OTAHCS-29, Washington, DC, 1984.
[85] R. Vinet, Y. Lozac'h, N. Beaudry, G. Drouin, Design methodology for a multifunctional hand prosthesis, Journal of Rehabilitation Research and Development 32 (4) (November 1995) 316–324.
[86] H. Sears, David Ford telephone interview, 21/6/10.
[87] G. Smit, R.M. Bongers, C.K. van der Sluis, D.H. Plettenburg, Efficiency of voluntary opening hand and hook prosthetic devices, 24 years of development?, Journal of Rehabilitation Research and Development 49 (4) (2012) 523–534.
[88] A. Kargov, C. Pylatiuk, J. Martin, S. Schulz, L. Döderlein, A comparison of the grip force distribution in natural hands and in prosthetic hands, Disability and Rehabilitation 26 (12) (2004) 705–711.
[89] G. Smit, D.H. Plettenburg, F.C.T. van der Helm, The lightweight Delft Cylinder Hand, IEEE Transactions on Neural Systems and Rehabilitation Engineering 23 (3) (August 2015) 431–440.
[90] J-I. Throndsen, Characterisation of shape memory alloys for use as an actuator, MSc thesis, Engineering Science, Oxford University, 1996.
[91] A.B. Soares, Shape Memory Alloy Actuators for Upper-Limb Prostheses, PhD thesis, Engineering Department, University of Edinburgh, 1997.
[92] Z. Wang, G. Hang, J. Li, Y. Wang, K. Xiao, A micro-robot fish with embedded SMA wire actuated flexible biomimetic fin, Sensors and Actuators. A, Physical 144 (2) (2008) 354–360.
[93] C.M. Light, An Intelligent Hand Prosthesis and Evaluation of Pathological and Prosthetic Hand Function, PhD thesis, Electrical Engineering Department, University of Southampton, 2000.
[94] W. Hill, Ø. Stavdahl, L.N. Hermansson, P. Kyberd, S. Swanson, S. Hubbard, Functional outcomes in the WHO-ICF model: establishment of the upper limb prosthetic outcome measures group, in: Special Issue, Journal of Prosthetics and Orthotics 21 (2) (April 2009) 115–119.
[95] Official Findings of the State-of-the-Science Conference, October 2009.
[96] L.A. Miller, S. Swanson (Eds.), Upper Limb Prosthetic Outcome Measures (ULPOM): a Working Group and Their Findings, Vol. 9, American Academy of Orthotists and Prosthetists, Lippincott, October 2009.
[97] World Health Organization, ICF International Classification of Functioning, Disability and Health, Dummy, 2001.

[98] World Health Organization, Towards a common language for Functioning, Disability and Health: ICF Beginner's guide - (who/eip/gpe/cas/01.3), The International Classification of Functioning, Disability and Health, 2002.
[99] V. Wright, Measurement of functional outcomes with individuals who use upper extremity prosthetic devices: current and future directions, Journal of Prosthetics and Orthotics 18 (2) (2006) 46–56.
[100] A.P. Shimamura, Muybridge in motion: travels in art, psychology and neurology, History of Photography 26 (4) (2002) 341–350.
[101] B. Abernathy, Dual-task methodology and motor skills research: some applications and methodological constraints, Journal of Human Movement Studies 14 (1988) 101–132.
[102] S. Deeny, C. Chicoine, L. Hargrove, T. Parrish, A. Jayaraman, A simple ERP method for quantitative analysis of cognitive workload in myoelectric prosthesis control and human-machine interaction, PLoS ONE 9 (11) (2014).
[103] J.B. Pelz, R. Canosa, Oculomotor behavior and perceptual strategies in complex tasks, Vision Research 41 (25–26) (2001) 3587–3596.
[104] M. Sobuh, L.P.J. Kenney, A. Galpin, S. Thies, J. McLaughlin, J. Kulkarni, P. Kyberd, Visuomotor behaviours when using a myoelectric prosthesis, Journal of NeuroEngineering and Rehabilitation 11 (17) (2014) 72–83.
[105] F.A. Popa, P.J. Kyberd, Identification of patterns in upper limb prosthetic usage by analysis of visual attention to areas of interest, in: "MEC'11 Symposium - Raising the Standard", Proceedings of the 2011 MyoElectric Controls/Powered Prosthetics Symposium, Fredericton, New Brunswick, Canada, August 14–19, Institute of Biomedical Engineering, 2011, pp. 226–229.
[106] P.J. Kyberd, D.J. Beard, J.J. Davey, D.J. Morrison, Survey of upper limb prostheses users in Oxfordshire, Journal of Prosthetics and Orthotics 10 (4) (Fall 1998) 85–91.
[107] P. Kyberd, W. Hill, Survey of upper limb prosthesis users in Sweden, the United Kingdom and Canada, Prosthetics and Orthotics International 35 (2) (June 2011) 234–241.
[108] J-D. Bartoe, Future international cooperation on space stations, Acta Astronautica 24 (1991) 377–385.
[109] E.A. Biddiss, T.T. Chau, Upper limb prosthesis use and abandonment: a survey of the last 25 years, Prosthetics and Orthotics International 31 (3) (September 2007) 236–257.
[110] A. Chadwell, The Reality of Myoelectric Prostheses: How do EMG skill, unpredictability of prosthesis response, and delays impact on user functionality and everyday prosthesis use?, PhD thesis, School of Health Sciences, University of Salford, 2018.
[111] A. Chadwell, L. Kenney, M. Granat, S. Thies, A. Galpin, J. Head, Upper limb activity of twenty myoelectric prosthesis users and twenty healthy anatomically intact adults, Scientific Data 6 (199) (2019).
[112] A. Chadwell, telephone interview, 20/9/19.
[113] Institution of Mechanical Engineers, Big ideas: the future of engineering in schools, 2014.
[114] A. Chadwell, L. Kenney, M. Granat, S. Thies, J. Head, A. Galpin, R. Baker, J. Kulkarni, Upper limb activity in myoelectric prosthesis users is biased towards the intact limb and appears unrelated to goal-directed task performance, Scientific Reports 8 (11084) (2018).
[115] A. Spiers, telephone interview, 4/10/19.
[116] T. Feix, I.M. Bullock, A.M. Dollar, Analysis of human grasping behavior: object characteristics and grasp type, IEEE Transactions on Haptics 7 (3) (July-September 2014) 311–323.
[117] J.C. Cochran, A.J. Spiers, A.M. Dollar, Analyzing exfordance use by unilateral upper-limb amputees, in: 7th IEEE International Conference on Biomedical Robotics and Biomechatronics (Biorob), IEEE, 2018, pp. 86–93.
[118] M. Jury, telephone interview, 10/10/19.
[119] E.B. Marsolais, R. Kobetic, Functional electrical stimulation for walking in paraplegia, JBJS 69 (5) (1987) 728–733.
[120] Y. Hara, Rehabilitation with functional electrical stimulation in stroke patients, International Journal of Physics and Medical Rehabilitation 1 (147) (2013) 2.

[121] R. Reiner, telephone interview, 18/11/19.
[122] Herodotus, Calliope IX, George Routledge and Sons Ltd, 1891.
[123] Pliny the Younger, Natural History, Vol. VII, xxvii, W. Heinwemann, 1942.
[124] J. Finch, The Durham Mummy: deformity and the concept of perfection in the ancient world, Bulletin of the John Rylands Library 89 (1) (2013) 111–132.
[125] American Academy of Orthopaedic Surgeons, Orthopaedic Appliances Atlas, Vol. 2: Artificial Limbs, JW Edwards-Ann, Arbor, Michigan, 1960.
[126] R. Benhamou, The artificial limb in preindustrial France, Technology and Culture 35 (4) (1994) 835–845.
[127] R.H. Meier, History of arm amputation, prosthetic restoration, and arm amputation rehabilitation, in: R.H. Meier, D.J. Atkins (Eds.), Functional Restoration of Adults and Children with Upper Extremity Amputation, Demos Medical Publishing, New York, 2004, pp. 1–7.
[128] Henry J. Cohn, Götz von Berlichingen and the Art of Military Autobiography, in: War, Literature and the Arts in Sixteenth-Century Europe, Springer, 1989, pp. 22–40.
[129] Science Museum Group Collections, Iron artificial arm, Europe, 1560-1600, https://collection.sciencemuseumgroup.org.uk/objects/co126844/.
[130] P. Berton, Flames Across the Boarder, Random House, Canada, 1981.
[131] Shadrach Byfield, A Narrative of a Light Company Soldier's Service in the Forty-First Regiment of Foot (1807-1814), W. Abbatt, 1910.
[132] W.D. Penrose, The discourse of disability in ancient Greece, Classical World 108 (4) (2015) 499–523.
[133] K.W. Alt, Evidence for stone age cranial surgery, Nature 387 (May 1997) 360.
[134] M.C. Lillie, Cranial surgery dates back to Mesolithic, Nature 391 (February 1998) 854.
[135] I. Micarelli, R. Paine, C. Giostra, M.A. Tafuri, A. Profico, M. Boggioni, F. Di Vincenzo, D. Massani, A. Papini, G. Manzi, Survival to amputation in pre-antibiotic era: a case study from a Longobard necropolis (6th-8th centuries ad), Journal of Anthropological Sciences 96 (2018) 1–16.
[136] B. Bryson, At Home, Doubleday, Canada, 2010.
[137] Hanger Incorporated, The J.E. Hanger Story, http://www.hanger.com/history/Pages/The-J.E.-Hanger-Story.aspx.
[138] W.T. Carnes, (Artificial Limb) - Patent No. 760,102, 1904.
[139] W.T. Carnes, (Artificial Arm) - Patent No. 1,046,966, 1912.
[140] S. Zahedi, telephone interview, 19/2/15.
[141] E.M. Little, Artificial Limbs and Amputation Stumps; a Practical Handbook, Blakiston, Philadelphia, 1922.
[142] W.T. Carnes, (Artificial Arm and Hand) - Patent No. 349,375 (2,287,781), 1942.
[143] Billings Gazette, Artificial Arm - Australian Re-Engineers Century-Old Artificial Limb, Billings Gazette, 12 2014.
[144] S. Jacobsen, David Ford telephone interview, 28/6/10.
[145] L.F. Lucaccini, P.K. Kaiser, J. Lyman, The French electric hand: some observations and conclusions, Bulletin of Prosthetics Research 10 (6) (1966) 30–51.
[146] A. Bennett Wilson (Ed.), Atlas of Limb Prosthetics: Surgical and Prosthetic Principles, American Academy of Orthopaedic Surgeons, Mosby Co., Saint Louis, 1981.
[147] D.S. Childress, M.V. Podlusky, Myoelectric control – letter to the editor, Medical and Biological Engineering 7 (3) (1969) 345.
[148] A. Muzumdar (Ed.), Powered Upper Limb Prostheses: Control, Implementation and Clinical Application, Springer, Berlin, New York, 2004.
[149] L. Boldon, A History of Myoelectric Control, University of New, Brunswick, Fredericton, Canada, 1983.
[150] G. Weltman, H. Groth, J. Lyman, An analysis of bioelectrical prosthesis control, Technical Report 59–49, Department of Engineering, University of California, Los Angeles, 1959.
[151] D.S. Childress, Historical aspects of powered limb prostheses, Clinical Prosthetics and Orthotics 9 (1) (1985) 2–13.

[152] C.K. Battye, A. Nightingale, J. Whillis, The use of myo-electric currents in the operation of prostheses, Journal of Bone and Joint Surgery. British Volume 37-B (3) (1955) 506–510.

[153] A.E. Kobrinski, S.V. Bolkovitin, L.M. Voskoboinikova, Problems of bioelectric control, in: J. Coles (Ed.), Proceedings of the 1st IFAC International Congress of Automatic and Remote Control, Moscow, 1960, pp. 619–629.

[154] A.H. Bottomley, Progress with the British Myoelectric Hand, in: Advances in External Control of Human Extremities, 1967, pp. 114–124.

[155] B. Popov, The bio-electrically controlled prosthesis, Journal of Bone and Joint Surgery. British Volume 47-B (3) (1965) 421–424.

[156] E. Sherman, G. Gingras, A.L. Lippay, New trends in externally powered upper extremity prostheses, World Medical Journal 15 (5) (September-October 1968) 121–125.

[157] G. Gingras, M. Mongeau, E.D. Sherman, A.L. Lippay, J. Hutchison, C. Corriveau, Bioelectric upper extremity prosthesis developed in Soviet Union: preliminary report, Archives of Physical Medicine and Rehabilitation 47 (4) (April 1996) 232–237.

[158] A.H. Bottomley, A.B. Kinnier-Wilson, A. Nightingale, Muscle substitutes and myoelectric control, Journal of the British Institute of Radio Engineers 26 (6) (December 1963) 429–448.

[159] US Department of Health, Progress in prosthetics, Technical Report 2, Department of Education and Welfare, Office of Vocational Rehabilitation, Washington, 1962.

[160] S.W. Alderson, The electric arm, in: P.E. Klopsteg, P.D. Wilson (Eds.), Human Limbs and Their Substitutes, Hafner, New York, 1968, pp. 359–408, Chapter 13.

[161] S.W. Alderson, Electrically operated artificial arm for above-the-elbow amputees, US Patent 2,580,987, 1952.

[162] H.R. Luce (Ed.), Foot Controlled Arm, Life Magazine, August 7 1950.

[163] S. Zahedi, The Intelligent Prosthesis – the first 6 years and the outlook for the future, Orthopädie Technik 12 (98) (1998) 952–957.

[164] P. Parker, K. Englehart, B. Hudgins, Control of powered upper limb prostheses, in: R. Merletti, P. Parker (Eds.), Electromyography: Physiology, Engineering, and Noninvasive Applications, IEEE Press, New York, 2004, pp. 453–475.

[165] I.D. Swain, J.M. Nightingale, An adaptive control system for a complete hand/arm prosthesis, Journal of Biomedical Engineering 2 (July 1980) 163–166.

[166] M.J. Topping, J.K. Smith, The development of Handy 1 A robotic system to assist the severely disabled, Technology and Disability 10 (2) (1999) 95–105.

[167] D. Foord, P. Kyberd, From design to research: upper limb prosthetic research and development in Canada, 1960-2000, Scientia Canadensis: Canadian Journal of the History of Science, Technology and Medicine (Scientia Canadensis: revue canadienne d'histoire des sciences des techniques et de la médecine) 38 (1) (2015) 50–71.

[168] D. Foord, P. Kyberd, Ideas and networks: the rise and fall of research bodies for powered artificial arms in America and Canada, 1945-1977, Scientia Canadensis: Canadian Journal of the History of Science, Technology and Medicine (Scientia Canadensis: revue canadienne d'histoire des sciences, des techniques et de la médecine) 38 (2) (2016) 35–56.

[169] Defense Advanced Research Projects Agency (DARPA), Revolutionizing prosthetics, BAA05-26, 2005.

[170] Defense Advanced Research Projects Agency (DARPA), Press release - DARPA initiates revolutionary prosthetics programs, February 2006.

[171] J.M. Burck, J.D. Bigelow, S.D. Harshbarger, Revolutionizing prosthetics: systems engineering challenges and opportunities, Johns Hopkins APL Technical Digest 30 (3) (2011) 186–197.

[172] E. Biden, telephone interview, 2/4/15.

[173] S. Meek, telephone interview, 23/2/15.

[174] E. Iversen, telephone interview, 12/11/14.

[175] L. Resnik, M. Borgia, G. Latlief, N. Sasson, L. Smurr-Walters, Self-reported and performance-based outcomes using DEKA arm, Journal of Rehabilitation Research and Development 51 (3) (2014) 351–362.
[176] J. Lafo, S. Correia, M. Borgia, F. Acluche, L. Resnik, Cognitive characteristics associated with device adoption, skill retention, and early withdrawal from a study of an advanced upper limb prosthesis, American Journal of Physical Medicine & Rehabilitation 98 (10) (2019) 879.
[177] Glaze Prosthetics, Glaze Prosthetics website, https://glazeprosthetics.com/.
[178] National Health Service, Your pregnancy and baby guide, http://news.bbc.co.uk/1/hi/uk/6215847.stm.
[179] T. Stephens, R. Brynner, Dark Remedy - the Impact of Thalidomide and Its Revival as a Vital Medicine, Persus Publishing, 2001.
[180] A.L. Florence, Is Thalidomide to blame?, British Medical Journal 2 (1960) 1954.
[181] S. Hubbard, David Ford telephone interview, 22/4/10.
[182] R. Cooper, telephone interview, 6/3/15.
[183] C. Tantibanchachai, US Regulatory Response to Thalidomide (1950-2000), Embryo Project Encyclopedia (2014-04-01), The Embryo Project, http://embryo.asu.edu/handle/10776/7733, 2014.
[184] Thalidomide Victims Association of Canada, The Canadian Tragedy, https://thalidomide.ca/en/the-canadian-tragedy/.
[185] T.L. Wellersen, Historical development of upper extremity prosthetics, Orthopaedic and Prosthetic Appliance Journal 11 (3) (1957) 73–77.
[186] R.W. Bulliet, The Wheel: Inventions and Reinventions, Columbia University Press, New York, 2016.
[187] D.W. Dorrance, (Artificial Hand) - US Patent No. US1042413, 29 October 1912.
[188] C. Antfolk, M. D'Alonzo, B. Rosén, G. Lundborg, F. Sebelius, C. Cipriani, Sensory feedback in upper limb prosthetics, Expert Review Medical Devices 10 (1) (2013) 45–54.
[189] R.N. Scott, R.H. Brittain, R.R. Caldwell, A.B. Cameron, V.A. Dunfield, Sensory-feedback system compatible with myoelectric control, Medical & Biological Engineering & Computing 18 (1980) 65–69.
[190] C. Cipriani, F. Zaccone, S. Micera, M.C. Carrozza, On the shared control of an EMG-controlled prosthetic hand: analysis of user-prosthesis interaction, IEEE Transactions on Robotics 24 (1) (2008) 170–184.
[191] M. Hichert, A.N. Vardy, D. Plettenburg, Fatigue-free operation of most body-powered prostheses not feasible for majority of users with trans-radial deficiency, Prosthetics and Orthotics International 42 (1) (2018) 84–92.
[192] D. Latour, Anchoring system for prosthetic and orthotic devices, US Patent app. 11/787,161, October 2007.
[193] J. Sensinger, Design and evaluation of voluntary opening and voluntary closing prosthetic terminal device, Journal of Rehabilitation Research and Development 52 (1) (2015) 63–76.
[194] M. Hichert, D.A. Abbink, P.J. Kyberd, D.H. Plettenburg, High cable forces deteriorate pinch force control in voluntary-closing body-powered prostheses, PLoS ONE (18 January 2017).
[195] J. Uellendahl, C. Heckathorne, Creative prosthetic solutions for bilateral upper extremity amputation, in: R. Meier, D. Atkins (Eds.), Functional Restoration of Adults and Children with Upper Extremity Amputation, Demos Publications, 2004, Chapter 21.
[196] H. Sears, private correspondence, August 2020.
[197] Motion Control Inc, Motion Control website, http://www.utaharm.com/.
[198] D. Latour, telephone interview, 5/8/20.
[199] C. White, L. Boucke, The UnDutchables: An Observation of the Netherlands: Its Culture and Its Inhabitants, White Boucke Pub, 2006.
[200] D. Plettenburg, telephone interview, 20/2/15.
[201] D.H. Plettenburg, The WILMER passive hand prosthesis for toddlers, Journal of Prosthetics and Orthotics 21 (2) (2009) 97–99.
[202] M. LeBlanc, Y. Setoguchi, J. Shaperman, L. Carlson, Mechanical work efficiencies of body-powered prehensors for young children, Children's Prosthetic and Orthotic Clinics 3 (27) (1992) 70–75.
[203] G. Smit, D.H. Plettenburg, Comparison of mechanical properties of silicone and PVC cosmetic gloves for articulating hand prostheses, Journal of Rehabilitation Research and Development 50 (5) (2013) 723–732.

[204] P. Steeper, telephone interview, 5/6/15.
[205] D. Plettenburg, telephone interview, 30/10/19.
[206] G. Smit, B. Maat, D. Plettenburg, P. Breedveld, A self-grasping hand prosthesis, in: "MEC '17 A Sense of What's to Come", Proceedings of the 2017 MyoElectric Controls/Powered Prosthetics, Symposium, Institute of Biomedical Engineering, Fredericton, New Brunswick, Canada, August 2017, pp. 15–18.
[207] S. Schulz, C. Pylatiuk, G. Bretthauer, A New Ultralight Anthropomorphic Hand, Proceedings 2001 ICRA. IEEE International Conference on Robotics and Automation (Cat. No. 01CH37164), vol. 3, IEEE, 2001, pp. 2437–2441.
[208] H. Dietl, telephone interview, 5/3/15.
[209] L. Milde (Ed.), Otto Bock Prosthetic Compendium - Upper Limb Prostheses, second edition, Hans Georg Näder, Duderstadt, Germany, 2011.
[210] D.T. Barry, K.E. Gordon, G.G. Hinton, Acoustic and surface EMG diagnosis of pediatric muscle disease, Muscle and Nerve: Official Journal of the American Association of Electrodiagnostic Medicine 13 (4) (1990) 286–290.
[211] B. Frauscher, A. Iranzo, C. Gaig, V. Gschliesser, M. Guaita, V. Raffelseder, L. Ehrmann, N. Sola, M. Salamero, E. Tolosa an, W. Poewe, J. Santamaria, Birgit Högl, Normative EMG values during REM sleep for the diagnosis of rem sleep behavior disorder, Sleep 35 (6) (2012) 835–847.
[212] J.J. Fudema, J.A. Fizzell, E.M. Nelson, Electromyography of experimentally immobilized skeletal muscles in cats, American Journal of Physiology 200 (5) (1961) 963–967.
[213] D.A. Jones, J.M. Round, Skeletal Muscle in Health and Disease, Manchester University Press, Oxford Road, Manchester, 1990.
[214] D.F. Lovely, The origins and nature of the myoelectric signal, in: A. Muzumdar (Ed.), Powered Upper Limb Prostheses, Springer-Verlag, 2004, pp. 17–33, Chapter 2.
[215] A. Fougner, M. Sæther, Ø Stavdahl, P. Kyberd, J. Blum, Cancellation of force induced artefacts in surface EMG using fsr measurements, in: MEC '08 Myoelectric Controls Symposium, Fredericton, New Brunswick, Canada, 13-15 August 2008, pp. 146–149.
[216] Ø. Stavdahl, P.J. Kyberd, A. Fougner, Measurement of bio-signals, US Patent no: 13057904, 2012, EU no: 2009784901 2011, 2012.
[217] P.A. Parker, Optimum signal processing for a multifunction myoelectric communication channel, Phd, Department of Electrical Engineering, University of New Brunswick, 1975.
[218] P.A. Parker, J.A. Stuller, R.N. Scott, Signal processing for the multistate myoelectric channel, in: Proceedings of the IEEE, 1977, pp. 662–674.
[219] R.N. Scott, J.E. Paciga, P.A. Parker, Operator error in multistate myoelectric control systems, Medical & Biological Engineering & Computing 16 (1978) 296–301.
[220] M. Ortiz-Catalan, B. Hakansson, R. Brånemark, Osseointegrated human-machine gateway for long-term sensory feedback and motor control of artificial limbs, Science Translational Medicine 6 (257) (8 October 2014) 6.
[221] D.W. Tan, M.A. Schiefer, M.W. Keith, J.R. Anderson, J. Tyler, D.J. Tyler, A neural interface provides long-term stable natural touch perception, Science Translational Medicine 6 (257) (8 October 2014) 257ra138.
[222] H.T. Lancashire, Y.A. Al Ajam, R.P. Dowling, C.J. Pendegrass, G.W. Blunn, Hard-wired epimysial recordings from normal and reinnervated muscle using a bone-anchored device, Plastic and Reconstructive Surgery Global Open 7 (9) (2019).
[223] K. Englehart, B. Hudgins, A.D.C. Chan, Continuous multifunction myoelectric control using pattern recognition, Technology and Disability 15 (2) (2003) 95–104.
[224] C. Almström, P. Herberts, L. Körner, Experience with Swedish multifunctional prosthetic hands controlled by pattern recognition of multiple myoelectric signals, International Orthopaedics (SICOT) 5 (1981) 15–21.
[225] I. Kamon, T. Flash, S. Edelman, Learning to Grasp Using Visual Information, Proceedings of IEEE International Conference on Robotics and Automation, vol. 3, IEEE, 1996, pp. 2470–2476.

[226] J. Pauli, Learning to recognize and grasp objects, Autonomous Robots 5 (3–4) (1998) 407–420.
[227] E. Arruda, J. Wyatt, M. Kopicki, Active vision for dexterous grasping of novel objects, in: 2016 IEEE/RSJ International Conference on Intelligent Robots and Systems (IROS), IEEE, 2016, pp. 2881–2888.
[228] Dr. Melvin J. Glimcher, Prosthetics Innovator, Dies at 88, New York Times, May 30 2014.
[229] N. Wiener, Cybernetics: The Science of Control and Communication in the Animal and the Machine, MIT Press, Cambridge, Mass, 1948.
[230] F. Conway, J. Siegelman, Dark Hero of the Information Age: Search of Norbert Wiener, The Father of Cybernetics, Basic Books, New York, 2005.
[231] P. Haggard, A.M. Wing, Coordinated responses following mechanical perturbation of the arm during prehension, Experimental Brain Research 102 (1995) 483–494.
[232] M. Amrich, telephone interview, 14/1/20.
[233] R. Alter, Bioelectric control of prostheses. - Technical report 446, Technical report, MIT, Cambridge, 1966.
[234] P.J. Kyberd, The Algorithmic Control of a multifunction hand prosthesis, PhD thesis, Electrical Engineering Department, Southampton University, 1990.
[235] I.L. Paul, Evaluation of Energy and Power Requirements for Externally Powered Upper-Extremity Prosthetic and Orthopedic Devices. Massachusetts Institute of Technology, Engineering Projects Laboratory, 1962.
[236] R. Hall, D. Shayler, Soyuz: A Universal Spacecraft, Springer Science and Business Media, 2003.
[237] R.W. Mann, Engineering design education and rehabilitation engineering, Journal of Rehabilitation Research and Development 39 (6) (2002) 23.
[238] Saturday Review, December 7 1963.
[239] Saturday Review, January 26 1964.
[240] N. Wiener, Dynamical systems in physics and biology, in: Nigel Calder (Ed.), The World in 1984, Vol. 1. The Complete New Scientist Series, Vol. 1, Pelican Books, 1965.
[241] T. Walley Williams III, telephone interview, 25/2/15.
[242] R.A. Rothchild, R.W. Mann, An EMG controlled, force sensing, proportional rate, elbow prosthesis, in: Proceedings of the Symposium of Biomedical Engineering, Vol. 1, 1966, pp. 106–109.
[243] R. Jerrard, telephone interview, 29/4/10.
[244] M.V. Podlusky, Forum, IEEE Spectrum 6 (2) (February 1969) 6.
[245] Energy Department for Business and Industrial Strategy UK Government, CE Marking, https://www.gov.uk/guidance/ce-marking.
[246] Robots of the future, The Salt Lake Tribune, January 9 2010.
[247] S.C. Jacobsen, D.F. Knutti, R.I. Johnson, H.H. Sears, Development of the Utah artificial arm, IEEE Transactions on Biomedical Engineering BME–29 (4) (April 1982) 249–269.
[248] E. Iversen, David Ford telephone interview, 21/6/10.
[249] P. Baker, The Story of Manned Space Stations: An Introduction, Springer, 2007.
[250] H. Sears, Evaluation and development of a hook type terminal device, PhD thesis, University of Utah, 1983.
[251] Harold H. Sears, Approaches to prescription of body-powered and myoelectric prostheses, Physical Medicine and Rehabilitation Clinics 2 (2) (1991) 361–371.
[252] E. Iversen, telephone interview, 4/3/20.
[253] T. Bertels, T. Schmalz, E. Ludwigs, Objectifying of functional advantages offered by wrist flexion, Journal of Prosthetics and Orthotics 21 (2) (2009) 74–78.
[254] P. Kyberd, The influence of passive wrist joints on the functionality of prosthetic hands, Prosthetics and Orthotics International 36 (1) (March 2012) 39–44.
[255] International Electrotechnical Commission, ANSI/IEC 60529-2004, 2004.
[256] E.D. Engeberg, S. Meek, Improved grasp force sensitivity for prosthetic hands through force-derived feedback, IEEE Transactions on Biomedical Engineering 55 (2) (February 2008) 817–821.
[257] D. Gow, private correspondence, August 2020.

[258] British Broadcasting Company, UK settles WWII debts to allies, http://news.bbc.co.uk/1/hi/uk/6215847.stm.
[259] A. Herman, How the Scots Invented the Modern World, Three Rivers Press, 2001.
[260] D.J. Gow, T.D. Dick, E.R.C. Draper, I.R. Loudon, P. Smith, The physiologically appropriate control of an electrically powered hand prosthesis, in: International Society of Prosthetics and Orthotics 4th World Congress, September 1983.
[261] J. Cole, Pride and a Daily Marathon, MIT Press, 1991.
[262] B. Stephens-Fripp, G. Alici, R. Mutlu, A review of non-invasive sensory feedback methods for transradial prosthetic hands, IEEE Access 6 (2018) 6878–6899.
[263] D.C. Simpson, J.G. Smith, An externally powered controlled complete arm prosthesis, Journal of Medical Engineering & Technology 1 (5) (1977) 275–277.
[264] J.A. Doubler, D.S. Childress, Design and evaluation of a prosthesis based on the concept of extended physiological proprioception, Journal of Rehabilitation Research and Development 21 (1) (1984) 19–31.
[265] D.C. Simpson, G. Kenworthy, The design of a complete arm prosthesis, Biomedical Engineering 8 (2) (February 1973) 56–59.
[266] D. Gow, David Ford interview, November 2010.
[267] T. Shibata, K. Inoue, R. Irie, Emotional robot for intelligent system-artificial emotional creature project, in: Proceedings 5th IEEE International Workshop on Robot and Human Communication. RO-MAN'96 TSUKUBA, IEEE, 1996, pp. 466–471.
[268] D. Gow, B. Marriot, W.B. Douglas, A modular prosthetic system for the upper-limb, in: International Society of Prosthetics and Orthotics, UK, Annual Scientific Meeting, Nottingham. International Society of Prosthetics and Orthotics, April 1989, pp. 12–14.
[269] P.J. Kyberd, N. Mustapha, F. Carnegie, P.H. Chappell, Clinical experience with a hierarchically controlled myoelectric hand prosthesis with vibro-tactile feedback, Prosthetics and Orthotics International 17 (1) (1993) 56–64.
[270] D. Gow, W. Douglas, C. Geggie, E. Monteith, D. Stewart, The development of the Edinburgh Modular Arm System, Proceedings of the Institution of Mechanical Engineers. Part H 215 (2001) 291–298.
[271] Field trials underway on revolutionary bionic arm, The Scotsman, August 1998.
[272] P.J. Kyberd, A.S. Poulton, L. Sandsjö, S. Jönsson, B. Jones, D. Gow, The ToMPAW modular prosthesis - a platform for research in Upper Limb prosthetics, Journal of Prosthetics and Orthotics 19 (1) (January 2007) 12–21.
[273] D.J. Gow, Motor drive system and linkage for hand prosthesis, US Patent no: 5,888,246, 1999.
[274] D.J. Gow, Upper limb prosthesis, US Patent no: 6,361,570 2002.
[275] Becker Mechanical Hand Co, Becker Mechanical Hand.
[276] Daniel B. Becker, Mechanical hand - US Patent 2,285,885, June 1942.
[277] E. Brown, N.L. Hancox, Reinforced plastics artificial limb component and method for making same, US Patent 4,397,048, August 1983.
[278] S. Mead, telephone interview, 5/3/15.
[279] G. McLearly, DG interview, 10/11/10.
[280] P. Kyberd, Assessment of functionality of multifunction prosthetic hands, Journal of Prosthetics and Orthotics 29 (1) (July 2017) 103–111.
[281] H. Dietl, telephone interview, 10/10/19.
[282] M. Hunter, telephone interview, 25/3/15.
[283] H. Dietl, private correspondence, August 2020.
[284] Ottobock, Ottobock website, http://www.ottobock.com/cps/rde/xchg/ob_com_en/hs.xsl/696.html.
[285] Otto Bock Healthcare Canada Acquires VASI, June 2005.
[286] M. Näder, Development of myoelectric upper limb prostheses: a glance at the past and a glimpse of the future, in: MEC '95 Myoelectric Controls Symposium, Fredericton, New Brunswick, Canada, 1995.

[287] H. Schmidl, The INAIL-CECA prostheses, Orthotics and Prosthetics 27 (1) (1973) 6–12.

[288] K. Englehart, B. Hudgins, P. Parker, Multifunction control of prostheses using the myoelectric signal, in: H.N.L. Teodorescu, L.C. Jain (Eds.), Intelligent Systems and Technologies in Rehabilitation Engineering, 1st edition, CRC Press, New York, NY, USA, 2000, pp. 153–208.

[289] D.S. Childress, Myoelectric control: brief history, signal origins, and signal processing, Capabilities of Northwestern University Prosthetics Research Laboratory and Rehabilitation Engineering Research Program 4 (6) (1995) 7.

[290] T. Sanderson, The evolution of Otto Bock Myoelectric Systems for the Paediatric Patient, ACPOC News 30 (4) (1996) 1–8.

[291] D.S. Childress, An approach to powered grasp, in: Proc. 4th Intl. Symp. on External Control of Human Extremities, Dubrovnik, 1972, pp. 159–167.

[292] R. Sörbye, Myoelectric prosthetic fitting in young children, Clinical Orthopaedics and Related Research 148 (May 1980) 34–40.

[293] Otto Bock, We offer the technology which the market demands, Dialog: the Magazine of Otto Bock, May 2010.

[294] T. Kuiken, D. Childress, W. Rymer, The hyper-reinnervation of rat skeletal muscle, Brain Research 676 (1) (1995) 113–123.

[295] T. Kuiken, L. Miller, R. Lipschulz, B. Lock, K. Stubblefield, P. Marasco, P. Zhou, G. Dumanian, Targeted reinnervation for enhanced prosthetic arm function in a woman with a proximal amputation: a case study, The Lancet 369 (9559) (3 February 2007) 371–380.

[296] Science 2.0, From The Laboratory To The Reality Of Daily Life, https://www.science20.com/newswire/laboratory_reality_daily_life.

[297] R. Todd, J.M. Nightingale, Adaptive prehension control for a prosthetic hand, in: D.B. Popovic (Ed.), Third International Symposium on External Control of Human Extremities, Yugoslav Committee for ETRAN, 1969, pp. 171–183.

[298] P. Steeper, telephone interview, 30/9/19.

[299] UK Department of Health and Social Security, Review of artificial limb and appliance services: (the report of an independent working party under the chairmanship of Professor Ian McColl), 1986.

[300] I. Swain, telephone interview, 10/7/20.

[301] L.L. Salisbury, A.B. Coleman, A mechanical hand with automatic proportional control of prehension, Medical Biological and Engineering 5 (5) (1967) 505–511.

[302] R.W. Todd, Adaptive control of a hand prosthesis, PhD thesis, Electrical Engineering Department, University of Southampton, 1969.

[303] R. Todd, telephone interview, 27/5/15.

[304] D. Jakšić, Mechanics of the Belgrade Hand, in: D.B. Popovic (Ed.), Proceedings of the 3rd Symposium on Advances in External Control of Human Extremities, Yugoslav Committee for ETRAN, 1969, pp. 145–150.

[305] R.D. Codd, J.M. Nightingale, R.W. Todd, An adaptive multi-functional hand prosthesis, Journal of Physiology (London) 232 (1973) 55–56.

[306] R.D. Codd, Development and evaluation of adaptive control for a hand prosthesis, PhD thesis, Electrical Engineering Department, University of Southampton, 1975.

[307] T. Sensky, A consumer's guide to "Bionic Arms", British Medical Journal 281 (6233) (1980) 126.

[308] D. Moore, Development of multifunctional adaptive hand prosthesis, PhD thesis, Electrical Engineering Department, University of Southampton, 1980.

[309] C.M. Light, P.H. Chappell, P.J. Kyberd, Establishing a standardized clinical assessment tool of pathologic and prosthetic hand function, Archives of Physical Medicine and Rehabilitation 83 (2002) 776–783.

[310] A. Murgia, P.J. Kyberd, P. Chappell, C.M. Light, The use of gait analysis techniques in the measure of hand function, The Journal of Hand Surgery 30B (Supplement 1) (June 2005) 83–84.

[311] S.B. Thies, L.P.J. Kenney, M. Sobuh, A. Galpin, P. Kyberd, R. Stine, M.J. Major, Skill assessment in upper limb myoelectric prosthesis users: validation of a clinically feasible method for characterising upper limb temporal and amplitude variability during the performance of functional tasks, Medical Engineering & Physics 47 (2017) 137–143.
[312] N. Storey, telephone interview, 5/3/15.
[313] I. Swain, telephone interview, 9/4/15.
[314] D.J. Brown, interview, 9/8/19.
[315] R.H. Jebsen, N. Taylor, R.B. Trieschmann, M.J. Trotter, L.A. Howard, An objective and standardized test of hand function, Archives of Physical Medicine and Rehabilitation 50 (6) (1969) 311–319.
[316] T.G. Bond, C.M. Fox, Applying the Rasch Model: Fundamental Measurement in the Human Sciences, Psychology Press, 2001.
[317] H. Bouwsema, P.J. Kyberd, W. Hill, C.K. van der Sluis, R.M. Bongers, Determining skill level in myoelectric prosthesis use with multiple outcome measures, Journal of Rehabilitation Research and Development 49 (9) (2012) 1331–1348.
[318] P. Herberts, telephone interview, 22/1/15.
[319] P. Herberts, Myoelectric signals in control of prostheses: studies on arm amputees and normal individuals, Acta Orthopaedica Scandinavica 40 (sup124) (1969) 1–83.
[320] P. Herberts, R. Kadefors, E. Kaiser, I. Petersén, Implantation of micro-circuits for myo-electric control of prostheses, Journal of Bone and Joint Surgery. British Volume 50 (4) (1968) 780–791.
[321] C. Almström, telephone interview, 26/5/15.
[322] K. Cain Winterberg, telephone interview, 3/3/15.
[323] F. Montagnani, M. Controzzi, C. Cipriani, Is it finger or wrist dexterity that is missing in current hand prostheses?, IEEE Transactions on Neural Systems and Rehabilitation Engineering 23 (4) (2015) 600–609.
[324] U. Bengtner, telephone interview, 21/5/15.
[325] R. Brånemark, interview, 27/4/15.
[326] R. Brånemark, L-O. Öhrnell, R. Skalak, L. Carlsson, P-I. Brånemark, Biomechanical characterization of osseointegration: an experimental in vivo investigation in the beagle dog, Journal of Orthopaedic Research 16 (1) (1998) 61–69.
[327] R. Brånemark, telephone interview, 26/5/15.
[328] R. Brånemark, P-I. Brånemark, B. Rydevik, R.R. Myers, Osseointegration in skeletal reconstruction and rehabilitation: a review, Journal of Rehabilitation Research and Development 38 (2) (2001) 175–182.
[329] I. Klineberg, G. Murray, Osseoperception: sensory function and proprioception, Advances in Dental Research 13 (1) (1999) 120–129.
[330] Per-Ingvar Brånemark, private communication.
[331] C.J. Pendegrass, A.E. Goodship, J.S. Price, G.W. Blunn, Nature's answer to breaching the skin barrier: an innovative development for amputees, Journal of Anatomy 209 (1) (2006) 59–67.
[332] N. Fitzpatrick, T.J. Smith, C.J. Pendegrass, R. Yeadon, M. Ring, A.E. Goodship, G.W. Blunn, Intraosseous transcutaneous amputation prosthesis (ITAP) for limb salvage in 4 dogs, Veterinary Surgery 40 (8) (2011) 909–925.
[333] J. Sensinger, telephone interview, 23/4/15.
[334] Northwestern University Prosthetics-Orthotics Center, History of NUPOC, 2020.
[335] K. Englehart, David Ford telephone interview, 23/4/10.
[336] T.A. Kuiken, L.A. Miller, R.D. Lipschutz, K.A. Stubblefield, G.A. Dumanian, Prosthetic command signals following targeted hyper-reinnervation nerve transfer surgery, in: 2005 IEEE Engineering in Medicine and Biology 27th Annual Conference, 2006, pp. 7652–7655.
[337] T.A. Kuiken, G. Li, B.A. Lock, R.D. Lipschutz, L.A. Miller, K.A. Stubblefield, K.B. Englehart, Targeted muscle reinnervation for real-time myoelectric control of multifunction artificial arms, JAMA 301 (6) (2009) 619–628.

[338] L.A. Miller, K.A. Stubblefield, R.D. Lipschutz, B.A. Lock, T.A. Kuiken, Improved myoelectric prosthesis control using targeted reinnervation surgery: a case series, IEEE Transactions on Neural Systems and Rehabilitation Engineering 16 (1) (2008) 46–50.
[339] J.B. Bowen, C.E. Wee, J. Kalik, I.L. Valerio, Targeted muscle reinnervation to improve pain, prosthetic tolerance, and bioprosthetic outcomes in the amputee, Advances in Wound Care 6 (8) (2017) 261–267.
[340] Guardian News and Media Limited, Robotic arm man Christian Kandlbauer dies in hospital after crash, http://www.guardian.co.uk/world/2010/oct/22/christian-kandlbauer-arm-dies-crash.
[341] T. Kuiken, telephone interview, 26/2/20.
[342] D.S. Childress, A tribute to Colin A McLaurin 1922-1997: build, don't talk, Journal of Rehabilitation Research and Development 35 (1) (1998) VII.
[343] W. Woods, Rehabilitation (a Combined Operation): Being a History of the Development and Carrying Out of a Plan for the Re-Establishment of a Million Young, Veterans of World War II, Queen's Press, Ottawa, 1953.
[344] Ontario Crippled Children's Centre, Annual Report, 1964.
[345] W.F. Sauter, The use of electric elbows in the rehabilitation of children with upper limb deficiencies, Prosthetics and Orthotics International 15 (2) (1991) 93–95.
[346] R. Reiter, Eine neue elektrokunsthand, Grenzgebiete der Medizin 1 (4) (1948) 133–135.
[347] R. Reiter, J. Kampik, Neue Ergebnisse der Klimatologie und Biophysik: Beiträge zur Wirkung meteorologischer, kosmischer und geophysikalischer Reize, Verlag Die Egge, 1948.
[348] Ontario Crippled Children's Centre, Annual Report, 1965.
[349] Ontario Crippled Children's Centre, Annual Report, 1967.
[350] Ontario Crippled Children's Centre, Annual Report, 1968.
[351] Ontario Crippled Children's Centre, Annual Report, 1969.
[352] Ontario Crippled Children's Centre, Annual Report, 1970.
[353] Ontario Crippled Children's Centre, Annual Report, 1973.
[354] M. Milner, David Ford telephone interview, 25/6/10.
[355] E. Sangster Morris, Milner on technology, Features 132 (2) (2004).
[356] M.M. Milner, On the odyssey: a personal journey, Assistive Technology 13 (1) (2001) 59–65.
[357] Ontario Crippled Children's Centre, Annual Report, 1981.
[358] Ontario Crippled Children's Centre, Annual Report, 1987.
[359] I. Timmen, David Ford telephone interview, 30/6/10.
[360] R.N. Scott, David Ford telephone interview, 12/5/10.
[361] E. Biden, David Ford telephone interview, 5/5/10.
[362] Department of National Health and Welfare, The report of the expert committee of the habilitation of congenital anomalies associated with Thalidomide, Technical Report 1, Department of National Health and Welfare, 1963.
[363] E.D. Shermand, A Russian bioelectric-controlled prosthesis: report of a research team from the Rehabilitation Institute of Montreal, Canadian Medical Association Journal 91 (24) (1964) 1268.
[364] P. Parker, telephone interview, 9/4/15.
[365] R.N. Scott, private communication.
[366] B. Hudgins, telephone interview, 5/5/10.
[367] B.S. Hudgins, A Novel Approach to Multifunction Myoelectric Control of Prostheses, PhD thesis, Department of Electrical Engineering, University of New Brunswick, 1991.
[368] B. Hudgins, P. Parker, R.N. Scott, A new strategy for multifunction myoelectric control, IEEE Transactions on Biomedical Engineering 40 (1) (1993) 82–94.
[369] A. Fougner, E. Scheme, A.D.C. Chan, K. Englehart, Ø. Stavdahl, Resolving the limb position effect in myoelectric pattern recognition, IEEE Transactions on Neural Systems and Rehabilitation Engineering 19 (6) (2011) 644–651.

[370] B. Lock, A.M. Simon, K. Stubblefield, L.J. Hargrove, Prosthesis-guided training for practical use of pattern recognition control of prostheses, in: MEC '95 Myoelectric Controls Symposium, Fredericton, New Brunswick, Canada, 2011, pp. 61–64.
[371] B. Lock, telephone interview, 8/11/19.
[372] B. Lock, M.K. Cava, E.W. Karlson, Real-world use of myoelectric pattern recognition: successes and challenges, in: TIPS '19 Trent International Prosthetics Symposium, Salford, United Kingdom, 2019, pp. 76–77.
[373] Q. Wang, B. Mao, S.I. Stoliarov, J. Sun, A review of lithium ion battery failure mechanisms and fire prevention strategies, Progress in Energy and Combustion Science 73 (2019) 95–131.
[374] S. Schulz, C. Pylatiuk, M. Reischl, J. Martin, R. Mikut, G. Bretthauer, A hydraulically driven multifunctional prosthetic hand, Robotica 23 (3) (2005) 293.
[375] Immanuel Nicolas Gaiser, Christian Pylatiuk, Stefan Schulz, Artem Kargov, Reinhold Oberle, Tino Werner, The fluidhand iii: a multifunctional prosthetic hand, Journal of Prosthetics and Orthotics 21 (2) (2009) 91–96.
[376] S. Schulz, First experiences with the Vincent Hand, in: Proceedings of the MyoElectric Controls/Powered Prosthetics Symposium Fredericton, Canada, 2011, 2011, pp. 14–19.
[377] Vincent Systems GmbH, Vincent Systems website, https://vincentsystems.de/en/.
[378] J. Gibbard, telephone interview, 9/3/20.
[379] J. Poirters, telephone interview, 4/6/20.
[380] H. Bouwsema, C.K. van der Sluis, R.M. Bongers, Effect of feedback during virtual training of grip force control with a myoelectric prosthesis, PLoS ONE 9 (5) (2014).
[381] T. Kuiken, Todd Kuiken private correspondence, 16/11/20.
[382] Evening Standard, 23 February 2009.

Index

Q

R